FLUID-INDUCED SEISMICITY

Serge A. Shapiro

Freie Universität Berlin

UNIVERSITY PRESS

CAMBRIDGE
UNIVERSITY PRESS

University Printing House, Cambridge CB2 8BS, United Kingdom

One Liberty Plaza, 20th Floor, New York, NY 10006, USA

477 Williamstown Road, Port Melbourne, VIC 3207, Australia

314-321, 3rd Floor, Plot 3, Splendor Forum, Jasola District Centre, New Delhi-110025, India

79 Anson Road, #06-04/06, Singapore 079906

Cambridge University Press is part of the University of Cambridge.

It furthers the University's mission by disseminating knowledge in the pursuit of
education, learning and research at the highest international levels of excellence.

www.cambridge.org
Information on this title: www.cambridge.org/9781108447928

First published 2015
First paperback edition 2017

A catalogue record for this publication is available from the British Library

Library of Congress Cataloging in Publication data
Shapiro, S. A.
Fluid-induced seismicity / Serge A. Shapiro, Earth Science Department,
Freie Universität Berlin.
pages   cm
Includes bibliographical references and index.
ISBN 978-0-521-88457-0
1. Rock mechanics.   2. Hydraulic fracturing.   3. Induced seismicity.
4. Reservoir-triggered seismicity.   5. Oil field flooding.   I. Title.
QE431.6.M4S53   2015
551.22–dc23
2014043958

ISBN 978-0-521-88457-0 Hardback
ISBN 978-1-108-44792-8 Paperback

FLUID-INDUCED SEISMICITY

The characterization of fluid-transport properties of rocks is one of the most important, yet difficult, challenges of reservoir geophysics, but is essential for optimal development of hydrocarbon and geothermal reservoirs. Production of shale oil, shale gas, heavy oil and geothermal energy, as well as carbon-dioxide sequestration, are relatively recent developments where borehole fluid injection is often employed to enhance fluid mobility. Unlike active seismic methods, which present fundamental difficulties for estimating the permeability of rocks, microseismicity induced by fluid injection in boreholes provides the potential to characterize physical processes related to fluid mobility and hydraulic-fracture growth in rocks.

This book provides a quantitative introduction to the underlying physics, application, interpretation, and hazard aspects of fluid-induced seismicity with a particular focus on its spatio-temporal dynamics. It presents many real-data examples of microseismic monitoring of hydraulic fracturing at hydrocarbon fields and of stimulations of enhanced geothermal systems. The author also covers introductory aspects of linear elasticity and poroelasticity theory, as well as elements of seismic rock physics and of the mechanics of earthquakes, enabling readers to develop a comprehensive understanding of the field. *Fluid-Induced Seismicity* is a valuable reference for researchers and graduate students working in the fields of geophysics, geology, geomechanics and petrophysics, and a practical guide for petroleum geoscientists and engineers working in the energy industry.

SERGE A. SHAPIRO is Professor of Geophysics at the Freie Universität Berlin, and since 2004, Director of the PHASE (PHysics and Application of Seismic Emission) university consortium project. From 2001 to 2008 he was one of the coordinators of the German Continental Deep Drilling Program. His research interests include seismogenic processes, wave phenomena, exploration seismology and rock physics. He received the SEG Virgil Kauffman Gold Medal in 2013 for his work on fluid-induced seismicity and rock physics, and in 2004 was elected a Fellow of the Institute of Physics.

Contents

Color plate section between pages 82 and 83

Preface

Characterization of fluid-transport properties of rocks is one of the most important, yet one of most challenging, goals of reservoir geophysics. However, active seismic methods have low sensitivity to rock permeability and mobility of pore fluids. On the other hand, it would be very attractive to have the possibility of exploring hydraulic properties of rocks using seismic methods because of their large penetration range and their high resolution. Microseismic monitoring of borehole fluid injections is exactly the tool that can provide us with such a possibility. Borehole fluid injections are often applied for stimulation and development of hydrocarbon and geothermal reservoirs. Production of shale gas and heavy oil as well as CO_2 sequestration are relatively recent technological areas that require broad applications of this technology. The fact that fluid injection causes seismicity has been well established for several decades (see, for example, Pearson, 1981, and Zoback and Harjes, 1997). Current ongoing research is aimed at quantifying and controlling this process. Understanding and monitoring of fluid-induced seismicity is necessary for hydraulic characterization of reservoirs and for assessments of reservoir stimulations.

Fluid-induced seismicity covers a wide range of processes between the two following limiting cases. In liquid-saturated hard rocks with low to moderate permeability (10^{-5}–10^{-2} darcy) and moderate bottom hole injection pressures (as a rule, less than the minimum absolute value of the principal compressive tectonic stress) the phenomenon of microseismicity triggering is often caused by the process of linear relaxation of pore-pressure perturbations (Shapiro *et al.*, 2005a,b). Note that we speak here about the linearity in the sense of corresponding differential equations. In porodynamics this process corresponds to the Frenkel–Biot slow wave propagation (see Biot, 1962, and a history review by Lopatnikov and Cheng, 2005, as well as an English translation of Frenkel, 2005). In the porodynamic low-frequency range (hours or days of fluid-injection duration) this process reduces to a linear pore-pressure diffusion. Then, the linear pore-pressure diffusion defines features of the rate of spatial growth, geometry and density of clouds of microearthquake hypocenters (Shapiro *et al.*, 2002, 2003, 2005a,b; Parotidis *et al.*, 2004). In some cases, spontaneously triggered natural seismicity, like earthquake swarms, also shows similar diffusion-like signatures (Parotidis *et al.*, 2003, 2004, 2005; Hainzl *et al.*, 2012; Shelly *et al.*, 2013).

Another extreme case is a strong non-linear fluid–solid interaction related to the hydraulic fracturing of sediments like a tight sandstone or a shale with extremely low permeability (10^{-9}–10^{-5} darcy). In this case a fluid injection leads to a strong enhancement of the permeability. Propagation of a hydraulic fracture is accompanied by opening of a new fracture volume, fracturing fluid loss and its infiltration into reservoir rocks, as well as diffusion of the injection pressure into the pore space of surrounding formations and inside the hydraulic fracture (Economides and Nolte, 2003). Some of these processes can be seen from features of spatio-temporal distributions of the induced microseismicity (Shapiro *et al.*, 2006b; Fischer *et al.*, 2008; Dinske *et al.*, 2010). The initial stage of fracture volume opening as well as the back front of induced seismicity (propagating after termination of the fluid injection) can be observed. Evaluation of spatio-temporal dynamics of induced microseismicity can help to estimate physical characteristics of hydraulic fractures, e.g. penetration rate of the fracture, its permeability as well as the permeability of the reservoir rock. Therefore, understanding and monitoring of fluid-induced seismicity by hydraulic fracturing can be useful for describing hydrocarbon and geothermal reservoirs and for estimating the results of hydraulic fracturing.

Seismicity induced by borehole fluid injections is a central topic of this book. It describes physical fundamentals of interpretation of fluid-induced seismicity. The first two chapters of the book provide readers with an introduction to the theoretical background of concepts and approaches useful for understanding fluid-induced seismicity. An application-interested reader can probably skip these two chapters and just go directly to Section 1.4 and then Chapters 3–5, using Chapters 1 and 2 mainly as reference material.

In Chapter 1 the book starts with a brief introduction to the theory of elasticity and seismic-wave propagation. This chapter also includes elements of fracture mechanics and of the geomechanics of faulting. Then there is an introductory description of earthquake sources of the seismic wavefield. Finally, the chapter contains a brief schematic description of methodical approaches of microseismic monitoring. Many important processing-related methodical aspects of microseismic monitoring remain outside of the scope of this book.

Chapter 2 provides a detailed introduction to the theory of poroelasticity. The main physical phenomena responsible for fluid-induced seismicity and discussed in this book in detail are fluid filtration and pore-pressure relaxation. They are closely related to slow waves in porous fluid-saturated materials. The dynamics of slow wavefields is the focus of this chapter. The chapter also includes a discussion of some non-linear effects related to deformations of the pore space. They are relevant for characterizing poroelastic coupling and for formulating models of the pressure-dependent permeability. Such models will be used for the considerations of non-linear pressure diffusion in subsequent chapters. The topic of thermo-poroelastic interaction is not discussed in the book.

In Chapters 3–5 of this book we describe the main quantitative features of different types of fluid-induced microseismicity. Different properties of induced seismicity related to reservoir characterization and hydraulic fracturing are addressed, along with the magnitude distribution of seismicity induced by borehole fluid injections. Evidently, this is an important question closely related to seismic hazard of injection sites. Many corresponding aspects of the book are also applicable to induced tectonic seismicity.

This book attempts to contribute to further elaboration of the seismicity-based reservoir characterization approach (see also Shapiro, 2008).

Acknowledgments

This book contains results of research funded by different institutions in different time periods. This includes the German Federal Ministry for the Environment, Nature Conservation and Nuclear Safety (BMU), and the German Federal Ministry of Education and Research (BMBF) supported the section of Geophysics of the Freie Universität Berlin for their projects MAGS and MEPRORISK. This also includes the Deutsche ForschungsGemeinschaft (DFG) in whose Heisenberg research program I started to work, in 1997, with interpretation of microseismic monitoring of fluid injection at the German KTB. At that time I worked at the Geophysical Institute of the Karlsruhe University and also spent a short period at the GeoForschungsZentrum Potsdam. I continued this research at the Geological School of Nancy, France, in close cooperation with Jean-Jacques Royer and Pascal Audigane, where our work was significantly supported by the GOCAD consortium project led by Professor Jean-Laurent Mallet. In 1999 I moved to the Freie Universität Berlin. Many of the results reported here were then obtained in our common work with Elmar Rothert, Jan Rindschwentner, Miltiadis Parotidis, Robert Patzig, Inna Edelman, Nicolas Delepine and Volker Rath. Corresponding research works were funded to a significant extent by the Wave Inversion Technology (WIT) university consortium project led by Professor Peter Hubral of Karlsruhe University.

Starting in 2005, the research reported here was to a large extent funded by the PHysics and Application of Seismic Emission (PHASE) university consortium project at the Freie Universität Berlin. Susanne Rentsch, Carsten Dinske, Jörn Kummerow, Stefan Buske, Erik Saenger, Stefan Lüth, Cornelius Langenbruch, Oliver Krüger, Nicolas Hummel, Anton Reshetnikov, Antonia Oelke, Radim Ciz, Maximilian Scholze, Changpeng Yu, Sibylle Mayr and Karsten Stürmer contributed strongly to the work performed in this project. Corresponding results are of especial importance for this book. I express my sincere thanks to all these colleagues and friends, and I would like to thank the sponsors of the PHASE consortium project and of two other consortium projects mentioned above as well as the DFG, the BMU and the BMBF for their generous support of my work related to this book.

I am also indebted to colleagues and institutions who helped me to access different microseismic data sets used here. These are Hans-Peter Harjes (Bochum University),[1] André Gérard and Roy Baria (SOCOMINE), Andrew Jupe (EGS Energy), Michael Fehler and James Rutledge (LANL), Shawn Maxwell (Pinnacle Technology), Kenneth Mahrer (USBR), Ted Urbancic, Adam Baig and Andreas Wuesterfeld (ESG), Hideshi Kaieda (Central Research Institute of Electric Power Industry, Tohoku), Takatoshi Ito (Institute of Fluid Science, Tohoku), Günter Asch (GFZ-Potsdam), Martin Karrenbach (P-GSI), Ulrich Schanz and Markus Häring (Geothermal Explorers), Sergey Stanchits and Georg Dresen (Deutsches GeoForschungsZentrum, GFZ-Potsdam).

I am deeply grateful to colleagues who provided me with their comments on the book manuscript. These are Boris Gurevich from Curtin University and CSIRO (Perth) and Robert Zimmerman from Imperial College London (the first two chapters). These are also my colleagues from the Freie Universität Berlin: Carsten Dinske (Chapter 4), Jörn Kummerow (various sections), Cornelius Langenbruch (Chapter 3), Oliver Krüger (Chapter 5). Of course, I have sole responsibility for the complete book content.

I sincerely acknowledge the Society of Exploration Geophysicists (SEG) for extending permission to use materials from a series of publications coauthored by me in *Geophysics* and *The Leading Edge*.

---

[1] Here, and in the following, the affiliations are given for the time periods during which the access to the data was made possible.

1

Elasticity, seismic events and microseismic monitoring

By "seismic events" we understand earthquakes of any size. There exists a broad scientific literature on earthquakes and on the processing of seismologic data. We refer readers interested in a detailed description of these subjects to corresponding books (see, for example, Lay and Wallace, 1995, and Shearer, 2009). We start this book with an introductory review of the theory of linear elasticity and of the mechanics of seismic events. The aim of this chapter is to describe classical fundamentals of the working frame necessary for our consideration of induced seismicity. We conclude this chapter with a short introduction to methodical aspects of the microseismic monitoring.

1.1 Linear elasticity and seismic waves

Deformations of a solid body are motions under which its shape and (or) its size change. Formally, deformations can be described by a field of a displacement vector $\mathbf{u}(\mathbf{r})$. This vector is a function of a location $\mathbf{r}$ of any point of the body in an initial reference state (e.g., the so-called unstrained configuration; see, for example, Segall, 2010). Initially we accept here the so-called Lagrangian formulation, i.e. we observe motions of a given particle of the body.

However, the field of displacements describes not only deformations of the body but also its possible rigid motions without changes of its shape and its size, such as translations and/or rotations.

In contrast to rigid motions, under deformations, distances (some or any) between particles of the body change. Therefore, to describe deformations, a mathematical function of the displacement field is used that excludes rigid motions of a solid and describes changes of distances between its particles only. This function is the strain tensor ϵ, which is a second-rank tensor with nine components ϵ_{ij}. Here the indices i and j can accept any of values 1, 2 and 3 denoting the coordinate directions of a Cartesian coordinate system in which the vectors $\mathbf{u}$ and $\mathbf{r}$ have been defined.

1.1.1 Strain

In the case of small deformations (i.e. where absolute values of all spatial derivatives of any components of the vector $\mathbf{u}(\mathbf{r})$ are much smaller than 1) the strain tensor has the form of a 3×3 symmetric matrix with the following components:

$$\epsilon_{ij} = \frac{1}{2}\left(\frac{\partial u_i}{\partial x_j} + \frac{\partial u_j}{\partial x_i}\right). \tag{1.1}$$

This form of the strain tensor describes deformations within a small vicinity of a given location. This form remains the same also by consideration of small deformations in the Eulerian formulation (see Segall, 2010), where instead of a motion of a given particle of the body (i.e. the Lagrangian approach) rather a motion at a given coordinate location (i.e. at a given point of the space) is considered. In this book we accept the small-deformation approximation and do not distinguish between the Lagrangian and Eulerian approaches.

Strains ϵ_{ij} can be arbitrary (small) numbers. However, because of their definition (1.1) they cannot be arbitrarily distributed in space. Spatial derivatives of strains must be constrained by the following compatibility equations (see Segall, 2010):

$$\frac{\partial^2 \epsilon_{ij}}{\partial x_k \partial x_l} + \frac{\partial^2 \epsilon_{kl}}{\partial x_i \partial x_j} = \frac{\partial^2 \epsilon_{ik}}{\partial x_j \partial x_l} + \frac{\partial^2 \epsilon_{jl}}{\partial x_k \partial x_i}. \tag{1.2}$$

Deformations of a body results from applications of loads to it. Deformations that will disappear completely if the loads are released are called elastic. Bodies that can have elastic deformations are called elastic bodies.

1.1.2 Stress

Elastic bodies resist their elastic deformations by means of elastic forces. Elastic forces in a solid body are analogous to a pressure in an ideal fluid. They occur due to mutual interactions of elastically deformed parts of the body. These interactions in turn take place on surfaces where the parts of the body are contacting each other (see also Landau and Lifshitz, 1987).

Let us consider an elementary part of a body under deformation (see Figure 1.1). Other parts of the body act by means of elastic forces onto this elementary part over its surface S. Let us consider a differentially small element of this surface at its arbitrary point $\mathbf{r}$. Such a surface element can be approximated by a differentially small part of a plane of area dS tangential to S at point $\mathbf{r}$ with a unit normal $\mathbf{n}$ directed outside this part of the surface. Owing to elastic deformations an elastic force $d\mathbf{F}(\mathbf{r}, \mathbf{n})$ (also called a stress force) acts on the plane element with the normal $\mathbf{n}$. The following limit defines a traction vector:

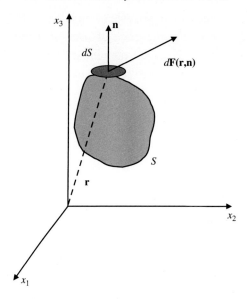

Figure 1.1 A sketch for defining a traction.

$$\boldsymbol{\tau}(\mathbf{r}, \mathbf{n}) = \lim_{dS \to 0} \frac{d\mathbf{F}(\mathbf{r}, \mathbf{n})}{dS}.\qquad(1.3)$$

Note that the traction has the same physical units as a pressure in a fluid (e.g. Pa in the SI system). Note also that the traction is a function of a location $\mathbf{r}$ and of an orientation of the normal $\mathbf{n}$.

Let us consider three plane elements parallel to coordinate planes at a given location. We assume also that their normals point in the positive directions of coordinate axes, which are perpendicular to the plane elements. Therefore, the corresponding three normals coincide with the unit basis vectors $\hat{\mathbf{x}}_1$, $\hat{\mathbf{x}}_2$ and $\hat{\mathbf{x}}_3$ of the Cartesian coordinate system under consideration. Tractions acting on these plane elements are $\boldsymbol{\tau}(\mathbf{r}, \hat{\mathbf{x}}_1)$, $\boldsymbol{\tau}(\mathbf{r}, \hat{\mathbf{x}}_2)$ and $\boldsymbol{\tau}(\mathbf{r}, \hat{\mathbf{x}}_3)$, respectively. A 3×3 matrix composed of nine coordinate components of these tractions defines the stress tensor, $\boldsymbol{\sigma}$. Its element σ_{ij} denotes the ith component of the traction acting on the surface with the normal $\hat{\mathbf{x}}_j$:

$$\sigma_{ij} = \tau_i(\hat{x}_j).\qquad(1.4)$$

Let us consider a differentially small elastic body under an elastic strain and assume for all deformation processes enough time to bring parts of this body into an equilibrium state. From the equilibrium conditions for the rotational moments (torques) of elastic forces it follows that the stress tensor is symmetric:

$$\sigma_{ij} = \sigma_{ji}.\qquad(1.5)$$

Note that if the body torques are negligible (which is usually the case) this relation is valid even in the case of the presence of rotational motions. This is because of the fact that, in the limit of a small elementary volume, the inertial forces are decreasing faster than the elastic force torques (see Auld, 1990, volume 1, section 2, for more details).

Similarly, a consideration of forces (elastic forces, body forces and inertial forces) acting on a volume element in the limit of its vanishing volume shows that elastic forces applied to the surface of such a volume must be in balance (see Auld, 1990, volume 1, section 2, for more details). It then follows that a traction $\tau(\mathbf{r}, \mathbf{n})$ acting on an arbitrarily oriented plane surface element can be computed by using the stress tensor:

$$\tau_i(\mathbf{n}) = \sigma_{ij} n_j. \tag{1.6}$$

Note that here and generally in this book (if not specially mentioned) we accept the agreement on summation on repeated indices, e.g. $a_i b_i = a_1 b_1 + a_2 b_2 + a_3 b_3$.

Definition (1.4) of the stress tensor corresponds to a common continuum mechanics sign convention that tensile stresses are positive and compressive stresses are negative (see, for example, a thin elementary volume and tractions acting on its outer surface with normals pointing outside this volume; Figure 1.2).

1.1.3 Stress–strain relations

The strain-tensor and stress-tensor notations give a general form of an observational fact, known as Hooke's law, that small elastic deformations are proportional to elastic forces:

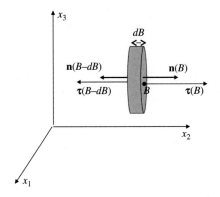

Figure 1.2 A sketch illustrating positiveness of tensile stresses. Indeed, equation (1.6) requires that the components σ_{22} in the both points, B and $B - dB$ must be positive. Note that the point B is shown as a dot on the right-hand side of the disc. The point denoted as $B - dB$ is not seen. It is on the left-hand side of the disc; dB denotes the width of the disc.

$$\epsilon_{ij} = S_{ijkl}\sigma_{kl}, \tag{1.7}$$

where the fourth-rank tensor **S**, with components S_{ijkl}, is the tensor of elastic compliances. Note that their physical units are inverse to the unit of stress: $1/\mathrm{Pa}$. Owing to the symmetry of the strain and stress tensors, the tensor of elastic compliances has the following symmetries:

$$S_{ijkl} = S_{jikl} = S_{ijlk}. \tag{1.8}$$

Another fourth-rank tensor **C**, with components C_{ijkl}, called the tensor of elastic stiffnesses, yields an alternative formulation of Hooke's law:

$$\sigma_{ij} = C_{ijkl}\epsilon_{kl}. \tag{1.9}$$

From this equation it is clear that the tensor of elastic stiffnesses also has the symmetry:

$$C_{ijkl} = C_{jikl} = C_{ijlk}. \tag{1.10}$$

Both the tensor of elastic stiffnesses and the tensor of elastic compliances are physical characteristics of a given elastic body.

Often both forms of Hooke's law (1.7) and (1.9) are written symbolically as (see Auld, 1990):

$$\epsilon = \mathbf{S} : \sigma, \quad \sigma = \mathbf{C} : \epsilon. \tag{1.11}$$

Here the double-dot (or double scalar) products denote summations over pairs of repeating indices in (1.7) and (1.9), respectively.

A deformed elastic body possesses an elastic strain energy. At zero strain this energy is equal to zero. With increasing strain by an increment $d\epsilon_{kl}$ due to the stress σ_{kl}, the volumetric density of this energy (energy per unit volume) must increase by the increment $dE = \sigma_{kl}d\epsilon_{kl}$ (see Landau and Lifshitz, 1987). The tensor of elastic stiffnesses can then be used to define the density of the elastic strain energy (by integration of the increment dE) as a positive quadratic function of non-zero strains:

$$E = \frac{1}{2}C_{ijkl}\epsilon_{ij}\epsilon_{kl} = \frac{1}{2}\sigma_{kl}\epsilon_{kl} = \frac{1}{2}S_{klij}\sigma_{kl}\sigma_{ij}, \tag{1.12}$$

where in the two last expressions the two forms of Hooke's law (1.7) and (1.9) have been used. The product $\epsilon_{ij}\epsilon_{kl}$ remains unchanged if the index pair ij is replaced by kl and kl is replaced by ij, respectively. Thus, the tensor of elastic stiffnesses as well as the the tensor of compliances must also have the following symmetry:

$$C_{ijkl} = C_{klij}, \quad S_{ijkl} = S_{klij}. \tag{1.13}$$

Elasticity, seismic events and microseismic monitoring

Symmetries (1.8), (1.10) and (1.13) of the stiffness and compliance tensors reduce the number of their independent components. From 81 possible components of a tensor (tensors' indices can be equal to 1, 2 or 3) only 21 components are mutually independent. These components are also called elastic moduli or elastic constants (the latter notation neglects such effects as pressure dependence and temperature dependence of these quantities). The requirement that the elastic strain energy must be a positive-definite quadratic form of arbitrary strain/stress components (called also the stability condition) provides additional restrictions on allowed values of elastic moduli.

1.1.4 Elastic moduli

The tensors C_{ijkl} and S_{ijkl} are inverse to each other so that (see Cheng, 1997):

$$C_{ijkl} S_{klmn} = \frac{1}{2}(\delta_{im}\delta_{jn} + \delta_{in}\delta_{jm}),\qquad(1.14)$$

where quantity δ_{kl} is the so-called Kronecker matrix, with components $\delta_{kl} = 1$, for $k = l$, and $\delta_{ij} = 0$ in other cases.

The tensors of stiffnesses and compliances can be expressed in convenient matrix forms by using their 21 independent components, respectively. For this, one uses the so-called contracted notation (or the Voigt notations). Let us introduce capital indices (e.g. I, J, etc.), which can take values $1, 2, 3, 4, 5$ and 6. The following correspondence between the capital indices and the pairs of the usual indices (ij) is assigned: $1 \rightarrow 11, 2 \rightarrow 22, 3 \rightarrow 33, 4 \rightarrow 23, 5 \rightarrow 13$, and $6 \rightarrow 12$. In these notations Hooke's law has the following forms (Jaeger et al., 2007; Auld, 1990):

$$
\begin{bmatrix} \epsilon_{11} \\ \epsilon_{22} \\ \epsilon_{33} \\ 2\epsilon_{23} \\ 2\epsilon_{13} \\ 2\epsilon_{12} \end{bmatrix}
=
\begin{bmatrix}
S_{11} & S_{12} & S_{13} & S_{14} & S_{15} & S_{16} \\
S_{12} & S_{22} & S_{23} & S_{24} & S_{25} & S_{26} \\
S_{13} & S_{23} & S_{33} & S_{34} & S_{35} & S_{36} \\
S_{14} & S_{24} & S_{34} & S_{44} & S_{45} & S_{46} \\
S_{15} & S_{25} & S_{35} & S_{45} & S_{55} & S_{56} \\
S_{16} & S_{26} & S_{36} & S_{46} & S_{56} & S_{66}
\end{bmatrix}
\begin{bmatrix} \sigma_{11} \\ \sigma_{22} \\ \sigma_{33} \\ \sigma_{23} \\ \sigma_{13} \\ \sigma_{12} \end{bmatrix},\qquad(1.15)
$$

$$
\begin{bmatrix} \sigma_{11} \\ \sigma_{22} \\ \sigma_{33} \\ \sigma_{23} \\ \sigma_{13} \\ \sigma_{12} \end{bmatrix}
=
\begin{bmatrix}
c_{11} & c_{12} & c_{13} & c_{14} & c_{15} & c_{16} \\
c_{12} & c_{22} & c_{23} & c_{24} & c_{25} & c_{26} \\
c_{13} & c_{23} & c_{33} & c_{34} & c_{35} & c_{36} \\
c_{14} & c_{24} & c_{34} & c_{44} & c_{45} & c_{46} \\
c_{15} & c_{25} & c_{35} & c_{45} & c_{55} & c_{56} \\
c_{16} & c_{26} & c_{36} & c_{46} & c_{56} & c_{66}
\end{bmatrix}
\begin{bmatrix} \epsilon_{11} \\ \epsilon_{22} \\ \epsilon_{33} \\ 2\epsilon_{23} \\ 2\epsilon_{13} \\ 2\epsilon_{12} \end{bmatrix}.\qquad(1.16)
$$

In these two equations the contracted notation is used in the two symmetric 6×6 matrices of components s_{IK} and c_{IK}, where I corresponds to a pair of normal indices, e.g. ij, and K corresponds to another their pair, e.g. kl. It is clear that these matrices are inverse to each other, i.e. their matrix product gives a 6×6 unit matrix. Their components are also called elastic compliances and elastic stiffnesses, respectively. The relations between the contracted-notation stiffness matrix components and corresponding components of the fourth-rank tensor of elastic stiffnesses is simple: $c_{IK} = C_{ijkl}$. This correspondence for the compliances is a bit more complicated: $s_{IK} = S_{ijkl}$ if $I, K = 1, 2, 3$, $s_{IK} = 2S_{ijkl}$ if $I = 1, 2, 3$ and $K = 4, 5, 6$, and $s_{IK} = 4S_{ijkl}$ if $I, K = 4, 5, 6$.

The higher the physical symmetry of the elastic medium, the smaller is the number of non-vanishing independent elastic constants. For mineral crystals, different symmetries are of relevance (see Auld, 1990, for a comprehensive description). In the most general case of triclinic crystals the elastic properties are characterized by 21 independent compliances (or, equivalently, 21 independent stiffnesses). This situation corresponds to equations (1.15) and (1.16), respectively. If the medium has a single symmetry plane (the monoclinic symmetry) then the number of independent constants will be reduced to 13 (for example, if we assume the xy coordinate plane as the plane of symmetry, this will result in the invariant coordinate transformation $z \to -z$ and thus, all elastic constants with odd numbers of index 3 must be equal to zero). This situation corresponds, for example, to a layered medium with a single system of plane cracks oblique to the lamination plane.

One of most relevant symmetries for rocks is the orthorhombic one. It can be applied to describe different geological situations, like rocks with three mutually perpendicular systems of cracks or horizontally layered rocks permeated by a single system of aligned vertical fractures. An orthorhombic medium has three mutually perpendicular symmetry planes. This means that in such a medium under corresponding coordinate transformations (reflections across symmetry planes) the tensors of elastic constants must remain unchanged. In a coordinate system with axes normal to the symmetry planes it follows that all components C_{ijkl} and S_{ijkl} with odd numbers of any index must be equal to zero. This leads to the following forms of the compliance and stiffness matrices, respectively:

$$
\begin{bmatrix}
s_{11} & s_{12} & s_{13} & 0 & 0 & 0 \\
s_{12} & s_{22} & s_{23} & 0 & 0 & 0 \\
s_{13} & s_{23} & s_{33} & 0 & 0 & 0 \\
0 & 0 & 0 & s_{44} & 0 & 0 \\
0 & 0 & 0 & 0 & s_{55} & 0 \\
0 & 0 & 0 & 0 & 0 & s_{66}
\end{bmatrix}
;
\begin{bmatrix}
c_{11} & c_{12} & c_{13} & 0 & 0 & 0 \\
c_{12} & c_{22} & c_{23} & 0 & 0 & 0 \\
c_{13} & c_{23} & c_{33} & 0 & 0 & 0 \\
0 & 0 & 0 & c_{44} & 0 & 0 \\
0 & 0 & 0 & 0 & c_{55} & 0 \\
0 & 0 & 0 & 0 & 0 & c_{66}
\end{bmatrix} . \quad (1.17)
$$

We see that nine independent constants are enough to completely describe the elastic properties of an orthorhombic medium. The compliances can be obtained from stiffnesses by the matrix inversion and vice versa. In the case of an arbitrary coordinate orientation, three additional constants (corresponding to three rotational angles) are required.

A useful and geologically relevant subset of orthorhombic symmetry is transverse isotropy. Layered sedimentary rocks can frequently be described by this symmetry. The plane of lamination is then the symmetry plane. If one of the coordinate planes coincides with the symmetry plane, then a coordinate axis normal to the symmetry plane will be an axis of an arbitrary-angle rotational symmetry. This symmetry results in four additional relations between the elastic constants, reducing the number of independent ones to five. If the symmetry axis coincides with the direction of the axis x_3, then in equations (1.17) additional relations will be (Auld, 1990): $s_{22} = s_{11}$, $s_{23} = s_{13}$, $s_{55} = s_{44}$ and $s_{66} = 2(s_{11} - s_{12})$. Correspondingly, $c_{22} = c_{11}$, $c_{23} = c_{13}$, $c_{55} = c_{44}$ and $c_{66} = (c_{11} - c_{12})/2$.

Finally, in the case of an elastic isotropic medium (all coordinate axes are arbitrary-angle rotational symmetry axes and any plane is a plane of symmetry), two constants remain independent only: $s_{22} = s_{33} = s_{11}$, $s_{23} = s_{13} = s_{12}$, $s_{66} = s_{55} = s_{44}$ and $s_{44} = 2(s_{11} - s_{12})$. Correspondingly, $c_{22} = c_{33} = c_{11}$, $c_{23} = c_{13} = c_{12}$, $c_{66} = c_{55} = c_{44}$ and $c_{44} = (c_{11} - c_{12})/2$. The independent elastic stiffnesses are usually denoted as the elastic moduli λ and μ, so that $c_{44} = \mu$ and $c_{12} = \lambda$. Inverting the matrix c_{ij} we obtain compliances of an isotropic medium:

$$s_{11} = \frac{\lambda + \mu}{\mu(3\lambda + 2\mu)}, \quad s_{12} = -\frac{\lambda}{2\mu(3\lambda + 2\mu)}, \quad s_{44} = \frac{1}{\mu}. \tag{1.18}$$

Let us consider a volumetric strain (dilatation) of an elementary volume V of an arbitrary anisotropic elastic medium:

$$\epsilon \equiv \frac{dV}{V}. \tag{1.19}$$

We can choose such an elementary volume to be a cuboid with side lengths l_x, l_y and l_z. Thus we see that

$$\epsilon = \frac{d(l_x l_y l_z)}{l_x l_y l_z} = \frac{dl_x}{l_x} + \frac{dl_y}{l_y} + \frac{dl_z}{l_z} = \epsilon_{11} + \epsilon_{22} + \epsilon_{33}. \tag{1.20}$$

Let us further assume that this dilatation is a result of a hydrostatic stress, $\sigma_{kl} = -p\delta_{kl}$, applied to the medium, where p is the pressure loading the medium. A general relation between the dilatation and the pressure can be obtained by taking a double-dot product (the scalar product) of Hooke's law (1.7) with the δ_{ij} (i.e. multiplying the both sides with δ_{ij} and summing up over repeating indices):

$$\epsilon = -S_{iikk} p. \tag{1.21}$$

The proportionality coefficient here is a bulk compressibility C^{mt} of the elastic material:

$$C^{mt} \equiv S_{iikk} = S_{1111} + S_{2222} + S_{3333} + 2(S_{1122} + S_{1133} + S_{2233})$$
$$= s_{11} + s_{22} + s_{33} + 2(s_{12} + s_{13} + s_{23}). \tag{1.22}$$

It follows from (1.19)–(1.21) that the bulk compressibility of a sample has the following relation to its bulk density ρ:

$$C^{mt} = -\frac{dV}{V\,dp} = -\frac{d(1/\rho)}{(1/\rho)dp} = \frac{1}{\rho}\frac{d\rho}{dp}. \tag{1.23}$$

In the case of an isotropic elastic material we obtain (see equations (1.22) and (1.18)) $C^{mt} = 3s_{11} + 6s_{12} = 1/(\lambda + 2\mu/3)$. Therefore,

$$K = \lambda + 2\mu/3 \tag{1.24}$$

is a bulk modulus describing the stiffness of the material to volumetric deformations.

The following representation of the stiffness tensor of an isotropic medium is useful (Aki and Richards, 2002):

$$C_{ijkl} = \lambda\delta_{ij}\delta_{kl} + \mu(\delta_{ik}\delta_{jl} + \delta_{il}\delta_{jk}). \tag{1.25}$$

In the same terms, Hooke's law for isotropic elastic media can be written in the following form:

$$\sigma_{ij} = \lambda\delta_{ij}\epsilon + 2\mu\epsilon_{ij}. \tag{1.26}$$

From this equation it follows that μ is a shear modulus of the material, describing its stiffness to shear deformations (under which $i \neq j$). It follows also that under uniaxial stress conditions (for example $\sigma_{33} \neq 0$ and $\sigma_{11} = \sigma_{22} = 0$) the ratio ν of the transverse strain to the longitudinal strain, $-\epsilon_{11}/\epsilon_{33}$, is equal to

$$\nu = \frac{\lambda}{2(\lambda + \mu)}. \tag{1.27}$$

This quantity is called Poisson's ratio. For an isotropic elastic solid the stability condition requires that both bulk and shear moduli must be positive. For Poisson's ratio this yields the restriction $-1 \leq \nu \leq 0.5$. For realistic rocks this coefficient is positive. Its upper limit of 0.5 corresponds to fluids. Frequently, its values for stiff tight isotropic rocks are close to 0.25 (corresponding to $\lambda \approx \mu$).

All elastic moduli introduced above will usually be assumed to be isothermal ones, if static deformations or processes being very slow in respect to the thermal diffusion are considered. In this book we consider processes that are faster than the temperature equilibration (e.g. wave propagation and pore-pressure equilibration). We will assume that these processes are approximately adiabatic. Thus, we assume that the elastic moduli introduced above are adiabatic. Note that the adiabatic and isothermal moduli of hard materials (e.g. rocks) differ by a small amount (see also Landau and Lifshitz, 1987).

In this book we will frequently assume that the elastic properties of the medium are isotropic. This simplifying assumption is often too rough for problems of seismic event location and imaging (which are not the main subject of our consideration). For such problems velocity models should take into account seismic anisotropy at least in the weak anisotropy approximation (Thomsen, 1986; Tsvankin, 2005; Grechka, 2009). For describing dominant effects responsible for the triggering of induced microseismicity the assumption of elastic isotropy seems to be adequate at least as the first approximation. For such effects hydraulic anisotropy of rocks is much more important. Elastic anisotropy in rocks is usually below 10% and seldom exceeds 30%, in respect to the velocity contrast between the slowest and fastest wave propagation directions. In shale the elastic anisotropy can be even higher. However, usually it is much smaller than a possible anisotropy of the hydraulic permeability, which can reach several orders of magnitude.

1.1.5 Dynamic equations and elastic waves

By an elastic deformation, a transfer of an elastic solid from one equilibrium state to another equilibrium state occurs by means of propagation of elastic waves. Elastic waves in rocks in the frequency range between 10^{-3} and 10^4 Hz are usually referred to as seismic waves. Resulting elastic forces acting on an elementary volume of the elastic medium define its acceleration vector. Owing to Hooke's law and the definition of the strain tensor, the second Newtonian law (i.e. the momentum conservation) takes the form of the following dynamic equation (Lamé equation):

$$\frac{\partial}{\partial x_j} C_{ijkl} \frac{\partial u_k}{\partial x_l} = \rho \frac{\partial^2 u_i}{\partial t^2}. \tag{1.28}$$

This equation describes the propagation of elastic waves in the most general case of a heterogeneous anisotropic elastic medium. Note that this is a system of three equations for three unknown components of the displacement vector. A plane-wave analysis (see also our later discussion of poroelastic waves) is instructive for investigating modes of propagation of elastic perturbations.

Let us consider the case of a homogeneous arbitrary anisotropic elastic medium. Then equation (1.28) simplifies to:

$$C_{ijkl} \frac{\partial^2 u_k}{\partial x_j x_l} = \rho \frac{\partial^2 u_i}{\partial t^2}. \tag{1.29}$$

In linear systems any wavefield can be decomposed into a superposition of independently propagating time-harmonic plane waves. We will designate their angular frequencies by ω. Thus, we can look for a solution of equation (1.29) in the following form:

$$\mathbf{u} = \mathbf{u}_0 e^{i(\omega t - \boldsymbol{\kappa} \mathbf{r})}, \tag{1.30}$$

Here $\boldsymbol{\kappa}$ is a wave vector of an arbitrary direction and an unknown length and $\mathbf{u}_0$ is an unknown polarization vector of a plane wave. Substituting this expression into equation (1.29) we obtain

$$C_{ijkl} u_{0k} \kappa_j \kappa_l = \rho \omega^2 u_{0i}. \tag{1.31}$$

Using the Kronecker matrix we can rewrite this equation in a more instructive way:

$$(C_{ijkl} \kappa_j \kappa_l - \rho \omega^2 \delta_{ik}) u_{0k} = 0. \tag{1.32}$$

This matrix equation is called the Christoffel equation. This is a homogeneous system of three algebraic linear equations for the three components of the polarization vector $\mathbf{u}_0$. Therefore, this equation system has non-zero solutions only if the following determinant is equal to zero:

$$\det(C_{ijkl} \kappa_j \kappa_l - \rho \omega^2 \delta_{ik}) = 0. \tag{1.33}$$

Equation (1.33) defines a relationship between the wave vector and the frequency for any possible plane wave in an elastic medium with the density ρ and the stiffness tensor C_{ijkl}. Thus, it is called a dispersion equation. It is clear that (1.33) is an equation for the following quantity:

$$c = \frac{|\omega|}{|\boldsymbol{\kappa}|}. \tag{1.34}$$

This quantity is the phase velocity of the corresponding plane wave. The condition (1.33) is a cubic equation in respect to c^2. Therefore, this equation describes dispersion relations of three different types of elastic waves. In weakly anisotropic media these waves correspond to an independently propagating quasi P- (nearly longitudinal) and two different quasi S- (nearly shear) waves. Substituting any of the three possible solutions for c^2 into equation (1.32) one obtains solutions for the corresponding polarization vectors $\mathbf{u}_0$.

Equation (1.30) shows that the phase velocity c controls how quickly the argument (i.e. the phase) of a given time-harmonic plane wave changes in space and time. In contrast, the equation

$$g_i = \frac{\partial \omega}{\partial \kappa_i} \tag{1.35}$$

provides components g_i of the group velocity vector of the wave type corresponding to the solution of equation (1.33) for c^2. The group velocity describes propagation of a spatial envelope of a group of plane waves with differentially close wave vectors. It is indeed a propagation velocity of wavefronts radiated by finite sources in their far fields in anisotropic elastic media. In elastic media the group velocity describes wave signal propagation along rays.

In the case of a homogeneous isotropic medium the phase and the group velocities coincide. In this case equation (1.33) provides two different solutions for c^2 only. They correspond to seismic P- and S-waves. Also a decomposition into rotational and dilatational parts of the displacement field (an example of such a decomposition will be shown later in our discussion of poroelastic wavefields) shows that equation (1.28) describes these two independently propagating elastic waves: the longitudinal P-wave and the shear S-wave with propagation velocities c_p and c_s, respectively:

$$c_p = \sqrt{\frac{\lambda + 2\mu}{\rho}}, \quad c_s = \sqrt{\frac{\mu}{\rho}}. \tag{1.36}$$

Note that similarly to K and μ, the so-called P-wave modulus, $c_{11} = \lambda + 2\mu = K + \frac{4}{3}\mu$, is always positive.

1.1.6 Point sources of elastic waves

Above we considered plane waves as elementary components of elastic wavefields. Solutions of the linear equation system (1.28) can also be obtained as a superposition of waves radiated by point sources. A rather general formulation of the superposition principle for a linear system (including the linear equation system of poroelasticity considered in the next chapter) is as follows. Let us assume that we know a (possibly tensorial) solution $G(\mathbf{r}, \mathbf{r_0}, t - t_0)$ for a given field quantity satisfying such a system with a source term equal to the Dirac function $\delta(\mathbf{r_0})\delta(t_0)$ (i.e. a singular source concentrated in point r_0 and time moment t_0). Such a solution is called a Green's function. Let us further assume that, in reality, the source terms are distributed in space and time, as a known source-density function $f_s(\mathbf{r}, t)$ describes. Then the corresponding field quantity produced by this source will be equal to the following superposition:

$$\int_{-\infty}^{\infty} G(\mathbf{r}, \mathbf{r_0}, t - t_0) f_s(\mathbf{r_0}, t_0) d^3\mathbf{r_0} dt_0. \tag{1.37}$$

Sometimes it is convenient to work with source terms having Heaviside-type temporal distributions $h(t)$ (where $h(t) = 0$ if $t < 0$ and $h(t) = 1$ if $t \geq 0$). Note that such Green's functions require their differentiation in respect to time (or alternatively, temporal differentiation of the function $f_s(\mathbf{r}, t)$) before applying the superposition (1.37).

Let us consider point-like distributions of instantaneous body-force densities in an isotropic homogeneous elastic medium. To complete the force balance a body-force source must be added to the left-hand part of the elastodynamic equation (1.28). Such a singular source $\mathbf{F}$ acting in direction $\hat{\mathbf{x}}_j$ can be expressed in the following form:

$$\mathbf{F}(\mathbf{r}, t) \propto \hat{\mathbf{x}}_j \delta(\mathbf{r}) \delta(t). \tag{1.38}$$

In the far field of this source (at the distance larger than several wavelengths) the wavefield is given by the following approximate Green's function (see chapter 4 of Aki and Richards, 2002):

$$G_{ij} \equiv u_i(\mathbf{r}, t) \propto \gamma_i \gamma_j \frac{\delta(t - r/c_p)}{4\pi \rho c_p^2 r} + (\delta_{ij} - \gamma_i \gamma_j) \frac{\delta(t - r/c_s)}{4\pi \rho c_s^2 r}, \tag{1.39}$$

where $r = |\mathbf{r}|$, and the quantities $\gamma_i = r_i/r$ are the direction cosines. The first term on the right-hand part describes a spherical P-wave. The second one describes a spherical S-wave.

1.1.7 Static equilibrium

Let us assume that elastic waves have had enough time to establish an equilibrium state in a given body. Then displacement vectors and strains will become independent of time. Therefore, an equation of elastic equilibrium can be obtained from equation (1.28) by removing the temporal derivatives of the displacements (i.e. by taking the static limit):

$$\frac{\partial}{\partial x_j} C_{ijkl} \frac{\partial u_k}{\partial x_l} = 0. \tag{1.40}$$

Note that in terms of the stress tensors

$$\frac{\partial}{\partial x_j} \sigma_{ij} = 0. \tag{1.41}$$

Using equation (1.26) for homogeneous isotropic media we obtain:

$$\lambda \frac{\partial \epsilon}{\partial x_i} + 2\mu \frac{\partial \epsilon_{ij}}{\partial x_j} = 0. \tag{1.42}$$

The second term on the left-hand side of this equation can be written explicitly as follows:

$$2\mu \frac{\partial \epsilon_{ij}}{\partial x_j} = \mu \frac{\partial(\partial u_i/\partial x_j + \partial u_j/\partial x_i)}{\partial x_j} = \mu \left(\frac{\partial^2 u_i}{\partial x_j^2} + \frac{\partial^2 u_j}{\partial x_i \partial x_j} \right). \tag{1.43}$$

The third part of this equation is equal to the ith component of the vector $\mu(\nabla^2\mathbf{u} + \nabla(\nabla u))$, where we have introduced a differential vector operator

$$\nabla \equiv (\partial/\partial x_1; \partial/\partial x_2; \partial/\partial x_3). \tag{1.44}$$

Note that with this notation:

$$\epsilon = \nabla\mathbf{u}. \tag{1.45}$$

Here and later we use also the Laplace operator

$$\nabla^2 \equiv \frac{\partial}{\partial x_i}\frac{\partial}{\partial x_i} \equiv \frac{\partial^2}{\partial x_1^2} + \frac{\partial^2}{\partial x_2^2} + \frac{\partial^2}{\partial x_3^2}. \tag{1.46}$$

Therefore, equation (1.42) has the following vectorial form:

$$(\lambda + \mu)\nabla\epsilon + \mu\nabla^2\mathbf{u} = 0, \tag{1.47}$$

Further we apply a known relation of the vector calculus:

$$\nabla(\nabla u) = \nabla^2\mathbf{u} + \nabla \times (\nabla \times \mathbf{u}). \tag{1.48}$$

Using this we obtain

$$(\lambda + 2\mu)\nabla\epsilon - \mu\nabla \times (\nabla \times \mathbf{u}) = 0. \tag{1.49}$$

Applying here once more the divergence operator (defined by equation 1.44) we obtain

$$\nabla^2\epsilon = 0. \tag{1.50}$$

Thus, we see that, in homogeneous isotropic media under equilibrium without body forces, the dilatation is a harmonic function (i.e. it is a solution of the Laplace equation). Also applying the Laplace operator directly to (1.47) yields:

$$\nabla^2\nabla^2\mathbf{u} = 0. \tag{1.51}$$

This is the so-called biharmonic equation for the displacement vector under static equilibrium conditions for an elastic medium (see Landau and Lifshitz, 1987).

1.2 Geomechanics of seismic events

Earthquakes are usually assumed to occur due to spontaneous quick shear motions (shear failure) along tectonic faults and cracks. A rather simple and practical formulation of conditions of such a brittle failure of rocks is based on Anderson's theory of faulting published in 1905. Anderson's theory was in turn based on earlier studies of Amontons in 1699 and Coulomb in 1773 on rock mechanics (Lay and Wallace, 1995). This formulation follows from a consideration of tectonic stresses.

1.2.1 Faults and principal stresses

Let us consider a tensor of tectonic stresses in its principal coordinate system. In such a coordinate system the traction vectors acting on the coordinate planes are directed along the normals to these planes. In other words, these traction vectors are parallel to the coordinate axes normal to the corresponding coordinate planes. These traction vectors are usually called principal tectonic stresses.

Frequently in tectonics and geomechanics the following sign notation is used for stresses: compressive stresses are positive, and a pore pressure is positive. However, in continuum mechanics it is usually assumed that compressive stresses are negative, and the pore pressure is positive. As was noted in the previous sections, in this book we accept the standard continuum mechanic sign-notation system, i.e. compressive stresses (a usual tectonic load is compressive) must be negative.

We will denote the projections of the principal stresses onto the corresponding (parallel to them) coordinate axes (note that each of the principal stresses has only one such non-vanishing projection) by $-\sigma_1$, $-\sigma_2$ and $-\sigma_3$, respectively. In this notation, in the case of compressive stresses, the algebraic quantities σ_1, σ_2 and σ_3 are positive. We will order the principal stresses so that the quantities σ_1, σ_2 and σ_3 will be in the following relations:

$$\sigma_1 \geq \sigma_2 \geq \sigma_3. \tag{1.52}$$

In the case of compressive stresses the quantities σ_1, σ_2 and σ_3 are equal to the absolute values of the maximum, the intermediate and the minimum compressive tectonic stresses, respectively.

Correspondingly, in the principal coordinate system, from the six independent components of the stress tensor σ_{ij} only three (diagonal) ones are non-vanishing: σ_{xx}, σ_{yy} and σ_{zz}. It is usually observed that one of the principal tectonic stresses is vertical. Then, x, y and z denote the two horizontal and one vertical coordinate axes, respectively. For example, in the normal faulting regime the maximum compressive stress σ_1 is vertical, and thus:

$$\sigma_{zz} = -\sigma_1, \quad \sigma_{yy} = -\sigma_2, \quad \sigma_{xx} = -\sigma_3. \tag{1.53}$$

In the strike-slip faulting regime the maximum compressive stress $\boldsymbol{\sigma_1}$ and the minimum compressive stress $\boldsymbol{\sigma_3}$ are both horizontal:

$$\sigma_{zz} = -\sigma_2, \quad \sigma_{yy} = -\sigma_3, \quad \sigma_{xx} = -\sigma_1. \tag{1.54}$$

Finally, in the thrust faulting regime the minimum compressive stress $\boldsymbol{\sigma_3}$ is vertical:

$$\sigma_{zz} = -\sigma_3, \quad \sigma_{yy} = -\sigma_2, \quad \sigma_{xx} = -\sigma_1. \tag{1.55}$$

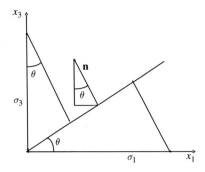

Figure 1.3 A fault and the principal stresses. The fault is normal to the plane of the maximum and minimum compressive stresses σ_1 and σ_3. It is shown as the line moving out at about 35° northeast from the origin. The compressive stresses σ_1 and σ_3 are shown as segments of the coordinate axes x_1 and x_3, respectively. The three right triangles are mutually similar and have the same angle θ.

In these three stress regimes potential fault surfaces (planes) are parallel to the direction of the intermediate principal compressive tectonic stress σ_2. Such a potential fault plane is normal to the plane of the maximum and minimum compressive stresses (σ_1, σ_3). Let us consider a potential fault plane making an angle θ with the coordinate axis of the maximum compression σ_1 (see Figure 1.3). A unit normal vector to this plane has the following components (note that we work in the principal-stress coordinate system):

$$n_1 = -\sin\theta, \quad n_2 = 0, \quad n_3 = \cos\theta. \tag{1.56}$$

Then a traction with the following components acts on this plane (we use equation (1.6) and our notations for the projection of principal stresses):

$$\tau_1 = \sigma_1 \sin\theta, \quad \tau_2 = 0, \quad \tau_3 = -\sigma_3 \cos\theta. \tag{1.57}$$

The component of this traction acting normal to the potential fault plane (we denote it as $-\sigma_n$, see below) is given by a scalar product between the plane's normal (1.56) and the traction (1.57):

$$-\sigma_n = -\sigma_1(\sin\theta)^2 - \sigma_3(\cos\theta)^2 = -\frac{\sigma_1+\sigma_3}{2} + \frac{\sigma_1-\sigma_3}{2}\cos(2\theta). \tag{1.58}$$

This quantity, $-\sigma_n$, is a normal stress acting onto the fault plane. Note that if principal stresses are compressive then the normal stress is compressive too. Thus, it is negative, and $\sigma_n > 0$. This explains our choice of the sign in front of σ_n in equation (1.58).

A unit vector $\mathbf{e}$ located in plane (x; z) and directed along the fault plane has the following components:

$$e_1 = \cos\theta, \quad e_2 = 0, \quad e_3 = \sin\theta. \tag{1.59}$$

Thus, the traction component τ_t tangential to the fault plane is

$$\tau_t = \frac{\sigma_1 - \sigma_3}{2} \sin(2\theta). \tag{1.60}$$

This is a shear stress acting on the fault plane.

Note that from equations (1.58) and (1.60) it follows that

$$\left(\sigma_n - \frac{\sigma_1 + \sigma_3}{2}\right)^2 + \tau_t^2 = \frac{(\sigma_1 - \sigma_3)^2}{4}. \tag{1.61}$$

This is the equation of a circle in the coordinate system (σ_n, τ_t) with the center at the point $(\sigma_m; 0)$, where

$$\sigma_m \equiv \frac{1}{2}(\sigma_1 + \sigma_3) \tag{1.62}$$

is the mean compressive stress. Equation (1.61) defines the so-called Mohr circle (see Figure 1.4). The diameter of Mohr's circle is equal to the differential stress:

$$\sigma_d \equiv \sigma_1 - \sigma_3. \tag{1.63}$$

The angle between a radius of the Mohr's circle and the negative direction of the σ_n-axis is equal to 2θ. The maximum possible shear stress on a fault plane is equal to $\sigma_d/2$. It corresponds to the angle $\theta = 45°$.

On the other hand, a friction-like force (per unit surface) acting along a potential fault plane and keeping such a fault stable (i.e. non-sliding) is given by Amontons' law:

$$\tau_f = \mu_f \sigma_n, \tag{1.64}$$

where μ_f is the friction coefficient. Note that, in the case of brittle failure of an intact rock, this coefficient is addressed as a coefficient of internal friction. In

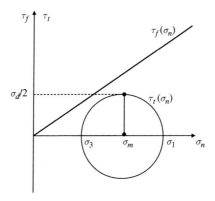

Figure 1.4 The shear stress and the friction force per unit surface of a fault plane as functions of the normal stress. Note that the center of the circle has coordinates $(\sigma_m; 0)$. Its top point has coordinates $(\sigma_m; \sigma_d/2)$.

this case it is usually higher than the friction coefficient on a pre-existing fault surface. Observation indicate that the induced seismicity occurs mainly on pre-existing defects of rocks (see, for example, Zoback, 2010, and further chapters of this book). Such defects (i.e. faults and cracks) will be destabilized if the absolute value of the shear stress $|\tau_t|$ exceeds the friction τ_f.

Frequently one expects that even in the case of a vanishing normal stress such a destabilization still requires a finite amount of shear stress C_c. Thus, the failure criterion (also called the Mohr–Coulomb failure criterion) for an arbitrary angle θ can be formulated as follows:

$$|\tau_t| = C_c + \tau_f. \tag{1.65}$$

The quantity C_c is called the cohesive strength (or just the cohesion). In the case of a failure of a pre-existing fault surface the cohesion is frequently neglected (see chapter 4 of Zoback, 2010). Finally, in the presence of fluids in pores of rocks the normal stress will be effectively reduced due to the pore pressure P_p and equation (1.65) takes the following form (see also our discussion of the Terzaghi's effective stress in the next chapter after equation (2.280)):

$$|\tau_t| = C_c + \mu_f(\sigma_n - P_p). \tag{1.66}$$

The pore pressure equally reduces all three principal stresses to corresponding effective principal stresses. Thus, it is convenient to consider Mohr's circle in the new coordinate system $(\sigma_n - P_p, \tau_t)$. Note that this coordinate transformation does not change differential and shear stresses.

Let us firstly consider the half-space $(\sigma_n - P_p, \tau_t)$ with positive shear stresses. Equation (1.66), expressed in geometric terms, is shown in Figure 1.5. There are three possible relative locations of the straight line

$$\tau_{fc} = C_c + \mu_f(\sigma_n - P_p) \tag{1.67}$$

expressing the sum of the cohesion stress with the friction force and of the semi-circle

$$|\tau_t| = \sqrt{(\sigma_1 - \sigma_n)(\sigma_n - \sigma_3)} \tag{1.68}$$

(see equation (1.61)) expressing the shear stress. They can have no intersections. In this case the friction force is too large and no faults will be destabilized. Another possibility is that the straight line and the semi-circle are intersect at two points. In this case the shear stress is more than sufficient to destabilize faults having orientations corresponding to the circle arc above the straight line. Finally, a situation is possible where the straight line just touches the circle in a single point. This means that the shear stress is exactly sufficient to destabilize faults of one special (optimal) orientation. The corresponding angle θ_{opt} can be obtained from a simple

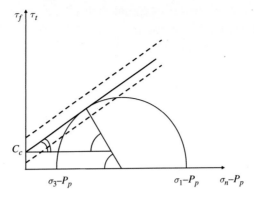

Figure 1.5 A friction–stress diagram taking into account a cohesion and a pore pressure acting on faults. The solid lines show force and angle relations at optimally oriented faults. The dashed lines show friction forces that are too large or too small in respect to the one acting on the optimally oriented fault.

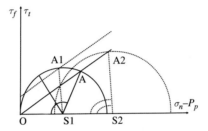

Figure 1.6 A friction–stress diagram for compressive stresses. The solid lines show force and angle relations for faults oriented at the lock-up angle. A doubled lock-up angle is given by the angle OS1A. The dashed lines show deviating situations. They indicate that corresponding angles between the largest compressive stress direction and faults (doubled these angles are given by the angles OS1A1 and OS2A2) are smaller than the lock-up angle.

geometric consideration. Recalling that in Figure 1.5 the friction coefficient μ_f gives a tangent of the friction straight line (i.e. $\mu_f = \tan\phi_f$) we obtain:

$$\theta_{opt} = \frac{1}{2}\arctan\frac{1}{\mu_f}. \tag{1.69}$$

Note that an equivalent consideration of the half-space ($\sigma_n - P_p, \tau_t$) with negative shear stresses will provide the same angle but with the negative sign.

Usually the friction coefficient on a pre-existing fault surface is between 0.6 and 1. Thus, $\theta_{opt} = 30° - 22.5°$. Figure 1.6 shows that for any compressive stress state the faults with $\theta > 2\theta_{opt}$ cannot slip. The quantity $2\theta_{opt}$ is called the lock-up angle. It is reported that faults with θ exceeding the lock-up angle are extremely rare (see Kanamori and Brodsky, 2004).

At the optimal orientations of potential fault planes the normal and tangential stresses can be explicitly expressed in terms of the friction coefficient. Indeed, one can substitute the expressions $\cot(2\theta_{opt}) = \mu_f = \tan\phi_f$, $\sin(2\theta_{opt}) = 1/\sqrt{1+\mu_f^2} = \cos\phi_f$ and $\cos(2\theta_{opt}) = \mu_f/\sqrt{1+\mu_f^2} = \sin\phi_f$ into equations (1.58) and (1.60):

$$-\sigma_n = -\sigma_m + \frac{\sigma_d}{2}\sin\phi_f. \tag{1.70}$$

$$\tau_t = \frac{\sigma_d}{2}\cos\phi_f. \tag{1.71}$$

The Mohr–Coulomb failure criterion (1.66) at the optimal angle takes then the following form:

$$\frac{1}{2}\sigma_d = C_c\cos\phi_f + \sin\phi_f(\sigma_m - P_p). \tag{1.72}$$

The difference between the left- and right-hand sides of this equation represents a failure criterion stress, FCS:

$$FCS = \frac{1}{2}\sigma_d - \sin\phi_f(\sigma_m - P_p) - C_c\cos\phi_f. \tag{1.73}$$

If FCS is negative in the vicinity of a given location, then the rock will be stable at this location. If changes of a failure criterion stress ΔFCS are positive, the rock will become less stable there. If the failure criterion stress becomes positive, the corresponding fault plane becomes unstable and brittle failure occurs: the fragments of rocks contacting along this fault plane slip relative to each other.

1.2.2 Friction coefficient

The Mohr–Coulomb criterion of rock failure is an approximation and simplification of reality. For example, a time-independent coefficient of friction is a strong assumption. The coefficient of internal friction introduced in the previous section does not describe the friction force along an already (and possibly long geological time) existing interface between two rock units. Theoretically the internal friction coefficient addresses friction along a potential internal plane in intact rocks before the failure occurs on this plane (Jaeger *et al.*, 2007). On the other hand, realistically, intact rocks do not exist, and failures probably occur along pre-existing zones of weakness. Moreover, the empirically obtained mathematical forms of the failure criterion of a plane of weakness and of the Mohr–Coulomb criterion coincide (see Jaeger *et al.*, 2007, equations 3.20 and 4.5). Thus, friction effects in both situations must be at least similar.

A great deal of active research in modern seismology has been dedicated to investigating the nature of the friction along pre-existing interfaces in rocks. A comprehensive review on the subject can be found in the last two chapters of the book of Segall (2010). The most important results are usually summarized by a system of empirically derived equations (Dieterich, 1978; Ruina, 1983) describing the dynamic friction coefficient μ_{fd} along surfaces sliding with the slip velocity v (Segall, 2010):

$$\mu_{fd} = \mu_{f0} + a_f \ln \frac{v}{v_0} + b_f \ln \frac{\theta_f}{\theta_{f0}}, \tag{1.74}$$

where μ_{f0} is a friction coefficient (frequently assumed to be close to 0.6) at a reference slip velocity v_0 and at the quantity θ_f (a so-called state variable) equal to its reference value θ_{f0}. The quantities a_f and b_f are empiric dimensionless constants defining the behavior of the dynamic friction.

The state variable expresses the temporal behavior of the friction due to damage accumulation at the friction interface. Alternatively, one of the following two differential equations (they have also been empirically derived) describing the state variable as a function of time can be used. The first one is called the aging law:

$$\frac{d\theta_f}{dt} = 1 - \frac{v\theta_f}{d_c}, \tag{1.75}$$

where d_c stands for the critical slip required for the friction coefficient to obtain its value by steady-state sliding after an abrupt change of the slip velocity. This model proposes that the state variable is just linearly mapped to the time in the case of $v = 0$ (i.e. a non-slipping contact).

Another model frequently used for describing the state variable is called the slip law:

$$\frac{d\theta_f}{dt} = -\frac{v\theta_f}{d_c} \ln \left(\frac{v\theta_f}{d_c} \right). \tag{1.76}$$

Both of these models predict that, in steady-state sliding, $d\theta_f/dt = 0$ and $\theta_f = d_c/v$. Assuming that $\theta_{f0} = d_c/v_0$ they yield:

$$\mu_{fd} = \mu_{f0} + (a_f - b_f) \ln \frac{v}{v_0}. \tag{1.77}$$

Thus, if the friction-law parameters a_f and b_f are such that $a_f > b_f$, then increasing velocity will lead to increasing friction. The fault tends to become more stable. This situation is called velocity strengthening. In the case of $a_f < b_f$ an increasing sliding velocity will decrease the friction and contribute to fault destabilization. Such a situation is called velocity weakening.

Usually variations of the friction coefficient related to the state and rate processes are considered to be of the order of 10%. They can be well hidden by the elastic heterogeneity and stress variations in real rocks and contribute to the randomness of the FCS distribution in space (and also in time).

1.2.3 Growth of finite cracks: a sufficient condition

Another simplification of the Mohr–Coulomb failure criterion is an assumption of a homogeneous stress state in the whole continuum. Thus, a destabilization condition of a part of a fault plane will be equally valid for the whole infinite corresponding fault plane. However, in reality, sizes of rock failures are finite. A sufficient condition of the unstable crack growth would be that a critical tensile stress must be reached in the vicinity of the fracture tip.

Based on this assumption and considering a finite elliptical two-dimensional (2D) initially tensile crack in the plane of the maximum and minimum stresses, Griffith (1924) derived the following criterion (see the derivation given in section 10.9 of Jaeger *et al.*, 2007):

$$(\sigma_1 - \sigma_3)^2 - 8T_0(\sigma_1 + \sigma_3) = 0 \quad \text{if} \quad \sigma_1 > -3\sigma_3, \tag{1.78}$$

and

$$\sigma_3 + T_0 = 0 \quad \text{if} \quad \sigma_1 < -3\sigma_3, \tag{1.79}$$

where the rock property T_0 is called the uniaxial tensile strength.

In terms of centers and diameters of Mohr's circles (i.e. in terms of σ_m and σ_d, respectively) these equations take the following forms:

$$\sigma_d^2 - 16T_0\sigma_m = 0 \quad \text{if} \quad 4\sigma_m > \sigma_d, \tag{1.80}$$

and

$$\sigma_m - \frac{1}{2}\sigma_d + T_0 = 0 \quad \text{if} \quad 4\sigma_m < \sigma_d. \tag{1.81}$$

In these terms the Mohr's circle equation (1.61) can be written as

$$f_M(\sigma_m) \equiv (\sigma_n - \sigma_m)^2 + \tau_t^2 - \frac{1}{4}\sigma_d^2 = 0. \tag{1.82}$$

The envelope of the Mohr circles, $\tau_t(\sigma_n)$ satisfying criterion (1.80), can be found from this criterion and two additional equations (see section 10.9 of Jaeger *et al.*, 2007):

$$f_M(\sigma_m) = 0, \quad \frac{\partial f_M(\sigma_m)}{\partial \sigma_m} = 0. \tag{1.83}$$

This envelope provides a form analogous to equation (1.65) for expressing the failure criterion for optimally oriented cracks:

$$\tau_t^2 = 4T_0(\sigma_n + T_0). \tag{1.84}$$

Using equation (1.81) for the Mohr envelope provides just its point $(-T_0; 0)$ on the plane $(\sigma_n; \tau_t)$. This is a tip of parabola (1.84).

McClintock and Walsh (1962) included slip friction in the case of a crack closing under normal compressive stress with the absolute value σ_c. For the case of a compressive normal stress $\sigma_n > 0$ they derived a modified Griffith criterion (for tensile stresses the original Griffith criterion expressed by the equation above is still applied):

$$\sigma_1(\sqrt{1 + \mu_f^2} - \mu_f) - \sigma_3(\sqrt{1 + \mu_f^2} + \mu_f) = 4T_0\sqrt{1 + (\sigma_c/T_0)} - 2\mu_f\sigma_c. \quad (1.85)$$

In terms of parameters of the Mohr circles this criterion has the following form:

$$\sigma_d\sqrt{1 + \mu_f^2} - 2\mu_f\sigma_m = 4T_0\sqrt{1 + (\sigma_c/T_0)} - 2\mu_f\sigma_c. \quad (1.86)$$

Using this equation and conditions (1.83) leads to a further modification of the Mohr envelope (note that conditions $\tau_t > 0$ and $\sigma_n < \sigma_m$ must additionally be taken into account):

$$\tau_t = 2T_0\sqrt{(1 + \sigma_c/T_0)} + \mu_f(\sigma_n - \sigma_c). \quad (1.87)$$

For $\sigma_c = 0$ and $T_0 = C_c/2$ this equation reduces to the Mohr–Coulomb criterion (1.65). Note that this result is based on the sufficient condition of the unstable crack growth. Therefore, the result of McClintock and Walsh (1962) can be considered as a substantiation of the empiric Mohr–Coulomb criterion. To some extend this clarifies the nature of the constants μ_f and C_c (see Scholz (2002) and references therein; equation 1.42 of Scholz (2002) is analogous to our equation above; note also a misprint in the equation of Scholz (2002): an erroneous factor 2 in front of the friction coefficient on its right-hand side).

1.2.4 Necessary conditions of crack growth and some results of fracture mechanics

The mechanics of rock failure investigates conditions of their destabilization and dynamics. This is a broad field of scientific research. Here we provide a brief overview of some results that we use in this book. For a more comprehensive treatment of the subject the reader should consult such books as Segall (2010), Jaeger *et al.* (2007), Scholz (2002) and references therein.

We consider a plane penny-shaped crack of radius $X/2$ or a 2D crack infinite in the x_2 direction and of length X in the x_1 direction. Griffith (1921, 1924) formulated conditions for a growth of finite cracks based on an energy-balance consideration. For a crack of length X to be in an equilibrium in a rock sample under a loading stress τ_l (this stress must be specified correspondingly to the loading configuration) the following energy balance must be fulfilled:

$$\delta E_s = \delta E_e + \delta E_l. \tag{1.88}$$

Here $\delta E_s = \gamma_c \delta S$ is an energy increment needed to create a new crack surface δS with the surface energy per unit area γ_c. For the crack to grow this energy increment must be equal to the sum of energy contributions supplied by the rock sample itself and by a system loading the sample. These are a loss of the sample's elastic energy δE_e and a loss of the energy of the loading system δE_l, respectively. The loss of the elastic energy δE_e occurs due to a decrease of the sample's elastic stiffness G_e caused by the growth of the crack. The stiffness G_e must be specified correspondingly to the loading configuration and to the geometry of a crack. The loss of the energy δE_l of the loading system occurs due to the work of this system spent for deforming the sample during the growth of the crack. Both of these energy changes result from the work of the loading forces along the crack surface. Thus, together they should be proportional to the length of the crack X, to its surface increment δS, as well as to the density of the elastic energy in the sample, $\tau_l^2/(2G_e)$. Therefore, the right-hand side of equation (1.88) must be proportional to $X\delta S\tau_l^2/G_e$.

If the energy balance (1.88) is disturbed so that its left-hand side is larger than the right-hand side, then the crack will not receive sufficient energy to grow. On the other hand, if the right-hand side of equation (1.88) is larger than the left-hand side, then the crack will grow unstably and the sample will fail. The critical crack length $X = 2a_c$ and the critical stress $\tau_l = \tau_c$ are given by the balance condition (1.88):

$$\gamma_c \frac{G_e}{C_g} = a_c \tau_c^2, \tag{1.89}$$

where C_g is a proportionality coefficient. For different modes of cracks, i.e. a tensile crack (opening mode $J = \text{I}$), a longitudinal-shear crack (in-plane shear mode $J = \text{II}$) and a transverse-shear crack (anti-plane shear mode $J = \text{III}$), and different load configurations this coefficient is usually of the order of 1. For example (see Jaeger *et al.*, 2007, p. 309), for a tensile load of a thin sample with a tensile (2D) crack under plane strain conditions, $G_e = 2\mu(1 + \nu)$ is the initial (pre-fractured) Young's modulus of the rock, and $C_g = \pi(1 - \nu^2)/2$, with μ and ν being the rock initial shear modulus and Poisson's ratio, respectively. For the plane stress 2D loading configuration, $G_e = 2\mu(1+\nu)$ and $C_g = \pi/2$. For a tensile load of a sample with a three-dimensional (3D) penny-shaped crack, G_e is again the Young's modulus and $C_g = \pi(1 - \nu^2)/4$. For a 2D transverse-shear crack under anti-plane strain conditions, $G_e = \mu$ and $C_g = \pi/4$ (see Kanamori and Brodsky, 2004, and references therein).

The left-hand side of equation (1.89) contains a combination of rock physical properties only. The right-hand side of this equation contains a combination of the crack length and the loading stress. For an arbitrary load τ_l and crack length X the quantity

$$K_J = \tau_l \sqrt{C_J X / 2} \tag{1.90}$$

is called the stress intensity factor. Here C_J is a proportionality coefficient depending on the crack mode J. For example (Rice, 1980), $C_J = \pi$ for plane 2D cracks (such cracks are infinite in their symmetry plane along the direction normal to the propagation direction of their tips). However, this coefficient is more complex for a circular crack.

From the linear elastic fracture mechanics (see Rice, 1980; and Jaeger *et al.*, 2007, pp. 311–314 and further references therein) it follows that on the plane of the crack ahead of its tip the local stress perturbations produced by the fracture, σ_J, are given by

$$\sigma_J = \frac{K_J}{\sqrt{2\pi r}}, \tag{1.91}$$

where r is a distance from the crack tip.

Another result of this theory is that the crack-surface displacement Δu (absolute value of the relative displacement of points across the crack) behind the crack tip is proportional to $\tau_l \sqrt{(X^2/4) - x^2}/G_e$, where $|x| \leq X/2$ is the distance from the crack center along the crack plane. For example, for modes I and II of 2D cracks, Δu is given by the following expression (Rice, 1980; Scholz, 2002; Segall, 2010):

$$\Delta u = 2\frac{\tau_l(1 - \nu)}{\mu}\sqrt{(X^2/4) - x^2}, \tag{1.92}$$

and for mode III:

$$\Delta u = 2\frac{\tau_l}{\mu}\sqrt{(X^2/4) - x^2}. \tag{1.93}$$

From the integration of a product of Δu taken in the vicinity of the propagating crack tip with the stress (1.91), one obtains the energy increment Υ released in the process of the crack growth by the local stress forces per unit increment in the crack length and per unit thickness in the x_2 direction (see, for example, Jaeger *et al.*, 2007):

$$\Upsilon = 2\frac{C_g K_J^2}{C_J G_e}. \tag{1.94}$$

For the crack to grow this energy increment must be equal to, or exceed, the energy increment required for creation of the new free surface:

$$\Upsilon_c = 2\gamma_c \tag{1.95}$$

This will again give condition (1.89). This will also give a critical value for the stress intensity factor:

$$K_c = \sqrt{\gamma_c C_J G_e / C_g}. \tag{1.96}$$

Sometimes K_c is called the fracture toughness. For modes I and III values of K_c are from about 0.1 MPa $\times$ m$^{1/2}$ for coal up to about 3.5 MPa $\times$ m$^{1/2}$ for granite and gabbro (see Scholz, 2002, and Jaeger *et al.*, 2007, and references therein).

Griffith's energetic criterion is a necessary rather than a sufficient condition of unstable crack growth. On the other hand, the sufficient condition (we have considered it in the previous section) and the necessary condition are rather close. In other words, creating tensile stresses of the order of critical ones simultaneously leads to satisfying the thermodynamic Griffith criterion (see Scholz, 2002, pp. 6–7) and vice versa.

The fact that under sufficiently high stress the atomic bonds will be broken and the material in the vicinity of the crack tip must behave inelastically means that the presence of the stress singularity in the results of the linear elastic fracture mechanics of the type (1.91) is an artefact of the theory. There exist theoretical approaches taking into account inelastic yielding at the crack tip (see Kanamori and Brodsky, 2004; Segall, 2010, and references therein). However, results of the linear elastic fracture mechanics are applicable at distances $r \gg l_y$, where l_y is the scale of the region of inelastic deformations. In seismology the necessary criterion (1.95) is often generalized to take into account the energy required for creating damaged zones (Kanamori and Brodsky, 2004). This criterion does not take into account seismic radiation occurring by dynamic propagation of the rupture. In this sense it is a condition for a quasi-static growth of the crack. The criterion for dynamic crack extension must also take into account the energy required for radiating seismic wavefields depending on the rupture velocity (Rice, 1980; Kanamori and Brodsky, 2004). Moreover, recent studies (Bouchbinder *et al.*, 2014) show the existence of a length scale l_n, where non-linear elastic effects are of importance. Scale l_n is a function of the crack velocity. It is shown that $l_n > l_y$. Thus, results of linear elastic fracture mechanics are applicable for $r \gg l_n$.

1.2.5 Earthquake motions on faults

The considerations related to equations (1.88)–(1.95) assume that the loading stress is equal to τ_l and the stress behind the tip of the crack is equal to 0. For earthquake mechanics, situations similar to the one considered by McClintock and Walsh (1962) are especially interesting: the crack surface displacement leads not to the complete stress release but rather to a decrease of the stress to the level τ_f supported by friction forces. Such situations are more relevant for the crack modes II and III. In this case the stress τ_l must be replaced by the stress drop $\Delta\sigma$:

$$\Delta\sigma = \tau_l - \tau_f. \tag{1.97}$$

Linear elastic fracture mechanics shows that the mean relative displacement of points across the crack $\overline{\Delta u}$ (given by the surface averaging of expressions like (1.92) and (1.93)) has the following relation to the stress drop (Jaeger *et al.*, 2007):

$$\overline{\Delta u} = \frac{X\Delta\sigma}{2C'\mu}, \tag{1.98}$$

where C' is a constant of the order of unity. For example, for a 2D plane vertical strike-slip fault extending from depth X to the free surface, $C' = 2/\pi$. For a circular of radius $X/2$ rupture surface, $C' = 3\pi(2-v)/(16(1-v)) = 7\pi/16$, where the last part assumes $v = 0.25$.

The quantity Δu can be related to the strain distributed over a crack zone. Let us firstly consider the following integral over a surface Ψ_c closely surrounding the crack:

$$\eta_{ij}^{\Psi} = \int_{\Psi_c} \frac{1}{2}(u_i n_j + u_j n_i)d^2\mathbf{r}, \tag{1.99}$$

where $n_j(\mathbf{r})$ are components of the outward normal of Ψ_c and $u_i(\mathbf{r})$ are displacement components at a given point $\mathbf{r}$ of Ψ_c.

In the case of a continuous elastic body with the surface Ψ_c (i.e. a differentiable displacement is given at all its points), Gauss' theorem gives

$$\eta_{ij}^{\Psi} = \int_{V_{\Psi}} \frac{1}{2}\left(\frac{\partial u_i}{\partial x_j} + \frac{\partial u_j}{\partial x_i}\right)d^3\mathbf{r}. \tag{1.100}$$

The integrand here is the strain tensor, and V_{Ψ} is the volume of rocks inside the surface Ψ_c. Thus, $\eta_{ij}^{\Psi}/V_{\Psi}$ is the volume-averaged strain.

In the case of a displacement discontinuity on a crack surface, integral (1.99) still represents the volume-integrated strain in the crack zone. To show this one considers a zero-thickness limit of the crack zone (see Rice, 1980). Let us denote by $+$ and $-$ the sides of the crack. At the zero-thickness limit these sides will geometrically coincide with each other and with the corresponding sides of the surface Ψ_c. The normals to the sides will point in opposite directions. Then:

$$\begin{aligned}
\eta_{ij}^{\Psi} &= \int_{\Psi_{c+}} \frac{1}{2}(u_i^+ n_j^+ + u_j^+ n_i^+)d^2\mathbf{r} + \int_{\Psi_{c-}} \frac{1}{2}(u_i^- n_j^- + u_j^- n_i^-)d^2\mathbf{r} \\
&= \int_{\Psi_{c+}} \frac{1}{2}(u_i^+ n_j^+ + u_j^+ n_i^+)d^2\mathbf{r} - \int_{\Psi_{c-}} \frac{1}{2}(u_i^- n_j^+ + u_j^- n_i^+)d^2\mathbf{r} \\
&= \int_{\Psi_{c+}} \frac{1}{2}(\Delta u_i n_j^+ + \Delta u_j n_i^+)d^2\mathbf{r}.
\end{aligned} \tag{1.101}$$

Accordingly to conventional notation, the fault zone below a non-vertical fault (in our notation, the rock contacting the Ψ_{c-}-side of the fault) is called the foot wall.

The fault zone above the fault plane (in our notation it corresponds to the rock contacting the Ψ_{c+}-side of the fault) is called the hanging wall. Thus, the slip vector (and more generally, the fault-displacement vector, i.e. the components Δu_i) describes the displacement of the hanging wall relative to the foot wall.

An equivalence between quantities (1.101) and (1.100) in the zero-thickness limit of the crack zone results then from the following substitution (see Rice, 1980):

$$\epsilon_{ij}^{\Psi} \equiv \frac{1}{2}\left(\frac{\partial u_i}{\partial x_j} + \frac{\partial u_j}{\partial x_i}\right) = \frac{1}{2}(\Delta u_i n_j^+ + \Delta u_j n_i^+)\delta_D(\Psi_{c+}), \tag{1.102}$$

where $\delta_D(\Psi_c)$ is the surface Dirac function turning a volume integral into a surface integral over the discontinuity surface contained in the volume. This Dirac function has physical dimension inverse to length.

Substituting (1.102) into (1.100) and performing the volume integration yields

$$\int_{V_{\Psi}} \epsilon_{ij}^{\Psi} d^3\mathbf{r} = \int_{\Psi_{c+}} p_{ij}(\mathbf{r})d^2\mathbf{r}, \tag{1.103}$$

where the quantity

$$p_{ij} = \frac{1}{2}(\Delta u_i n_j^+ + \Delta u_j n_i^+) \tag{1.104}$$

is called the potency density tensor (Segall, 2010).

Note that, according to (1.101), the right-hand part of equation (1.103) coincides with η_{ij}^{Ψ}. Therefore, (1.99) and (1.100) become equivalent also in the case of a crack zone. Therefore, the quantity

$$\epsilon_{ij}^{\Psi} = p_{ij}\delta_D(\Psi_{c+}) \tag{1.105}$$

represents the strain in the crack zone.

The potency density tensor can be rewritten in the following form:

$$p_{ij} = \Delta u_0 \frac{1}{2}(\hat{u}_i n_j^+ + \hat{u}_j n_i^+), \tag{1.106}$$

where $\Delta u_0 = \sqrt{\Delta u_i \Delta u_i}$ is the absolute value of the relative displacement of points across the crack (also called the fault displacement). The unit vector with components $\hat{u}_i = \Delta u_i/\Delta u_0$ defines a direction of the fault-displacement vector.

If the fault-displacement vector represents just a slip along a planar crack surface then the vector $\hat{\mathbf{u}}$ will represent the slip direction of the Ψ_{c+}-side of the crack and the vectors $\hat{\mathbf{u}}$ and $\mathbf{n}^+$ will be orthogonal. Thus, $\hat{u}_i n_i^+ = 0$. Therefore, in the case of a pure slip event the trace of the potency density tensor is equal to zero. Further, $\hat{u}_i$ and n_i^+ are components of unit vectors. Finally, we know that vector $\mathbf{n}^+$ points into the upper half-space. These three conditions leave in the potency density tensor of a slip event four independent components only. They can be expressed over the

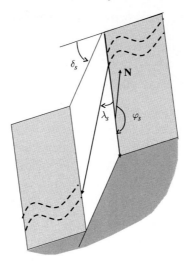

Figure 1.7 Angles defining fault and slip orientations. The rock shown to the right of the fault plane is the foot wall. The hanging wall is shown to the left of the fault plane. Note that the rake has a negative value for this situation.

absolute value of the slip Δu_0 and the following three angles. The first one (the strike ϕ_s) is the azimuth of the straight line representing the intersection of the free surface with the plane containing Ψ_{c+}. This angle is measured clockwise relative to due north toward the dipping direction of the fault. In other words, the strike direction is defined so that, for an observer looking into this direction, the fault surface is dipping to the right (see Figure 1.7).

The strike angle can take values from the range $0 \leq \phi_s \leq 2\pi$. The second one (the dip δ_s) is the angle between the plane Ψ_{c+} and the free surface ($0 \leq \delta_s \leq \pi/2$). The third one (the rake λ_s) is the angle measured from the strike direction of Ψ_{c+} to the direction of the slip vector. We follow here the notation accepted on pages 101–102 of Aki and Richards (2002) and assume that the rake takes values from the range $-\pi \leq \lambda_s \leq \pi$. Negative angles are measured clockwise from the strike direction. Note that for normal faults (such as the one shown in Figure 1.7) the rake has negative values. For reverse faults (thrust faults) it is positive. If slip vector is horizontal then one speaks about strike-slip faults.

Finally, in the case of a slip along the fault plane, the following useful relation for the components of the potency density tensor follows from (1.106):

$$\Delta u_0 = \sqrt{2 p_{ij} p_{ij}}. \tag{1.107}$$

Note that in a more general case of a fault displacement being not necessarily parallel to the surface Ψ_{c+} the tensor of potency density has five independent components. Thus, the four parameters mentioned before must be complemented by one more, for example, the angle between the vector of normal to Ψ_{c+} and the slip vector.

1.3 Elastic wavefields radiated by earthquakes

Growing cracks and slipping tectonic faults radiate elastic waves. A general approach to formulating these wavefields was proposed by Backus and Mulcahy (1976a,b). This approach was elegantly described by Rice (1980). Below we follow his description, slightly modifying it for further applications in this book.

Let us consider a tectonic domain (or a sample of a rock) of volume V and surface S with a fault (or a crack) inside. Elastic wavefields radiated by an earthquake on this fault contribute to the strain field $\epsilon_{ij}(\mathbf{r}, t)$ of the tectonic domain. During the earthquake, part of the fault suffers a deformation $\epsilon_{ij}^T(\mathbf{r}, t)$. One calls this deformation a transformation strain. Earthquakes release tectonic stress. Thus, the transformation strain is such that it will correspond to a zero-stress state ($\sigma_{ij} = 0$) of a fault vicinity, if it is made free (isolated) from the rest of the tectonic domain. Therefore, if a source region were isolated, it would be the case that:

$$\epsilon_{ij}(\mathbf{r}, t)|_{source}^{isolated} = \epsilon_{ij}^T(\mathbf{r}, t). \tag{1.108}$$

Using Hooke's law (1.7) we can write the following formulation for such a hypothetical zero-stress state:

$$0 = C_{ijkl}(\epsilon_{ij}(\mathbf{r}, t) - \epsilon_{ij}^T(\mathbf{r}, t))|_{source}^{isolated}. \tag{1.109}$$

This zero stress would be valid at any instant t because if the rupture starts to propagate the stress will be released. Therefore, the total strain would be equal to the transformation strain, as equation (1.108) states. Any additional strain would vanish.

However, an earthquake-source region is a part of a tectonic domain. Then, the right-hand part of equation (1.109) cannot be considered under conditions of dynamic isolation. Thus, it must represent not a zero stress but rather an altered stress state, where the strain is measured relative to the transformation strain:

$$\sigma_{ij} = C_{ijkl}\left(\epsilon_{kl}(\mathbf{r}, t) - \epsilon_{kl}^T(\mathbf{r}, t)\right). \tag{1.110}$$

Clearly, this stress state must satisfy the general elastodynamic equation (1.28):

$$\frac{\partial}{\partial x_j}\left(C_{ijkl}\left(\epsilon_{kl}(\mathbf{r}, t) - \epsilon_{kl}^T(\mathbf{r}, t)\right)\right) = \rho\frac{\partial^2 u_i}{\partial t^2}. \tag{1.111}$$

Let us compare this equation with the general elastodynamic equation (1.28) formulated for the case of a perturbation source in the form of a body-force density vector with components $F_i(\mathbf{r}, t)$ and vanishing transformation strain:

$$\frac{\partial}{\partial x_j}(C_{ijkl}\epsilon_{kl}(\mathbf{r},t)) + F_i(\mathbf{r},t) = \rho \frac{\partial^2 u_i}{\partial t^2}. \tag{1.112}$$

Thus, we conclude that, in the case of a perturbation source in the form of a transformation strain, the effective volumetric force density is given by the following expression:

$$F_i(\mathbf{r},t) = -\frac{\partial}{\partial x_j}m_{ij}(\mathbf{r},t), \tag{1.113}$$

where

$$m_{ij}(\mathbf{r},t) = C_{ijkl}\epsilon_{kl}^T(\mathbf{r},t) \tag{1.114}$$

is called the moment density tensor. Note that due to the symmetry of the tensor of elastic stiffnesses the moment density tensor is also symmetric: $m_{ij} = m_{ji}$.

Therefore, equation (1.113) allows us to compute the displacement field perturbation u_i caused by an earthquake using results of the elasticity theory derived for a perturbation source in the form of a body force. Thus, to obtain the displacement one integrates a product of the effective force (1.113) and the elastodynamic Green's function $G_{ij}(\mathbf{r},\mathbf{r}',t-t')$ over the volume V and time. Assuming that the altered stress (1.110) vanishes on the surface S of the tectonic domain and taking into account the symmetry of the moment density tensor this integration is reduced to the following result (Rice, 1980):

$$u_i(\mathbf{r},t) = \frac{1}{2}\int_{-\infty}^{t}\int_{V} H_{ijk}(\mathbf{r},\mathbf{r}',t-t')m_{jk}(\mathbf{r}',t')d^3\mathbf{r}'dt', \tag{1.115}$$

where the function

$$H_{ijk}(\mathbf{r},\mathbf{r}',t-t') = \frac{\partial G_{ij}(\mathbf{r},\mathbf{r}',t-t')}{\partial x_k'} + \frac{\partial G_{ik}(\mathbf{r},\mathbf{r}',t-t')}{\partial x_j'} \tag{1.116}$$

is a response of the medium to a perturbation source in the form of a singular (δ-function-like) concentration of the moment density. Taking into account that Green's function $G_{ij}(\mathbf{r},\mathbf{r}',t-t')$ is a displacement response u_i at the point $\mathbf{r}$ to a singular body-force density acting in the direction x_j at the location $\mathbf{r}'$, we see that $H_{ijk}(\mathbf{r},\mathbf{r}',t-t')$ is a displacement response u_i at $\mathbf{r}$ to two couples of singular body forces at $\mathbf{r}'$. Each of these force couples produces a response corresponding to one of the two terms on the right-hand side of equation (1.116). The first term corresponds to two forces acting in opposite directions along the axis x_j. However, they are applied at two differentially close points shifted in opposite directions from the point $\mathbf{r}'$ along the axis x_k as shown in Figure 1.8. In the second term the corresponding two forces act in opposite directions along the axis x_k. They are applied at two differentially close points spaced along the axis x_j. In the case of $j \neq k$ such a combination of the four forces (i.e. of the two couples of forces)

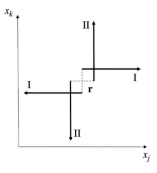

Figure 1.8 Double couple of forces corresponding to the Green function given by equation (1.116).

is called the double couple. In the case of $j = k$ the both force couples coincide and represent a dipole of doubled forces (note, however, the factor $1/2$ in front of the right-hand-part integral in equation (1.115)). Such a combination of forces is called the linear vector dipole. Note that all these combinations of two force couples have resulting rotational (angular) momentum (in respect to the source point $\mathbf{r}'$) exactly equal to zero. Thus we conclude that, in a general case (including also a generally anisotropic heterogeneous elastic medium), an earthquake radiates seismic wavefields as a combination of double couples and linear vector dipoles does. These double couples and linear vector dipoles can be distributed in a source region V of the earthquake and in a time t_s of the source process.

Using a standard approach of the linear system theory (e.g. forward and inverse temporal Fourier transforms with multiplication and division by the factor $i\omega$), equation (1.115) can be rewritten in an equivalent form with a response to a step function of time instead of the impulse response:

$$u_i(\mathbf{r}, t) = \frac{1}{2} \int_{-\infty}^{t} \int_{V} E_{ijk}(\mathbf{r}, \mathbf{r}', t - t') \frac{\partial m_{jk}(\mathbf{r}', t')}{\partial t'} d^3\mathbf{r}' dt', \qquad (1.117)$$

where

$$E_{ijk}(\mathbf{r}, \mathbf{r}', t) = \int_{0}^{t} H_{ijk}(\mathbf{r}, \mathbf{r}', t') dt' \qquad (1.118)$$

is the displacement response to a singular spatial distribution of double couples and linear vector dipoles being step functions of time.

In the low-frequency range, where wavelengths of radiated wavefields are much larger than the size of the source domain and at distances much larger than the source domain, we can neglect the r'-dependence of response $E_{ijk}(\mathbf{r}, \mathbf{r}', t)$ in integral (1.117) and obtain:

$$u_i(\mathbf{r}, t) = \frac{1}{2} \int_{-\infty}^{t} E_{ijk}(\mathbf{r}, \mathbf{r}', t - t') \frac{\partial M_{jk}(t')}{\partial t'} dt', \qquad (1.119)$$

where $\mathbf{r}'$ represents an (averaged) source location and

$$M_{ij}(t) = \int_V m_{ij}(\mathbf{r}, t)d^3\mathbf{r} \tag{1.120}$$

is a total moment tensor of the earthquake.

Equation (1.119) describes seismic wavefields radiated by a point earthquake source. Therefore, in the low-frequency limit, it is reasonable to assume that the duration time of the source process t_s is very short. It is very short also in respect to the wave propagation time from the source to observation points. Thus, in the low-frequency limit we are interested in observational times such that $t \gg t_s$. The moment tensor vanishes for times larger than t_s. Then the integral (1.119) becomes

$$u_i(\mathbf{r}, t) = \frac{1}{2}E_{ijk}(\mathbf{r}, \mathbf{r}', t)M_{jk}(t_s), \tag{1.121}$$

or, in the spectral domain:

$$\hat{u}_i(\mathbf{r}, \omega) = \frac{1}{2}\hat{E}_{ijk}(\mathbf{r}, \mathbf{r}', \omega)M_{jk}(t_s). \tag{1.122}$$

In other words, the low-frequency limit of the displacement spectrum is proportional to the total moment accumulated at the end of the source process. The Fourier transform of the response function E_{ijk} is a factor in the spectrum of u_i. It takes into account various propagation effects of seismic wavefields in rocks. These are effects like geometrical spreading, elastic anisotropy, reflection, refraction, diffraction and (more general) scattering on medium heterogeneities. There will also be effects of inelastic attenuation and velocity dispersion often caused by fluids in rocks, if the Green's functions take these effects into account.

If we further consider the observation distances and the wavelengths to be significantly larger than the source domain, but increase the frequency so that the wavelengths become significantly shorter than the observation distances, then the far-field approximation of the Green's functions can be applied in equation (1.119). In this approximation the elastic wavefields are equal to a sum of propagating elementary waves radiated by a point source. In elastic homogeneous isotropic media these are spherical P- and S-waves. The far-field Green's function is given by (1.39). The far-field displacements of the spherical P- and S-waves radiated by a point-like earthquake will be correspondingly given by the following expressions (see Rice, 1980):

$$u_i^P(\mathbf{r}, t) = \frac{1}{4\pi\rho c_p^3 r}\gamma_i\gamma_j\gamma_k\frac{\partial M_{jk}(t - r/c_p)}{\partial t}, \tag{1.123}$$

$$u_i^S(\mathbf{r}, t) = \frac{1}{4\pi\rho c_s^3 r}(\delta_{ij} - \gamma_i\gamma_j)\gamma_k\frac{\partial M_{jk}(t - r/c_s)}{\partial t}, \tag{1.124}$$

where $r = |\mathbf{r} - \mathbf{r}'|$, and the quantities $\gamma_i = r_i/r$ are the direction cosines.

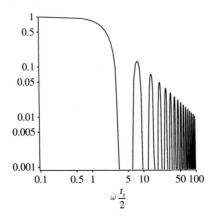

Figure 1.9 Normalized amplitude spectrum (1.125).

Results (1.123) and (1.124) help to understand some general features of the frequency spectra of earthquakes. Indeed, let us make a simple assumption that during the process time t_s the moment rate is constant and it is equal to zero outside of this time interval. Then the moment rate is a boxcar function of length t_s (i.e. the moment itself is a ramp function) and of height $M_{jk}(t_s)/t_s$. Its amplitude spectrum (the absolute value of the forward Fourier transform of the boxcar function) is then equal to

$$|M_{jk}(t_s)|\left|\frac{\sin(\omega t_s/2)}{t_s\omega/2}\right|. \tag{1.125}$$

This is an even frequency function. Thus, we consider it for positive frequencies only (see Figure 1.9). In the low-frequency limit this spectrum tends to a constant equal to $|M_{jk}(t_s)|$. Such behavior corresponds to equation (1.122). In the case of an isotropic homogeneous elastic medium the low-frequency limits of the displacement spectra of the radiated P- and S-waves will be equal to

$$\hat{u}_i^P(\mathbf{r},0) = \frac{1}{4\pi\rho c_p^3 r}\gamma_i\gamma_j\gamma_k M_{jk}(t_s), \tag{1.126}$$

$$\hat{u}_i^S(\mathbf{r},0) = \frac{1}{4\pi\rho c_s^3 r}(\delta_{ij}-\gamma_i\gamma_j)\gamma_k M_{jk}(t_s). \tag{1.127}$$

However, for frequencies higher than approximately

$$\omega_c \equiv \frac{2}{t_s}, \tag{1.128}$$

the envelope of this spectrum starts to decay as $1/\omega$. The amplitude spectrum (1.125) is proportional to an absolute value of a decreasing harmonic function.

It is equal to zero at the frequency values $\omega = 2\pi/t_s, 4\pi/t_s, 6\pi/t_s$. The quantity ω_c is called the corner frequency.

The results (1.123)–(1.125) describe wavefields radiated by a point source of duration t_s. If a rupture is finite in space then one can compute the radiated wavefields as a linear superposition of such point sources (at least approximately) distributed along the rupture surface. Such a superposition takes into account a finite rupture propagation velocity as well as directivity effects (similar to the Doppler effect, e.g. shorter stronger pulses are radiated in the direction of the rupture propagation and longer weaker pulses are radiated in the opposite direction). To account for a finite rupture time t_d one must convolve the boxcar moment rate function of duration t_s with a boxcar function of duration t_d (see Shearer, 2009). This is the so-called Haskell fault model (see Lay and Wallace, 1995). The corresponding spectrum will be given by a product of two functions of the form (1.125) with characteristic times t_s and t_d. Thus, the high-frequency decay of the spectrum will be given by $1/\omega^2$.

Similar spectra are actually observed. This fact gave rise to other similar models of the earthquake spectrum. One of the most influential models was proposed by Brune (1970):

$$|\hat{u}(\mathbf{r}, \omega)| = \frac{|\hat{u}(\mathbf{r}, 0)|}{1 + (\omega/\omega_c)^2}, \tag{1.129}$$

which has the same main features as the Haskell pulse. Note that here the corner frequency effectively corresponds to the intersection of the high- and low-frequency asymptotes of the displacement spectrum (Aki and Richards, 2002). This frequency (often modeled as an average over all radiation directions) is a function of the rupture geometry, size and velocity, and of the wave-signal speed. This intuitive understanding is summarized in the following numerically and analytically supported relation between the rupture length X and the corner frequency (Brune, 1970; Madariaga, 1976; Shearer, 2009):

$$X = \frac{4\pi k_r c_s}{\omega_c}, \tag{1.130}$$

where k_r is a model-dependent and rupture-velocity-dependent numerical coefficient usually assumed to be of the order of 0.2–0.6.

Let us consider a simplified model of an earthquake as a slip with a surface-average fault displacement $\overline{\Delta u}(t)$ along a plane rupture with the area A. Further we assume that the rupture is located in a homogeneous isotropic elastic medium. Then equations (1.102)–(1.106), (1.120), (1.114) and (1.25) give the following result for the moment tensor:

$$M_{ij}(t) = \int_{V_\Psi} C_{ijkl}\, p_{kl}(t)\, \delta_D(\Psi_{c+})\, d^3\hat{x}$$

$$= \overline{\Delta u}(t)\, A\, (\lambda \delta_{ij}\delta_{kl} + \mu(\delta_{ik}\delta_{jl} + \delta_{il}\delta_{jk}))\frac{1}{2}(\hat{u}_k n_l^+ + \hat{u}_l n_k^+) \tag{1.131}$$

$$= M_0(t)(\hat{u}_i n_j^+ + \hat{u}_j n_i^+),$$

where the quantity

$$M_0(t) = \mu \overline{\Delta u}(t) A \tag{1.132}$$

is called the scalar seismic moment. Combining this with equation (1.98) we obtain

$$\Delta\sigma = \frac{2C'\overline{\Delta u}\mu}{X} = \frac{8C'M_0}{\pi X^3} = \frac{C_0 M_0}{X^3}, \tag{1.133}$$

where $M_0 = M_0(t_s) = \mu \overline{\Delta u}(t_s) A = \mu \overline{\Delta u}\pi X^2/4$, and X stands for a characteristic length of the rupture. In the case of a circular rupture, X is the diameter of the rupture. Geometry-controlled constants $C_0 = 8C'/\pi$ and C' are of the order 1.

Further, using equation (1.130) to estimate the characteristic length of the rupture from the corner frequency, one obtains an equation often used to estimate the stress drop:

$$\Delta\sigma = M_0 \frac{C'}{\pi} \left(\frac{\omega_c}{2\pi k_r c_s}\right)^3, \tag{1.134}$$

where, for example, $C' = 7\pi/16$ for a circular rupture. Note that this equation is based on a series of critical assumptions about the rupture processes that may not be necessarily valid (see Beresnev, 2001). Moreover, this stress-drop estimate is very sensitive to the corner frequency, which is in turn a parameter influenced by many not completely determinable factors. These are, for example, the seismic attenuation, local site effects, heterogeneity of the medium and some instrumental features.

Finally, the scalar moment is related to the moment magnitude M_w by the following definition (Kanamori, 1977; Shearer 2009):

$$M_w = \frac{2}{3}(\lg M_0 - 9.1), \tag{1.135}$$

where the moment is taken in Nm (SI system). Note also that, in this book, we use the International Standard notation $\lg \equiv \log_{10}$ and $\ln \equiv \log_e$. We imply moment magnitudes when addressing earthquake magnitudes throughout this book.

Multicomponent records of P- and S-waveforms can be used to find focal mechanisms of earthquakes. Such solutions usually represent fault plane (double-couple) approximations of tectonic motions at earthquake faults. Moreover, more complete moment-tensor solutions can be found. They also include non-double-couple components of the moment tensor, i.e. isotropic and compensated-linear-vector-dipole (CLVD) ones (see Shearer, 2009, section 9.2).

1.4 Introduction to microseismic monitoring

Microseismic monitoring is a method of seismic investigation of the underground. It is based on detection and location of small-magnitude earthquakes (usually called microearthquakes or sometimes also called seismic emission) often occurring in rocks due to various artificial or natural processes. Microseismic monitoring is a powerful method of geophysical reservoir characterization. It can contribute to delineation and imaging of reservoirs and their heterogeneous structures, characterization of hydraulic and elastic properties of rocks, understanding of geomechanical features like tectonic stresses and pore-pressure distributions.

Results of microseismic monitoring might be further combined with four-dimensional (4D) reflection seismics, reservoir simulations, borehole measurements and stimulation and production data. Such integrated information provides a rather complete picture of relevant processes in underground targets. One of the most common applications of microseismic monitoring is the mapping of hydraulic fractures in hydrocarbon reservoirs (Urbancic and Baig, 2013). Maxwell (2014) gives a detailed description of many practical, historical, technical and environmental aspects of this application.

1.4.1 Detection of seismic events

To record seismic events, arrays of seismic receivers are required. Seismic receivers (also called geophones) can be placed on the Earth's surface (see Duncan and Eisner, 2010) and/or in several or in a single monitoring borehole. A schematic sketch of seismic monitoring is shown in Figure 1.10. Figures 1.11 and 1.12 show an example of real borehole-monitoring system for three hydraulic fracture stages at a shale-gas reservoir.

Multicomponent seismic receivers record components of vectorial seismic wavefields (i.e. components of motions of rock particles) as functions of time (i.e. seismic traces or seismograms; see Figures 1.13 and 1.14). Application of arrays of seismic receivers and their ability to record different components of seismic wavefields allow for detection of seismic waves radiated by earthquakes. This is usually done by different types of analysis of seismic traces, such as their energy analysis including ratios of energy averaged in short and long time windows of seismic traces, their statistical analysis, their integral transforms including Fourier

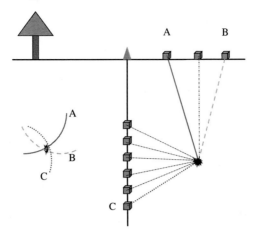

Figure 1.10 A sketch of possible acquisition geometry of microseismic monitoring. Multicomponent geophones can be placed in several boreholes or in a
single monitoring borehole. Multicomponent geophones also can be placed on
the Earth's surface. Geophones (denoted by the cubes) record seismic wavefields
radiated by a microearthquake (a spot to the right). A sketch of the triangulation
principle of event location using travel time information at geophones A, B and C
(the elements of differential isochrones are schematically shown by corresponding
arcs) is given on the left.

and wavelet analysis, analysis of particle motions, or combinations of those (see,
for example, the literature review in Rentsch *et al.*, 2007). Sometimes one also
applies the technique of master event (a matched-filter technique) originally proposed for the detection of global seismic events (Shearer, 1994). The data basis
of template earthquakes is enriched with time and additional events can be also
detected retrospectively.

Further processing of microseismic wavefields usually requires identification of
seismic phases (for example arrivals of the P- and S-waves) and the picking of their
arrival times. In spite of the fact that the picking of arrival times can be accom-
plished with automatic algorithms, manual picking is still frequently performed to
increase the accuracy of the travel-time information.

1.4.2 Seismic multiplets

The waveform similarity is important for event location. Seismic events characterized by a high similarity are referred to as a seismic multiplet. Figure 1.15
shows an example of seismic traces of a multiplet composed of three different
microearthquakes at Cotton Valley.

Such events probably occur on the same structures and have the same tectonic mechanisms. To quantify the waveform similarity one can use their

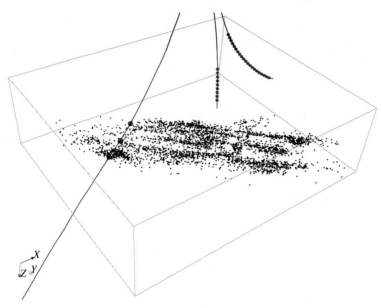

Figure 1.11 Hypocenters (dots) of microseismic events induced by three hydraulic-fracturing stages conducted from a horizontal section of a borehole (black line) in a shale-gas reservoir in Canada. The corresponding perforations are shown as small spheres. The distances between the perforations are approximately 100 m. The depth range of the microseismic clouds is approximately 1700–1800 m. The observation system was composed of two chains of three-component geophones (small cubes) located in two different monitoring boreholes (two other black lines with cubes). The asymmetry of the microseismic clouds in respect to the perforation locations is due to the asymmetric position of geophones in respect to the perforated borehole (see also Figure 1.12). The figure is courtesy of Anton Reshetnikov, Freie Universität Berlin.

cross-correlation coefficients. For example, Scholze *et al.*, (2010) used the following definitions. The normalized cross-correlation function $c_{ij}(t)$ between two traces $u_i(t)$ and $u_j(t)$ is defined as

$$c_{ij}(t) = \frac{\int u_i(t')\, u_j(t'-t)\, dt'}{\left(\int u_i^2(t')dt' \times \int u_j^2(t')dt'\right)^{\frac{1}{2}}}. \tag{1.136}$$

The cross-correlation coefficient is then defined as the absolute maximum of the cross-correlation function, $CC_{ij} = max[|c_{ij}(t)|]$. It can be calculated for the complete seismograms or separately for specified time windows containing selected wave phases (e.g. P- or S-waveforms).

Clustered multiplet events can be observed in the seismicity induced by artificial rock stimulations as well as in natural seismicity. Multiplet clusters can frequently

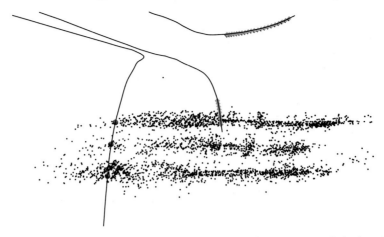

Figure 1.12 The microseismic clouds, the monitoring system and the boreholes shown in Figure 1.11 but now in the map view. The figure is courtesy of Anton Reshetnikov, Freie Universität Berlin.

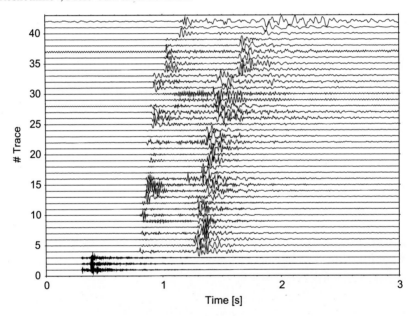

Figure 1.13 Seismic traces of the vertical and two horizontal components of the wavefield radiated by a microearthquake that occurred during a water-injection experiment in the pilot borehole of the German Continental Deep Drilling Project (KTB). The geophone array was composed of different types of seismic receivers. Five three-component 4.5 Hz seismic sensors were deployed around the KTB in shallow boreholes (25–50 m depth). Additionally, eight three-component 1 Hz sensors were deployed on the surface. The sampling rate of these 13 shallow geophones was 200 Hz. One more three-component 15 Hz geophone was installed at 3500 m depth in the main KTB borehole. Its sampling rate was 1000 Hz. Corresponding seismic traces of the vertical and two horizontal components have numbers 1, 2 and 3, respectively. The figure is courtesy of Jörn Kummerow, Freie Universität Berlin. (From Shapiro, 2008, EAGE Publications bv.)

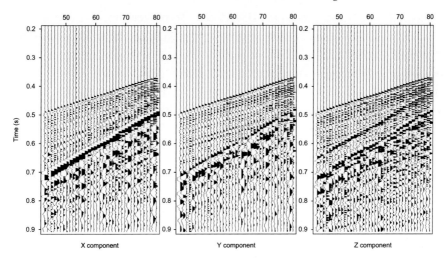

Figure 1.14 Seismograms of east, south and depth components (from the left, respectively) of the wavefield radiated by a tectonic microearthquake at the San Andreas Fault. The seismograms were recorded by a borehole array of the Paulsson Geophysical Services Inc. (P/GSI) in the main borehole of the San Andreas Fault Observatory at Depth. The array is composed of 80 three-component 15 Hz borehole geophones located at depths between 878 m and 1703 m below sea level. The horizontal axis represents the number of geophones (the traces shown were recorded by geophones with numbers 43–80). The vertical axis is time in seconds, and the sampling rate is 0.25 ms. (Modified after Reshetnikov *et al.*, 2010.)

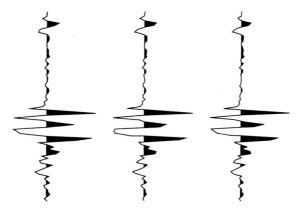

Figure 1.15 A multiplet of three different microseismic events recorded at the same borehole geophone during a hydraulic fracturing at Cotton Valley. The traces are 0.1 s long. The arrivals of the P- and S-waves are clearly seen at the top and in the middle of traces, respectively. The figure is courtesy of Karsten Stürmer, Freie Universität Berlin.

be located with high precision. Based on the processing of multiplet events, Kummerow (2013) proposed an automatic approach of enhancing consistency in determining arrival times of direct seismic P- and S-waves. He cross-correlated microseismic borehole array data for such events. This leads to a cross-linking of waveforms of different events in a multiplet recorded by several closely spaced receivers. The picked travel times are then iteratively corrected by forcing a consistent time alignment of such waveforms. Improved travel times lead finally to better event locations.

Following the work of Lin and Shearer (2007) on tectonic earthquakes, Kummerow *et al.*, (2012) proposed applying microseismic multiplets to estimate the spatial distribution of the velocity ratio of the longitudinal and shear waves, c_p/c_s. This ratio is directly related to the Poisson ratio. High-precision P and S arrival time differences must be calculated for pairs of microseismic events. Multiplets are highly suitable for this task. Then, the velocity ratios are determined from the slope of distributions of differential P- versus S-travel times. If done for all elementary spatial cells, this will provide a 3D c_p/c_s-ratio image. Such a procedure requires approximate event locations as an input.

1.4.3 Location of seismic events

Travel-time characteristics of P- and S-waves (e.g. their arrival times, curvatures of arrival time surfaces as functions of receiver locations, and differences in arrival times of longitudinal and shear waves) as well as their vectors of particle displacements (i.e. their polarizations) can be used for locating earthquake hypocenters. A hypocenter of an earthquake is the starting point of the corresponding rupture. To locate an earthquake means finding the spatial coordinates of its hypocenter and its origin time. It is one of the most important issues in earthquake seismology (Thurber and Rabinowitz, 2000) as well as in the microseismic monitoring (Rentsch *et al.*, 2007).

Most location procedures require P- and S-wave arrival times as well as a velocity model between the hypocenter and the receiver array. In some approaches, expected arrival times are calculated and compared with measured ones for every receiver. Then, event locations and sometimes velocity-model updates are found by grid-search-type optimization algorithms which minimize arrival-time residuals (Lomax *et al.*, 2009). Often, different simplified variations of such approaches are applied. For example, for each geophone of the monitoring array one uses differences between P- and S-wave arrival times and a given velocity model to calculate spatial surfaces (differential isochrones) of possible hypocenter locations. A hypocenter is then assigned to an intersecting region of these surfaces (of course, they are spheres, in the case of simple constant isotropic velocity models). Other location approaches use full waveform seismograms. These are migration-type location methods (Rentsch *et al.*, 2010) or methods using seismogram cross-correlations in combinations with arrival-time-based wavefield characteristics (Richards *et al.*, 2006).

Seismic multiplets and seismogram cross-correlations are extensively used in relative earthquake location methods such as the double-difference method of Waldhauser and Ellsworth (2000). In such approaches the cross-correlation of P- and S- wave traces are used for precise estimates of travel-time differences for pairs of earthquakes. Then residuals between observed travel-time differences and the velocity-model-based predictions of these differences are minimized. This leads to high-resolution images inside seismicity clusters. Some event-location procedures exploit the dependence of waveform similarity on event spacing. Given an initial set of precise absolute earthquake locations, the maximum waveform cross-correlation coefficients for all possible pairs of events are calculated, and a relation between event separation and cross-correlation coefficients is established. This is used to locate further events by calculating their waveform similarity relative to all located events. An advantage of this approach is that it can be applied to events registered even by a single seismic sensor. Kummerow (2010) proposed this method and applied it to microseismic data recorded during a hydraulic experiment at the Continental Deep Drilling Site (KTB) in Germany. The number of reliably determined hypocenters was increased by a factor of about eight compared with standard location methods. Another variation of this approach inverts both the measured arrival times and cross-correlation values of the waveforms for the hypocenter coordinates. Such algorithms have the potential to significantly better locate seismic events in comparison to the purely arrival-time-based approaches.

An important aspect of event location is a construction of the velocity model (Pavlis, 1986). The elastic anisotropy of rocks must be accounted for. Thus the velocity model must also include parameters characterizing the anisotropy. Elastic moduli C_{ijkl} or S_{ijkl} can be used for this task. Thomsen (1986) proposed a convenient way to describe seismic anisotropy using the so-called anisotropy parameters. In many relevant situations of microseismic monitoring in sedimentary rocks the transverse isotropy is a good approximation. For the transverse isotropy with the vertical (x_3) rotational symmetry axis, Thomsen (1986) introduced the following three anisotropy parameters:

$$\epsilon_a \equiv \frac{c_{11} - c_{33}}{2c_{33}}, \tag{1.137}$$

$$\gamma_a \equiv \frac{c_{66} - c_{55}}{2c_{55}}, \tag{1.138}$$

$$\delta_a \equiv \frac{(c_{13} + c_{55})^2 - (c_{33} - c_{55})^2}{2c_{33}(c_{33} - c_{55})}, \tag{1.139}$$

where, for elastic moduli, we used the contracted notation discussed in Section 1.1.4. We also added the index a to Thomsen's original notation for the anisotropy parameters in order to distinguish them from other notations used in the book. To define the velocity model for a homogeneous transversely isotropic medium with the vertical symmetry axis (the so-called VTI medium), these three parameters are usually complemented by the velocities of P- and S-waves propagating along the x_3 direction:

$$c_{p0} = \sqrt{\frac{c_{33}}{\rho}}, \tag{1.140}$$

$$c_{s0} = \sqrt{\frac{c_{55}}{\rho}}. \tag{1.141}$$

These two velocities and the three Thomsen parameters can be used to reconstruct five independent elastic moduli of a transverse isotropic medium. For such a reconstruction the sign of $(c_{13} + c_{55})$ must be additionally known. Usually it is assumed that $c_{13} + c_{55} > 0$ (Tsvankin, 2005). In Section 1.1.4 we saw that, in isotropic media, $c_{33} = c_{22} = c_{11} = \lambda + 2\mu$, $c_{23} = c_{13} = c_{12} = \lambda$ and $c_{66} = c_{55} = c_{44} = \mu$. Therefore, in isotropic rocks the anisotropy parameters vanish. This indicates that the anisotropy parameters can be considered as a measure of the strength of anisotropy. Owing to the stability conditions (see Section 1.1.4) the anisotropy parameters cannot be less than -0.5 (Grechka, 2009). They have no upper bounds. Shale is often characterized by a very strong anisotropy, with the anisotropy parameters above 0.3. Possible values of ϵ_a above 0.5 and the typical relation $\epsilon_a > \delta_a$ have been observed (Tsvankin, 2005). The case for Thomsen parameters smaller than 0.1 is referred to as weak anisotropy. Sandstones are often weakly anisotropic. Carbonates are frequently nearly isotropic. Crystalline rocks like basalt and granite often do not show significant elastic anisotropy. However, fractures, an oriented texture and tectonic stress can change this situation.

In Section 1.1.5 we saw that, in elastically anisotropic rocks, three different types of seismic body waves can be observed. This is often the case for the microseismic wavefieds in shale (see Figure 1.16).

Even in the case of weak seismic anisotropy, significant location errors can occur by neglecting this effect. Some inversion-based approaches to event location and macromodel construction have been proposed. For example, Grechka and Yaskevich (2013) proposed a simultaneous inversion and location approach for general

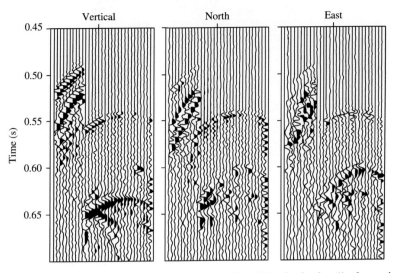

Figure 1.16 Traces of a microseismic event induced by the hydraulic fracturing in a shale-gas reservoir in Canada. The observation system is shown in Figures 1.11 and 1.12. Approximate directions of polarization components are indicated by the titles of the seismograms. Arrivals of the quasi P-wave are clearly recorded by the all 30 geophones. The arrivals are approximately seen in the time interval 0.48–0.51 s on the first nine geophones and as a curved moveout in the time interval 0.54–0.56 s on the last 21 geophones. The slower quasi S-wave (which corresponds to the qSV-wave in a VTI medium) is seen on all the last 21 geophones as a moveout in the time interval 0.63–0.65 s. The quicker quasi S-wave (which corresponds to the SH-wave in a VTI medium) is seen on the north- and east components of the last 21 geophones. Its first arrivals are approximately in the time interval 0.60–0.62 s. Both these qS-waves can be also seen on the first nine geophones. Their arrivals partially interfere and closely follow the qP-wave. A moveout of the slower qS-wave reflected by a layer interface is seen on the vertical component of the sixth to twelfth geophones (from the left) of the second array in the time interval 0.65–0.67 s. The complexity of the waveforms is due to an approximate tilted VTI symmetry of the medium, its heterogeneity, the related ray bending and not-exact geophone orientations. At this site and depth interval of this layered shale deposit Yu (2013) estimated values of Thomsen parameters in the range 0.15–0.4 for ϵ_a, in the range 0.15–0.2 for δ_a and in the range 0.15–0.7 for γ_a. The figure is courtesy of Anton Reshetnikov, Freie Universität Berlin.

triclinic anisotropic media. Yu *et al.* (2013) proposed a variation of this approach for transversely isotropic layered media. Velocity anisotropy is taken into account in an additional step after conventional event locating. In this step, the isotropic velocity model and corresponding locations serve as an initial model. Then, the anisotropy parameters as well as the resulting locations are obtained simultaneously. The algorithm uses ray theory for travel-time computations and known numerical approaches for non-linear optimization.

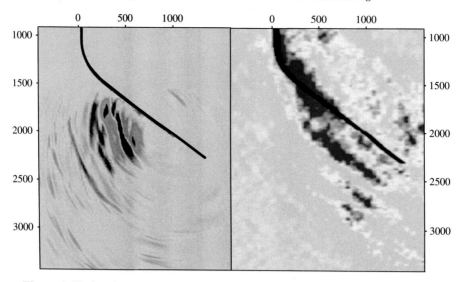

Figure 1.17 On the left: a result of microseismic reflection imaging at the main borehole of the San Andreas Fault Observatory at Depth. On the right: a result of reflection seismic imaging from the free surface at the same location. The borehole is shown by the black line. The scale is given in meters. Both images are similar; however, the microseismic image has significantly higher resolution. (Modified after Reshetnikov *et al.*, 2010.)

1.4.4 Microseismic reflection imaging

An interesting possibility of using microseismic information is given by treating microearthquakes as active sources. First, high-precision location of microseismic events is necessary. Second, one migrates the corresponding reflected wavefield using one of reflection seismic-imaging techniques. For example, Reshetnikov *et al.* (2010) used a Fresnel-volume-limited Kirchhoff summation (a directional migration algorithm) to obtain a high-resolution image of a subsurface region illuminated by microseismic wavefields. They applied the method to several micro-seismic events recorded by a borehole array in the SAFOD (San Andreas Fault Observatory at Depth) main hole and compared the findings with results of surface seismic reflection profiling (Figure 1.17).

Usually, the number of located microseismic events is much larger than the number of sensors. Using the principle of reciprocity, one can swap sources and receivers. Thus, by such a type of imaging one considers a monitoring system as several "sources" and a cloud of microearthquakes as an array of spatially distributed "receivers." In the case of the Basel geothermal experiment it was possible to obtain a high-resolution reflection image (Reshetnikov *et al.*, 2013). More-over, it seems that identification of singular large-scale fractures was possible. The consistency and extremely high resolution of the method can be enhanced in its multisource multireceiver configuration.

Frequently we observe significant microseismic reflections, indicating the presence of strong heterogeneities. Numerical and theoretical studies of waves reflected at thin fluid layers (see Oelke *et al.*, 2013) show that in the microseismic frequency range ($f = 50$–400 Hz) hydraulic fractures and thin fluid-filled cracks (modeled as fluid layers of thicknesses 10^{-3}–10^{-2} m) can produce strong reflection coefficients (0.2 and higher). In addition, the double-couple sources corresponding to microseismic events can produce rather complex reflection signatures due to non-radially symmetric radiation patterns.

2

Fundamentals of poroelasticity

Fluid-saturated rocks are multiphase media. Elastic stresses are supported there by the solid phase of rocks. Pore pressure is supported by saturating fluids. In such media the elastic stresses and the pore pressure are coupled.

The mechanics of poroelastic media was developed by Maurice Biot in a series of his seminal publications in the 1940s–1970s (see, for example, Biot 1956, 1962). This theory was further developed and reformulated by several scientists. A very incomplete list of corresponding works includes the publications of Gassmann (1951), Brown and Korringa (1975), Rice and Cleary (1976), Chandler and Johnson (1981), Rudnicki (1986), Zimmerman *et al.* (1986), Detournay and Cheng (1993) and Cheng (1997).

We consider induced microseismicity as a poromechanical phenomenon. Coussy stated in the preface to his book *Poromechanics* (2004): "We define *Poromechanics* as the study of porous materials whose mechanical behavior is significantly influenced by the pore fluid." In this book we use several approximations derived from the theory of poroelasticity, which is part of poromechanics.

We start our introduction to poroelasticity by considering elastic compliances of porous rocks. Here we follow and correspondingly modify the approach of Zimmerman *et al.* (1986) and Brown and Korringa (1975). Following Detournay and Cheng (1993) we consider a very small elementary sample of a porous fluid-saturated rock with a characteristic length dl, which is assumed to be more than ten times larger than the typical size of pores and grains. Such a sample can be also considered as a representative volume of a statistically homogeneous porous medium. Both solid and fluid phases are assumed to be connected in a 3D rock structure. The pore space of a rock sample is defined as the total space occupied by all connected voids (including fractures and pores) in the sample. Let us denote by V_p the volume of the pore space of a sample of volume V. The ratio

$$\phi = \frac{V_p}{V} \tag{2.1}$$

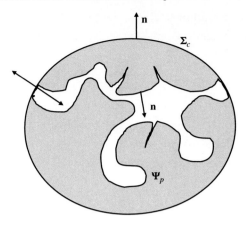

Figure 2.1 A sketch of external and internal surfaces of a porous system.

is the so-called connected porosity of the rock. Throughout the book we consider total porosity to mean connected porosity. Isolated pores are understood as part of the solid material (grain material).

Such a porous sample has two closed surfaces on the sample scale: its external surface Σ_c and surface Ψ_p of its pore space (see Figure 2.1). The surface Ψ_p of the connected pore space is the sample's internal surface.

2.1 Linear stress–strain relations in poroelastic materials

We consider a thought experiment where a uniform confining stress, σ_{ij} (also called total stress), is applied to the external surface Σ_c of the rock sample. Simultaneously, a pore pressure, P_p, is applied to the internal surface Ψ_p. Surface Σ_c is the external surface of the solid part of the rock. Simultaneously it also seals pores. At a given point $\mathbf{r}$ of the surface Σ_c the applied traction τ is then

$$\tau_i = \sigma_{ij} n_j(\mathbf{r}), \tag{2.2}$$

where $n_j(\hat{x})$ are components of the outward unit normal to Σ_c.

Let us assume that the confining stress and/or pore pressure has changed from the load state (σ_{ij}^0, P_p^0) to the load state (σ_{ij}, P_p). As a result, points of the external surface have been displaced by $u_i(\mathbf{r})$. The displacement is assumed to be very small in comparison with the size of the rock sample.

Following Brown and Korringa (1975), we introduce a symmetric tensor (note also the analogy to equation (1.99)):

$$\eta_{ij} = \int_{\Sigma_c} \frac{1}{2}(u_i n_j + u_j n_i) d^2\mathbf{r}. \tag{2.3}$$

In the case of a continuous elastic body replacing the porous rock (i.e. a differentiable displacement **u** is given at all its points) the Gauss' theorem gives

$$\eta_{ij} = \int_V \frac{1}{2}\left(\frac{\partial u_i}{\partial x_j} + \frac{\partial u_j}{\partial x_i}\right)d^3\mathbf{r}. \tag{2.4}$$

The integrand here is the strain tensor and V is the volume of the sample. Thus, $\epsilon_{ij} = \eta_{ij}/V$ is the volume averaged strain.

We introduce also another symmetric tensor,

$$\zeta_{ij} = \int_{\Psi_p} \frac{1}{2}(u_i n_j + u_j n_i)d^2\mathbf{r}, \tag{2.5}$$

where Ψ_p is the surface of the pore space, $\mathbf{r}$ is a point on this surface, u_i is a component of the displacement of points $\mathbf{r}$ on this surface slightly deformed by changing the load, and n_i is a component of the outward normal to this surface (the normal is directed into the space of pores, see Figure 2.1). At points where surface Σ_c seals the pores it coincides with surface Ψ_p; however, their normals have opposite directions.

If we assume that the pore space is filled with some material (e.g. a fluid, clay or cement) then, analogously with tensor η_{ij}, a quantity $-\zeta_{ij}$ will denote a volume-averaged strain of this material multiplied by its volume. Note that the minus sign is due to the direction of the normal of surface Ψ_p. Thus, the quantity $-\zeta_{ii}$ denotes a change of the volume of this pore-filling material.

In the linearized theory of poroelasticity the initial load state is not of importance. It is especially so if incremental small-strain effects are considered. For simplicity we will assume that the initial load state corresponds to $\sigma_{ij}^0 = P_p^0 = 0$. Later, in Section 2.9, we consider some effects of the initial stress state of rocks. In realistic experiments a uniformly distributed scalar pore pressure P_p can be applied to surface Ψ_p. However, on the following pages (until equation (2.19)) we will assume a more general, albeit abstract, possibility that a uniformly distributed stress tensor σ_{ij}^f is applied to the surface Ψ_p (note that the stress compressional with respect to the solid phase is negative). In spite of its quite artificial character, this assumption can be advantageous. For example, it will allow us to imagine a load configuration where the compliance tensor of the grain material can be measured (at least in an average sense) without destroing the porous sample (see our discussion of experiments with an unjacketed sample below). If the load on surface Ψ_p is hydrostatic, then $\sigma_{ij}^f = -P_p \delta_{ij}$. The effective stress is defined as:

$$\sigma_{ij}^e = \sigma_{ij} - \sigma_{ij}^f. \tag{2.6}$$

Note that in this equation we introduce only a notation, "effective stress." At this point we do not discuss any physical quantity for which this stress is indeed "effective." We will discuss this later in this chapter and especially in Section 2.9.6.

Further, for a completely hydrostatic load (i.e., also the confining stress is given by a confining pressure $\sigma_{ij} = -P_c\delta_{ij}$) we have:

$$\sigma_{ij}^e = \sigma_{ij} - \sigma_{ij}^f = (-P_c + P_p)\delta_{ij} = -(P_c - P_p)\delta_{ij}. \qquad (2.7)$$

By analogy with Brown and Korringa (1975) and Zimmerman *et al.* (1986) we introduce different compliances of an anisotropic porous body corresponding to different configurations of its loading. Indeed, it is possible to apply loads to both surfaces of a porous sample (see Figure 2.2). It is equally possible to measure displacements of these two surfaces by loading any one of them. Thus, there are four different combinations of loading a surface and measuring displacements of the same or of the other surface. We assume that the applied incremental load is small (the sample can be pre-stressed; in this case we discuss incremental stresses and strains). Equally we assume that the resulting incremental displacements are small (so that the strains of the sample and of its pore space are small). Such loading experiments lead to formulating four Hooke's laws relating the corresponding strain and stress tensors. This allows us to define four compliance tensors.

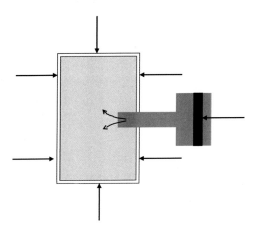

Figure 2.2 A jacketed porous sample (the doubled boundary of the sample symbolizes its jacketting envelope) provides a possibility to apply independent loads to its external and internal (pore-space) surfaces. Thin curved arrows inside the sample symbolize uniformally distributed loading of pore-space surface Ψ_p. Thick arrows outside the sample are applied to its external surface Σ_c. Note that in the case of a fluid-saturated sample this figure is a draft of a realistic laboratory experiment.

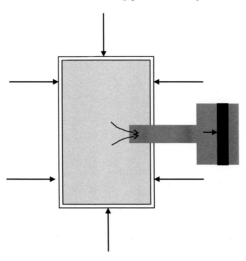

Figure 2.3 A sketch of a loading configuration, where a jacketed sample provides the possibility of establishing drained conditions.

Generally surfaces Σ_c and Ψ_p are isolated from each other and can be subjected to different loads. Let us first assume that only the surface Σ_c has been subjected to a changing load. Thus, the load σ_{ij}^f has been unchanged. In the case of fluid saturation this is a load configuration of a jacketed sample organized in such a way that the fluid can freely drain away from the sample (see Figure 2.3). The so-called drained compliance tensor can be then introduced analogously to Hooke's law for elastic media:

$$\frac{1}{V}\eta_{ij}|_{\sigma^f} = S_{ijkl}^{dr}\sigma_{kl}^e. \tag{2.8}$$

The compliance tensor introduced here characterizes the elastic properties of the drained skeleton of the rock. In equation (2.8), and later in other similar equations expressing different Hooke's laws for poroelastic systems, corresponding components of compliance tensors (here S_{ijkl}^{dr}) are defined as partial derivatives of strain components (here η_{ij}/V) over stress components (here σ_{kl}^e), respectively (see Brown and Korringa, 1975; and Shapiro and Kaselow, 2005). Note that in such partial derivatives the symmetric off-diagonal stress components (σ_{kl}^e and σ_{lk}^e) are formally assumed to be independent, in spite of the fact that in reality they are equal.

Another loading configuration (using an unjacketed sample) can be defined in the following way. Both surfaces Σ_c and Ψ_p are loaded by the same changing part of the stress (see Figure 2.4). Therefore, the effective stress remains unchanged. Such a homogeneous distribution of the stress on the internal and external surfaces of the sample is equivalent to a distribution of the stress in a homogeneous elastic

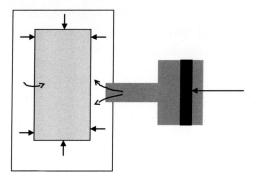

Figure 2.4 An unjacketed sample providing a possibility to establish constant-effective-stress conditions. For a changing part of the load (symbolized by thin arrows) we assume an equivalent access to the internal and the external surfaces of the sample. In the case of a fluid-saturated sample this can be achieved by filling the space around the sample by the saturating fluid. The changing part of the load is then the fluid pressure.

medium constructed from the same material as the skeleton of the rock. Thus, the following form of the Hooke's law introduces the tensor of compliance of the skeleton material. The skeleton material is also often called the grain material. In the case of a heterogeneous grain material the corresponding compliance tensor characterizes its elastic properties in an effective (averaged) sense:

$$\frac{1}{V}\eta_{ij}|_{\sigma^e} = S^{gr}_{ijkl}\sigma^f_{kl}. \tag{2.9}$$

The two compliance tensors introduced above can be found from measurements of displacements of the external surface Σ_c. Measurements of displacements of the internal surface Ψ_p provide two more compliance tensors. The first one can be called the compliance tensor of the pore space. It is obtained from the above-described experiment with an unjacketed sample (under a constant effective stress, see Figure 2.4):

$$-\frac{1}{V_p}\zeta_{ij}|_{\sigma^e} = S^p_{ijkl}\sigma^f_{kl}. \tag{2.10}$$

If the material of the solid skeleton of the rock (i.e. the grain material) is homogeneous and linear, then the condition of a constant, uniformly distributed loading stress will be equivalent to the replacement of the material in pores by the grain material. Thus, the porosity does not change. Moreover, in this case the volume-averaged strain is independent of the averaging domain. This yields the equivalence of equations (2.9) and (2.10) and leads to the following statement:

$$S^p_{ijkl} = S^{gr}_{ijkl}. \tag{2.11}$$

Finally, one more (but not-independent) compliance can be estimated from displacement measurements of the internal surface. If we assume that effective stress changes but the stress σ^f remains constant, then:

$$-\frac{1}{V}\zeta_{ij}|_{\sigma^f} = S'_{ijkl}\sigma^e_{kl}. \tag{2.12}$$

Using the reciprocity theorem analogously to Brown and Korringa (1975) we obtain (see the Appendix to this chapter):

$$S'_{ijkl} = S^{dr}_{klij} - S^{gr}_{klij}. \tag{2.13}$$

Let us consider a strain $\epsilon_{ij} = \eta_{ij}/V$ of a porous elastic sample due to a loading by changing both stresses, σ^e and σ^f. Note that we assume the strain to be small and neglect all non-linear terms (in respect to any displacement gradients). Under arbitrary (but small) changes of a load the strain will change due to applying incremental σ^e by keeping a constant stress σ^f plus an effect of applying incremental σ^f from inside and outside (i.e. keeping $\sigma^e = const.$). This is a consequence of a linear superposition of deformation effects of independent loads. Therefore, using equations (2.8) and (2.9) as well as the definition of the effective stress (2.6) we obtain

$$\epsilon_{ij} = S^{dr}_{ijkl}(\sigma_{kl} - \sigma^f_{kl}) + S^{gr}_{ijkl}\sigma^f_{kl}. \tag{2.14}$$

Analogously we can consider the strain of the pore space (later we will explicitly assume that the pore space is indeed saturated by a fluid). This strain will change due to σ^e by keeping a constant stress σ^f plus an effect of applying incremental σ^f from inside and outside of the porous sample (i.e. keeping $\sigma^e = const.$). Therefore, using equations (2.10), (2.12) and (2.13) we obtain:

$$-\frac{1}{V_p}\zeta_{ij} = S^P_{ijkl}\sigma^f_{kl} + \frac{1}{\phi}\left(S^{dr}_{klij} - S^{gr}_{klij}\right)\left(\sigma_{kl} - \sigma^f_{kl}\right). \tag{2.15}$$

Let us assume that the pore space is saturated by a fluid or filled with another solid material. The following strain of the infilling material ϵ^f_{ij} results just from the action of a homogeneously distributed stress σ^f_{kl} in it:

$$\epsilon^f_{ij} = S^f_{ijkl}\sigma^f_{kl}. \tag{2.16}$$

In the case of an undrained system, when the infilling material (including a fluid) has no possibility to migrate (at least on average) outside of the solid frame (i.e. the mass of the material infilling the pore space is constant), two strains (2.15) and (2.16) must be equal (see also the comments after equation (2.19)), and thus:

$$S^f_{ijkl}\sigma^f_{kl} = S^P_{ijkl}\sigma^f_{kl} + \frac{1}{\phi}\left(S^{dr}_{klij} - S^{gr}_{klij}\right)\left(\sigma_{kl} - \sigma^f_{kl}\right). \tag{2.17}$$

In this case the so-called undrained compressibility of the saturated porous rock S^u_{ijkl} can be calculated. For this the strain (2.14) must be rewritten by explicitly using the undrained compressibility and the confining stress:

$$S^u_{ijkl}\sigma_{kl} = S^{dr}_{ijkl}(\sigma_{kl} - \sigma^f_{kl}) + S^{gr}_{ijkl}\sigma^f_{kl}. \tag{2.18}$$

Expressing tensor σ^f_{kl} from equation (2.17), substituting it into equation (2.18) and eliminating from there tensor σ_{kl} we obtain the following result:

$$S^u_{ijkl} = S^{dr}_{ijkl} -$$
$$- [S^{gr} - S^{dr}]_{ijmn} [(\phi(S^f - S^P) + S^{dr} - S^{gr})^{-1}]_{mnos} [S^{gr} - S^{dr}]_{oskl}, \tag{2.19}$$

where $[\ldots]_{ijkl}$ denotes corresponding components of tensors obtained after applying tensor operations indicated in the square brackets. The derivation of equation (2.19) has not explicitly used the fact that the material infilling the pore space is a fluid. However, equations (2.16), (2.17) and (2.18) imply a possibility of using homogeneously distributed tensors σ^f_{ij} and S^f_{ijkl} for describing the strain–stress relations in this material. For a pore pressure in a homogeneous fluid infilling the pore space this is really the case. Ciz and Shapiro (2007) heuristically assumed that such uniformly distributed tensors can, on average, describe the strain–stress relations even in the case of a solid infilling material. Under such an assumption, equation (2.19) can be used to approximately compute compliance tensors in such situations, where the pore space is filled by a solid or nearly solid material like a cement, a gas hydrate or an oil with an extremely high viscosity (e.g. a heavy oil). In Section 2.2.2 we will pay more attention to undrained systems.

2.2 Linear stress–strain relations in fluid-saturated materials

In the rest of this book we will concentrate on the classical case, where the infill material of the pore space is a fluid. Therefore, from this point on we must set $\sigma^f_{ij} = -P_p\delta_{ij}$. Then, Hooke's laws (2.14) and (2.15) for the strains of the sample and of the pore space take the following forms, respectively:

$$\epsilon_{ij} = S^{dr}_{ijkl}\sigma_{kl} + (S^{dr}_{ijkk} - S^{gr}_{ijkk})P_p, \tag{2.20}$$

$$-\frac{1}{V_p}\zeta_{ij} = \frac{1}{\phi}(S^{dr}_{klij} - S^{gr}_{klij})\sigma_{kl} + \frac{1}{\phi}(S^{dr}_{ijkk} - S^{gr}_{ijkk} - \phi S^P_{ijkk})P_p. \tag{2.21}$$

2.2.1 Independent confining stress and pore pressure

Quantity $-\zeta_{ij}/V_p$ expresses the strain of the pore space as an independent body. However, the pore space along with the pore fluid, which is assumed to saturate the pore space completely, is a part of the porous sample. The following quantity is more natural for describing deformations of the pore space as a part of the rock sample:

$$\xi_{ij} = -\frac{\zeta_{ij}}{V}. \tag{2.22}$$

We further consider the quantity ξ_{ij} under conditions of a pore pressure acting on the pore-space surface and a confining stress acting on the outer surface of the sample. We also assume a homogeneous skeleton material. This last assumption (implying that equation (2.11) is valid) is often adequate. Elastic moduli of common minerals are of the same order of magnitude and at least an order of magnitude higher than the bulk moduli of common pore fluids. Thus, to a first approximation, spatial variations of grain bulk and shear moduli can be neglected. This allows us to introduce in this section several important relations. Hooke's law for the quantity ξ_{ij} is obtained from equation (2.21):

$$\xi_{ij} = (S_{klij}^{dr} - S_{klij}^{gr})\sigma_{kl} + (S_{ijkk}^{dr} - S_{ijkk}^{gr} - \phi S_{ijkk}^{gr})P_p. \tag{2.23}$$

On the other hand the following dilatational strain of the fluid ϵ^f results from a homogeneously distributed pressure in it:

$$\epsilon^f = -C^f P_p, \tag{2.24}$$

where C^f is the bulk compressibility of the fluid. In respect to the sample volume this is the following strain (note the difference in the strain notations ϵ^f and ε^f):

$$\varepsilon^f = -\phi C^f P_p. \tag{2.25}$$

Therefore, the dilatational characteristic of the pore space

$$\chi = \xi_{ii} - \varepsilon^f = \xi_{ii} + \phi C^f P_p \tag{2.26}$$

expresses volumetric deformations of the pore space due to additional fluid-mass migration only (because the effect of the fluid dilatation has been eliminated by the second terms of the right-hand sides of the equation above).

A demonstration of this fact follows from equations (7.50)–(7.53) of Jaeger *et al.* (2007). Let us consider a change of the fluid volume in a small sample of a porous fluid-saturated material. We denote a variable mass of the fluid in the sample by m_f. The fluid has a pressure-dependent mass density ρ_f. Generally, an incremental change of the fluid volume is produced by applying pore pressure P_p and confining stress σ_{kl}. It is defined by a change in the fluid mass δm_f and a change in the fluid density $\delta \rho_f$. The change of the fluid volume coincides with the change of the pore-space volume δV_p (we consider fully saturated materials). Then (see also equation (1.23))

$$\delta V_p = \delta \frac{m_f}{\rho_f} = \frac{\delta m_f}{\rho_f} - \frac{m_f \delta \rho_f}{\rho_f^2} = \frac{\delta m_f}{\rho_f} - V_p C^f P_p. \qquad (2.27)$$

Dividing this equation by the volume of the rock sample and taking into account (2.22) we obtain:

$$\xi_{ii} = \frac{\delta m_f}{V \rho_f} - \phi C^f P_p. \qquad (2.28)$$

Combining this with (2.26) we see that

$$\chi = \frac{\delta m_f}{V \rho_f}. \qquad (2.29)$$

Thus, χ is indeed equal to the relative change of the pore-space volume due to influent or effluent fluid mass only.

The corresponding Hooke's law is obtained from equations (2.23) and (2.26):

$$\chi = (S_{klii}^{dr} - S_{klii}^{gr}) \sigma_{kl} + (C^{dr} - C^{gr} - \phi C^{gr} + \phi C^f) P_p, \qquad (2.30)$$

where C^{dr} and C^{gr} are bulk compressibilities of the drained skeleton of the rock and of its grain material, respectively.

By inverting for the stress tensors, Hooke's law (2.20) can also be written in the stress–strain form:

$$\sigma_{ij} = C_{ijkl}^{dr} \epsilon_{kl} - \alpha_{ij} P_p. \qquad (2.31)$$

Here $C_{ijkl}^{dr} \equiv [(S^{dr})^{-1}]_{ijkl}$ denotes the components of the stiffness tensor of the drained rock skeleton. We have also introduced a new tensor with components α_{ij}:

$$\alpha_{ij} \equiv C_{ijmn}^{dr} (S_{mnkk}^{dr} - S_{mnkk}^{gr}). \qquad (2.32)$$

Note that the sample strain ϵ_{kl} can be expressed from equation (2.31) over the confining stress and the pore pressure:

$$\epsilon_{kl} = S_{klij}^{dr} (\sigma_{ij} + \alpha_{ij} P_p). \qquad (2.33)$$

Thus, quantity α_{ij} plays the role of a tensorial effective stress coefficient for the strain of the sample (see our discussion of the effective stress in Section 2.9.6).

In an isotropic medium for tensors of compliances and stiffnesses the following relations hold: $S_{mnkk} = \delta_{mn}/(3\lambda + 2\mu)$ and $C_{ijmm} = \delta_{ij}(3\lambda + 2\mu)$. Therefore, tensor α_{ij} simplifies further:

$$\alpha_{ij} = \alpha \delta_{ij}, \qquad (2.34)$$

where

$$\alpha \equiv 1 - \frac{C^{gr}}{C^{dr}} = 1 - \frac{K_{dr}}{K_{gr}} \qquad (2.35)$$

is called the Biot–Willis coefficient. Here K_{dr} and K_{gr} are the bulk moduli of the drained skeleton of the rock and of its grain material, respectively. Note that definition (2.35) of the scalar quantity α is meaningful for anisotropic media too.

In isotropic media, equation (2.31) takes the following form:

$$\sigma_{ij} = \lambda_{dr}\delta_{ij}\epsilon + 2\mu_{dr}\epsilon_{ij} - \alpha P_p\delta_{ij}. \tag{2.36}$$

A double-dot scalar product with δ_{ij} then gives

$$\epsilon = \frac{1}{K_{dr}}\left(\frac{1}{3}\sigma_{ii} + \alpha P_p\right). \tag{2.37}$$

The quantity $\sigma_{ii}/3$ is an average confining normal stress. Taking into account the form of equation (2.37) the combination $\alpha P_p + \sigma_{ii}/3$ is sometimes called an effective stress for the sample's dilatation $\epsilon \equiv \epsilon_{ii}$ (note the difference between the effective stress defined previously in equation (2.6) and the effective stress for the sample's dilatation; we will again discuss this subject in Section 2.9.6).

Let us further simplify the dilatational quantity χ using the notation introduced above. For this we substitute the confining stress from equation (2.31) into equation (2.30):

$$\chi = (S_{klii}^{dr} - S_{klii}^{gr})(C_{klmn}^{dr}\epsilon_{mn} - \alpha_{kl}P_p) + (C^{dr} - C^{gr} - \phi C^{gr} + \phi C^f)P_p$$
$$= \alpha_{mn}\epsilon_{mn} + M_a^{-1}P_p, \tag{2.38}$$

where in the last part of this equation we have introduced a new quantity M_a:

$$M_a \equiv [C^{dr}\alpha^2 + (\alpha - \phi)C^{gr} + \phi C^f - \alpha_{kl}\alpha_{mn}S_{klmn}^{dr}]^{-1}. \tag{2.39}$$

In an isotropic medium $M_a = M$, where the quantity

$$M \equiv \left[\frac{\phi}{K_f} + \frac{\alpha - \phi}{K_{gr}}\right]^{-1} \tag{2.40}$$

is frequently called the Biot modulus (Detournay and Cheng, 1993). From equation (2.38) we finally obtain:

$$P_p = M_a(\chi - \alpha_{mn}\epsilon_{mn}). \tag{2.41}$$

For isotropic systems this equation simplifies to the following important relation between the pore pressure and the dilatations:

$$P_p = M(\chi - \alpha\epsilon). \tag{2.42}$$

This equation and equation (2.37) show that, in elastically isotropic media, of the four dilatational- and pressure-type quantities ϵ, χ, P_p and σ_{ii}, only two are independent. Thus, two more relations between any three of these quantities can additionally be written down:

$$\chi = \left(\frac{1}{M} + \frac{\alpha^2}{K_{dr}}\right) P_p + \frac{\alpha}{3K_{dr}}\sigma_{ii} \equiv \frac{\alpha}{K_{dr}}\left(\frac{1}{B}P_p + \frac{1}{3}\sigma_{ii}\right), \quad (2.43)$$

and

$$\frac{1}{3}\sigma_{ii} = (K_{dr} + \alpha^2 M)\epsilon - \alpha M\chi \equiv K_u\epsilon - \alpha M\chi. \quad (2.44)$$

In the two equations above we have introduced the bulk modulus of an undrained rock K_u and a coefficient B (called also Skempton's coefficient). We will discuss these quantities in the next section. They are explicitly given by equations (2.51) and (2.54), respectively.

For some rock samples, drained and grain bulk moduli as well as Poisson's ratios and other poroelastic constants are given in table 7.2 of the book by Jaeger *et al.* (2007).

2.2.2 Linear stress–strain relations in undrained media

Let us again return to a quite general situation of a possibly heterogeneous anisotropic grain material, and consider a special case of undrained systems (see Figure 2.5). Later we will see that seismic body waves can usually be considered as propagating in undrained rocks.

In the case of an undrained system, when the fluid has no possibility of migrating (at least on average) relative to the solid frame (i.e. the mass of the material infilling

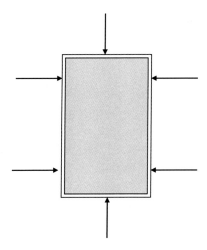

Figure 2.5 A jacketed porous sample representing an undrained system.

the pore space is constant), the strains ε^f and $-\zeta_{ii}/V_p$ must be equal (i.e. the quantity χ is vanishing in equation (2.26)), and thus equations (2.21) and (2.25) yield:

$$- \phi C^f P_p = (S^{dr}_{klii} - S^{gr}_{klii})\sigma_{kl} + (C^{dr} - C^{gr} - \phi C^p)P_p, \qquad (2.45)$$

where we recall that C^{dr}, C^{gr}, C^p are bulk compressibilities of the drained skeleton of the rock, of its grain material and of its pore space. Now the pore pressure can be calculated:

$$P_p = -\frac{1}{3}B_{kl}\sigma_{kl}, \qquad (2.46)$$

where we have introduced a tensor with components B_{kl}:

$$B_{kl} \equiv 3\frac{S^{dr}_{klii} - S^{gr}_{klii}}{C^{dr} - C^{gr} + \phi(C^f - C^p)}. \qquad (2.47)$$

This tensor relates pore-pressure changes in an undrained anisotropic sample to a confining stress tensor causing these changes.

The undrained compliance of the saturated porous rock S^u_{ijkl} can also be calculated. For this strain (2.20) must be rewritten by explicitly using the undrained compliance and the confining stress:

$$S^u_{ijkl}\sigma_{kl} = S^{dr}_{ijkl}\sigma_{kl} + (S^{dr}_{ijkk} - S^{gr}_{ijkk})P_p. \qquad (2.48)$$

Substituting pore pressure P_p from equation (2.46) into equation (2.48) and excluding tensor σ_{kl} we obtain the following result:

$$\begin{aligned}
S^u_{ijkl} &= S^{dr}_{ijkl} - \frac{(S^{dr}_{mmij} - S^{gr}_{mmij})(S^{dr}_{nnkl} - S^{gr}_{nnkl})}{\phi(C^f - C^p) + C^{dr} - C^{gr}} \\
&= S^{dr}_{ijkl} - \frac{1}{3}B_{ij}(S^{dr}_{nnkl} - S^{gr}_{nnkl}).
\end{aligned} \qquad (2.49)$$

This equation was first derived by Brown and Korringa (1975).

Let us assume an orthorhombic symmetry (including isotropy and transverse isotropy) of the drained skeleton and grain materials with coinciding symmetry planes. Then quantities of the type S_{mmij} will vanish if i and j do not coincide. Therefore, the following components of the tensors of compliances and stiffnesses of undrained rocks (in contracted notations) $s_{44}, s_{55}, s_{66}, c_{44}, c_{55}, c_{66}$ are not affected by the presence of fluids in pores. These components coincide with corresponding quantities of the drained skeleton. They describe dynamics of shear deformations of the sample, and thus logically are independent of the pore fluid. Note that owing to the same reasons in orthorhombic media, matrix B_{kl}, defined by equation (2.47), is also diagonal.

If all materials are homogeneous and isotropic, then a twofold double-dot product of the Brown–Korringa equation (2.49) with δ_{ij} and δ_{kl} will give the famous Gassmann equation (Gassmann, 1951):

$$C^u = C^{dr} - \frac{(C^{dr} - C^{gr})^2}{\phi(C^f - C^{gr}) + C^{dr} - C^{gr}}. \tag{2.50}$$

This equation is the basis of fluid-substitution calculations broadly applied in reservoir geophysics. Note that here we set $C^p = C^{gr}$. The Gassmann equation is frequently written in terms of the corresponding bulk moduli, i.e. the bulk modulus of the undrained (i.e. fluid-saturated) medium K_u, the bulk modulus of the drained medium (i.e. of the skeleton) K_{dr}, the bulk modulus of the grain material (i.e. of a material of which the skeleton "is made," frequently also called a solid material) K_{gr}, and the bulk modulus of the fluid K_f:

$$K_u = K_{dr} + \alpha^2 M, \tag{2.51}$$

where M is the Biot modulus defined by equation (2.40). For any type of saturating fluid the bulk modulus of a saturated rock K_u must be larger than the bulk modulus of the rock skeleton K_{dr}. Thus, the modulus M must be always positive. Therefore, the following restriction must be always valid: $\alpha \geq \phi$ (see equation (2.40)). An even more-strict limitation for α can be obtained from the upper bound of the quantity K_{dr} if the rock skeleton is considered as a composite material of porosity ϕ (see Zimmerman *et al.*, 1986; see also our discussion later in Section 2.9.4).

In an undrained homogeneous porous sample an applied confining pressure P_c causes changes of the pore pressure. The general equation (2.46) will describe this if we substitute $\sigma_{kl} = -P_c\delta_{kl}$ and make corresponding simplifications

$$P_p = BP_c = -\frac{1}{3}B\sigma_{kk}, \tag{2.52}$$

where the quantity

$$B \equiv \frac{C^{dr} - C^{gr}}{C^{dr} - C^{gr} + \phi(C^f - C^{gr})} = \frac{1}{3}B_{ii} \tag{2.53}$$

is called the Skempton coefficient. In isotropic rocks the Skempton coefficient can be further simplified:

$$B = \frac{\alpha M}{K_u}. \tag{2.54}$$

Note that we have already used this relation in equation (2.43).

Finally, in undrained systems $\chi = 0$, and equation (2.41) yields

$$P_p = -M_a\alpha_{mn}\epsilon_{mn}. \tag{2.55}$$

In isotropic systems this equation simplifies to

$$P_p = -M\alpha\epsilon. \qquad (2.56)$$

Substituting relation (2.55) into equation (2.31) yields

$$\sigma_{ij} = C^u_{ijkl}\epsilon_{kl}, \qquad (2.57)$$

where the stiffness tensor of an undrained rock C^u_{ijkl} is given by

$$C^u_{ijkl} = C^{dr}_{ijkl} + \alpha_{ij}\alpha_{kl}M_a. \qquad (2.58)$$

In undrained isotropic systems the substitution of (2.56) into (2.36) yields the following form of Hooke's law:

$$\sigma_{ij} = (\lambda_{dr} + \alpha^2 M)\delta_{ij}\epsilon + 2\mu_{dr}\epsilon_{ij}, \qquad (2.59)$$

which is consistent with the Gassmann equation (2.51) and with a well-known result (already discussed above in a general form for orthorhombic media) for the undrained shear modulus:

$$\mu_u = \mu_{dr}. \qquad (2.60)$$

2.3 Fluid flow and dynamic poroelasticity

Our consideration above was restricted to static poroelastic phenomena. Dynamic effects are given by a time-dependent balance of forces (in other words, momentum conservation or Newton's second law) acting on a representative volume of a porous fluid-saturated rock. Therefore, a description of the rock dynamics must also take into account processes related to fluid flows.

We continue to consider a differentially small (but representative) elementary volume of a fluid-saturated rock. We introduce a velocity vector $\mathbf{q}$ of an average displacement of the volume of the pore fluid with respect to the skeleton (this displacement is sometimes called a global flow). This vector is also called the fluid flux vector (see Jaeger *et al.*, 2007) or the vector of filtration velocity. Its component q_i is equal to a volume of the fluid crossing in a unit time a unit planar surface of the rock (attached to the solid skeleton) with a normal directed along axis $\hat{\mathbf{x}}_i$. Let us further denote a vector of the corresponding average displacement of the fluid volume (relative to the solid skeleton) as $\mathbf{w} = (w_1, w_2, w_3)$. Therefore,

$$\frac{\partial\mathbf{w}}{\partial t} = \mathbf{q}. \qquad (2.61)$$

We recall that $\mathbf{q}$ describes a motion of the volume of fluid in the pore space. Therefore, according to definition (2.61) we can also write

$$\mathbf{w} = \phi(\mathbf{u}^f - \mathbf{u}), \qquad (2.62)$$

where $\mathbf{u}^f$ is a displacement vector of an average (representative) fluid particle relative to its initial location (in a coordinate system common for the solid as well as the fluid). We recall also that $\mathbf{u}$ is a displacement of an average (representative) particle of the solid relative to its initial location. It follows from (2.61) that the dilatation $\partial w_i / \partial x_i$ is equal to a fluid-volume decrement in the corresponding elementary volume of the rock. Corresponding to the definition (2.26) of the quantity χ we must have:

$$\frac{\partial w_i}{\partial x_i} = -\chi. \tag{2.63}$$

Taking the time derivative at the left- and right-hand sides of this equation, we obtain

$$\frac{\partial \chi}{\partial t} + \frac{\partial q_i}{\partial x_i} = 0. \tag{2.64}$$

This equation expresses the fluid-mass conservation in terms of migration of an additional fluid volume per unit time necessary to compensate temporal changes of the pore-space volume per unit volume of the rock. This is a form of the so-called fluid-continuity equation (Jaeger *et al.*, 2007).

Taking into account the fluid-mass component permanently remaining in the pore space during a deformation process, the continuity equation can be rewritten in terms of masses explicitly (Coussy, 2004; Segall, 2010):

$$\frac{\partial \rho_f \phi}{\partial t} + \frac{\partial \rho_f q_i}{\partial x_i} = 0, \tag{2.65}$$

where the first term on the left-hand side describes temporal changes of a fluid mass per unit volume of the rock, and the second term describes divergence of the vector of the fluid-mass flux.

An experimental observation called Darcy's law relates a stationary filtration velocity of a laminar fluid flow in a porous system to a gradient of the pore pressure:

$$q_i = -\frac{k_{ij}}{\eta} \frac{\partial P_p}{\partial x_j}, \tag{2.66}$$

where η is the dynamic viscosity of the pore fluid. This equation defines a tensor with components k_{ij}, which is called the tensor of permeability. Here (and generally, if not otherwise specified) the gravity force is neglected. Darcy's law can be also written in a form explicitly expressing an equilibrium between elastic and friction forces acting on the fluid:

$$\eta [k^{-1}]_{ij} q_j = -\frac{\partial P_p}{\partial x_i}, \tag{2.67}$$

where $[k^{-1}]_{ij}$ denotes corresponding components of the tensor inverse to the tensor of permeability.

Summarizing, equation (2.31) describes the elastic stress acting on the rock. Its spatial derivatives represent elastic forces. Equation (2.41) describes the fluid pressure. Its gradient gives a fluid driving force. The friction force acting on the fluid participates in Darcy's law (2.67). Biot (1962) combined these equations into the following system of equations describing the balance of dynamic forces applied to an elementary representative volume of a fluid-saturated rock:

$$\frac{\partial}{\partial x_j} \left(\left(C^{dr}_{ijkl} + M_a \alpha_{ij} \alpha_{kl} \right) \epsilon_{kl} - \alpha_{ij} M_a \chi \right) = \rho \frac{\partial^2}{\partial t^2} u_i + \rho_f \frac{\partial}{\partial t} q_i, \qquad (2.68)$$

$$-\frac{\partial}{\partial x_i} (M_a (\chi - \alpha_{mn} \epsilon_{mn})) - \eta [k^{-1}]_{ij} q_j = \rho_f \frac{\partial^2}{\partial t^2} u_i + \rho_f m_{ij} \frac{\partial}{\partial t} q_j. \qquad (2.69)$$

Here the first equation expresses a force balance for a small (but still representative) element of a fluid-saturated rock as a whole. The second equation expresses a force balance for the fluid in this rock element. The left-hand sides of these equations express the elastic and friction forces. Their right-hand sides describe inertial forces. We have introduced here some new notations: ρ is the total density of the fluid-saturated rock; the matrix m_{ij} (here made dimensionless) was introduced by Biot (1962) as a volume-averaged matrix product $a_{ki} a_{kj}$, where $a_{ki}(\mathbf{r})$ is a matrix coefficient relating a vector field of the local micro-velocity of the fluid in pores (in respect to the skeleton) to the global (macro) filtration-velocity vector q_i.

System (2.68)–(2.69) is a system of six equations for all displacement and filtration-velocity components. It is restricted in its validity for a not too high frequency range, where the fluid flow in pores can be assumed to be of Poiseuille type (laminar viscous flow in a tube). Biot (1956) estimates that one quarter of the shear-wave length (the boundary layer) in a viscous fluid should not become shorter than the pore diameter d_p. This leads to the following restriction of the frequency range:

$$\omega < \omega_l \equiv \frac{\pi^2 \eta}{2 d_p^2 \rho_f}, \qquad (2.70)$$

where ω is an angular frequency of a wave motion and ω_l is a characteristic frequency of the Poiseuille flow. For example, assuming rather large pores of $d_p = 10^{-6}$ m and a saturating flud with physical properties close to those of the water, $\rho_f = 10^3$ kg/m^3 and $\eta = 10^{-3}$ Pa·s, we obtain $\omega_l \approx 5 \cdot 10^6$. Therefore, the characteristic frequency of the Poiseuille flow is very high, of the order of MHz. It becomes even higher for smaller pores. If condition (2.70) is not satisfied, the viscosity, the permeability and the matrix m_{ij} will become frequency dependent.

However, this fact is not significant for our further consideration, which will be restricted later to a much lower frequency range.

Equation system (2.68)–(2.69) describes four wavefields independently propagating in a homogeneous anisotropic poroelastic continuum (Carcione, 2007). Three of them are analogous to the three elastic waves in a general anisotropic solid. The fourth one corresponds to Biot's slow wave, which we will discuss later in detail.

To discuss the wave motions we first assume elastic and hydraulic isotropy and homogeneity of rocks. The hydraulic isotropy is given by statistical isotropy of the pore space. This leads to statistical isotropy of the fluid micro-velocity field. Therefore, the following relation holds: $k_{ij} = k\delta_{ij}$, where k is a scalar permeability of the rock. Further, $m_{ij} = T\delta_{ij}/\phi$, where $T \geq 1$ is a parameter (called tortuosity) of the order of a mean square of a ratio of two lengths: a trajectory length of a pore-fluid particle between two points of the pore space to a straight line distance between these two points (see Carcione, 2007, and references therein). Equations (2.68)–(2.69) then simplify to the following ones:

$$(\lambda_{dr} + M\alpha^2)\frac{\partial \epsilon}{\partial x_i} - \alpha M \frac{\partial \chi}{\partial x_i} + 2\mu_{dr}\frac{\partial \epsilon_{ij}}{\partial x_j} = \rho\frac{\partial^2}{\partial t^2}u_i + \rho_f\frac{\partial}{\partial t}q_i, \qquad (2.71)$$

$$-\frac{\partial}{\partial x_i}(M(\chi - \alpha\epsilon)) = \rho_f\frac{\partial^2}{\partial t^2}u_i + \rho_f\frac{T}{\phi}\frac{\partial}{\partial t}q_i + \frac{\eta}{k}q_i. \qquad (2.72)$$

This system of equations describes propagation of two dilatational waves and one shear (rotational) wave (see Biot, 1956). These equations explicitly contain a dilatation of a porous fluid-saturated elementary rock volume $\epsilon = \nabla\mathbf{u}$ (equal to the dilatation of the skeleton) and a dilatation of the pore space of this elementary rock volume caused by a fluid flow in or out of this volume: $\chi = -\nabla\mathbf{w}$.

2.4 Poroelastic wavefields

To analyze possible wave motions in more detail we apply the divergence operator to equations (2.71) and (2.72). Taking into account that

$$\frac{\partial^2\epsilon_{ij}}{\partial x_i\partial x_j} = \frac{\partial^2\epsilon}{\partial x_i^2} = \nabla^2\epsilon, \qquad (2.73)$$

and using also continuity equation (2.64), we obtain

$$\nabla^2[H\epsilon - \alpha M\chi] = \rho\frac{\partial^2}{\partial t^2}\epsilon - \rho_f\frac{\partial^2}{\partial t^2}\chi, \qquad (2.74)$$

$$\nabla^2(M\alpha\epsilon - M\chi) = \rho_f\frac{\partial^2}{\partial t^2}\epsilon - \rho_f\frac{T}{\phi}\frac{\partial^2}{\partial t^2}\chi - \frac{\eta}{k}\frac{\partial}{\partial t}\chi. \qquad (2.75)$$

We have introduced here a poroelastic modulus

$$H \equiv \lambda_{dr} + 2\mu_{dr} + M\alpha^2. \tag{2.76}$$

This quantity is equal to the P-wave modulus of isotropic undrained rocks. This can be seen from a comparison of equation (2.76) with equations (2.51) and (2.60).

Let us also consider rotational (shear) wave motions. For this we apply the **curl** operator (also called the rotor operator) to equations (2.71) and (2.72). This operator is equal to a vectorial product of the vector ∇ with the vectors on the left- and right-hand sides of these equations. Taking into account that $\nabla \times \nabla \epsilon = 0$, $\nabla \times \nabla \chi = 0$ and that $2\partial \epsilon_{ij}/\partial x_j$ is the ith component of the vector $\nabla^2 \mathbf{u} + \nabla(\nabla \mathbf{u})$ (see equations (1.43)–(1.46)), we obtain:

$$\mu_{dr} \nabla^2 \nabla \times \mathbf{u} = \rho \frac{\partial^2}{\partial t^2} \nabla \times \mathbf{u} + \rho_f \frac{\partial}{\partial t} \nabla \times \mathbf{q}, \tag{2.77}$$

$$0 = \rho_f \frac{\partial^2}{\partial t^2} \nabla \times \mathbf{u} + \rho_f \frac{T}{\phi} \frac{\partial}{\partial t} \nabla \times \mathbf{q} + \frac{\eta}{k} \nabla \times \mathbf{q}. \tag{2.78}$$

Comparing both equation systems, (2.74)–(2.75) and (2.77)–(2.78), we observe that they are mutually independent. Therefore, these two equation systems describe independent wavefields, respectively. Indeed, dilatational quantities are not involved in equations (2.77)–(2.78). Equally, rotational quantities are not involved in equations (2.74)–(2.75). Thus, dilatational and rotational motions propagate mutually independently.

To further analyze wave motions it is convenient to write equation systems (2.74)–(2.75) and (2.77)–(2.78) in terms of the displacement vector fields $\mathbf{u}$ and $\mathbf{w}$. The vector fields $\mathbf{u}$ and $\mathbf{w}$ can always be represented as sums of their dilatational and rotational vector-field parts $\mathbf{u_d}$, $\mathbf{u_r}$ and $\mathbf{w_d}$, $\mathbf{w_r}$, respectively. Thus:

$$\begin{aligned} \mathbf{u} &= \mathbf{u_d} + \mathbf{u_r}, \\ \mathbf{w} &= \mathbf{w_d} + \mathbf{w_r}, \\ \nabla \times \mathbf{u_d} &= 0, \\ \nabla \mathbf{u_r} &= 0, \\ \nabla \times \mathbf{w_d} &= 0, \\ \nabla \mathbf{w_r} &= 0. \end{aligned} \tag{2.79}$$

Taking into account these decompositions and the fact that, if both the vector product and the scalar product of the vector ∇ with a vector field are equal to zero everywhere in space, then such a vector field (which must be also equal to zero at infinite distances from its source) will be equal to zero too, we obtain two other equation systems, respectively:

$$\nabla^2 \left[H\mathbf{u_d} + \alpha M\mathbf{w_d} \right] = \rho \frac{\partial^2}{\partial t^2}\mathbf{u_d} + \rho_f \frac{\partial^2}{\partial t^2}\mathbf{w_d} \tag{2.80}$$

$$\nabla^2 (M\alpha\mathbf{u_d} + M\mathbf{w_d}) = \rho_f \frac{\partial^2}{\partial t^2}\mathbf{u_d} + \rho_f \frac{T}{\phi} \frac{\partial^2}{\partial t^2}\mathbf{w_d} + \frac{\eta}{k}\frac{\partial}{\partial t}\mathbf{w_d} \tag{2.81}$$

and

$$\mu_{dr} \nabla^2 \mathbf{u_r} = \rho \frac{\partial^2}{\partial t^2}\mathbf{u_r} + \rho_f \frac{\partial^2}{\partial t^2}\mathbf{w_r} \tag{2.82}$$

$$0 = \rho_f \frac{\partial^2}{\partial t^2}\mathbf{u_r} + \rho_f \frac{T}{\phi}\frac{\partial^2}{\partial t^2}\mathbf{w_r} + \frac{\eta}{k}\frac{\partial}{\partial t}\mathbf{w_r}. \tag{2.83}$$

We observe that in an infinite homogeneous isotropic porous fluid-saturated medium, dilatational and rotational parts of the displacement fields are described by completely independent equation systems. Thus, perturbations of these field parts propagate mutually independently. Because equation systems (2.80)–(2.81) and (2.82)–(2.83) describe the displacement fields completely, then they describe all possible wave motions. Therefore, an infinite homogeneous medium can independently propagate dilatational and rotational waves only. Note that additional boundary conditions (for example, a free surface) can lead to the emergence of additional independent wave motions (for example, surface waves), where the dilatational and rotational displacement fields can be coupled. Such wavefields also are solutions of the initial equation system (2.71) and (2.72). They are beyond the scope of this book.

2.4.1 Dispersion relations for poroelastic wavefields

Any wavefield can be decomposed (by Fourier transformation) into a superposition of time-harmonic plane waves. Thus, further analysis of possible wave motions is performed by substituting into equation systems (2.80)–(2.81) and (2.82)–(2.83) trial wavefield solutions in the forms:

$$\begin{aligned}
\mathbf{u_d} &= \mathbf{u_{d0}}e^{i(\omega t - \boldsymbol{\kappa_d}\mathbf{r})}, \\
\mathbf{w_d} &= \mathbf{w_{d0}}e^{i(\omega t - \boldsymbol{\kappa_d}\mathbf{r})}, \\
\mathbf{u_r} &= \mathbf{u_{r0}}e^{i(\omega t - \boldsymbol{\kappa_r}\mathbf{r})}, \\
\mathbf{w_r} &= \mathbf{w_{r0}}e^{i(\omega t - \boldsymbol{\kappa_r}\mathbf{r})}.
\end{aligned} \tag{2.84}$$

Here $\boldsymbol{\kappa_d}$ and $\boldsymbol{\kappa_r}$ are unknown wave vectors and $\mathbf{u_{d0}}$, $\mathbf{w_{d0}}$, $\mathbf{u_{r0}}$ and $\mathbf{w_{r0}}$ are unknown polarization vectors of the corresponding plane waves.

We restrict our consideration of wave motions to isotropic systems. Thus, all propagation directions are equivalent. For example, we can consider plane waves propagating along the x-direction (i.e. $\hat{\mathbf{x}}_1$-direction). Then the scalar products $\boldsymbol{\kappa_d}\mathbf{r}$

and $\kappa_r \mathbf{r}$ can be replaced by $\kappa_d x$ and $\kappa_r x$, respectively. Here κ_d and κ_r are corresponding wave numbers (i.e. x-projections of the wave vectors). Further, because of the conditions (2.79), the polarization vectors of the dilatational waves must be parallel to $\hat{\mathbf{x}}_1$ and can be replaced by the corresponding vector components, u_{d0} and w_{d0}. The rotational wave motions must be polarized orthogonally to the propagation direction. Moreover, equation (2.83) indicates that the vectors $\mathbf{u}_{r0}$ and $\mathbf{w}_{r0}$ must be parallel. We can choose the y-direction (i.e. $\hat{\mathbf{x}}_2$-direction) to be parallel to the polarization of the rotational waves and replace the rotational polarization vectors by their corresponding vector components u_{r0} and w_{r0}. Taking these simplifications into account we obtain:

$$0 = (\kappa_d^2 H - \omega^2 \rho)u_{d0} + (\kappa_d^2 \alpha M - \omega^2 \rho_f)w_{d0} \tag{2.85}$$

$$0 = (\kappa_d^2 M\alpha - \omega^2 \rho_f)u_{d0} + (\kappa_d^2 M - \omega^2 \rho_f \frac{T}{\phi} + i\omega\frac{\eta}{k})w_{d0} \tag{2.86}$$

and

$$0 = (\kappa_r^2 \mu_{dr} - \omega^2 \rho)u_{r0} - \omega^2 \rho_f w_{r0} \tag{2.87}$$

$$0 = \rho_f u_{r0} + (\rho_f \frac{T}{\phi} - i\frac{\eta}{k\omega})w_{r0}. \tag{2.88}$$

These are two linear algebraic equation systems for wavefield polarizations u_{d0}, w_{d0}, u_{r0} and w_{r0}. They will have non-zero solutions only if the corresponding determinants are equal to zero, respectively:

$$0 = (\kappa_d^2 H - \omega^2 \rho)(\kappa_d^2 M - \omega^2 \rho_f \frac{T}{\phi} + i\omega\frac{\eta}{k}) - (\kappa_d^2 \alpha M - \omega^2 \rho_f)^2 \tag{2.89}$$

and

$$0 = (\kappa_r^2 \mu_{dr} - \omega^2 \rho)(\rho_f \frac{T}{\phi} - i\frac{\eta}{k\omega}) + \omega^2 \rho_f^2. \tag{2.90}$$

These are dispersion equations relating wave numbers and frequencies of possible wave motions (note the analogy to equation (1.33)).

We consider first rotational wave motions. Solving equation (2.90) we obtain the following wave number:

$$\kappa_r^2 = \omega^2 \frac{\rho}{\mu_{dr}}(1 - \frac{\rho_f \phi}{T\rho}(1 - i\frac{\phi\eta}{\rho_f Tk\omega})^{-1}). \tag{2.91}$$

For frequencies significantly below the critical frequency (also called the characteristic frequency of the global flow)

$$\omega_c \equiv \frac{\phi\eta}{\rho_f Tk} \tag{2.92}$$

the absolute value of the quantity $(1 - i\eta\phi/(\rho_f T k\omega))^{-1}$ is much smaller than 1, and we will obtain $\kappa_r^2 = \omega^2\rho/\mu_{dr}$. This gives the following phase velocity:

$$c_s = \sqrt{\frac{\mu_{dr}}{\rho}}. \tag{2.93}$$

This is a velocity of a seismic shear wave in an undrained isotropic rock (note the difference in the density of an undrained medium and of its drained skeleton).

Usually for realistic rocks and fluids the critical frequency $\omega_c/(2\pi)$ is of the order of 10^5 Hz or higher. It can be even higher than the frequency limit of a laminar flow in pores (see equation (2.70)). Therefore, in such fields as borehole seismology, reflection seismology, earthquake seismology and hydrogeology, the system of poroelastic equations can be considered in the low-frequency (in respect to ω_c) range.

The low-frequency range has the following meaning in terms of forces acting on the fluid in the pore space. In the low-frequency range viscous forces dominate over inertial forces (acting on the fluid due to wave-motion-related variations of velocities of the solid skeleton). In the high-frequency range the situation is opposite. Accelerations of the solid skeleton become so high that they dominate over the viscous friction between the solid and the fluid.

In the low-frequency range, all terms of order ω^2 can be neglected with respect to terms of order ω and lower. Correspondingly, equation (2.89) simplifies to

$$0 = (\kappa_d^2 H - \omega^2\rho)(\kappa_d^2 M + i\omega\frac{\eta}{k}) - (\kappa_d^2\alpha M - \omega^2\rho_f)^2. \tag{2.94}$$

This is a bi-quadratic equation having two solutions for κ_d^2. Thus, we observe a possibility of two types of dilatational waves. The wave number can be of the order of ω or of a lower ω-order. Higher orders in ω are excluded in the low-frequency range. We first assume that $\kappa_d = O(\omega)$. Then the second term in the equation above is of the order $O(\omega^4)$. It can be neglected in respect to the first one (which is $O(\omega^3)$), and we obtain

$$0 = \kappa_d^2 H - \omega^2\rho. \tag{2.95}$$

Thus, the phase velocity of this first dilatational wave is:

$$c_{PI} = \sqrt{\frac{H}{\rho}}. \tag{2.96}$$

This is a velocity of a seismic longitudinal wave in an undrained isotropic rock (compare this expression to equations (2.76), (2.51) and (2.60)). Thus, the quantity H is indeed the modulus of seismic P-waves.

Further, we consider a possibility that the wave number can be of a lower order than $O(\omega)$. Then we can neglect terms $O(\omega^2)$ in respect to the terms $O(\omega)$ and $O(\kappa_d^2)$, and equation (2.94) simplifies to:

$$0 = \kappa_d^2 H(\kappa_d^2 M + i\omega\frac{\eta}{k}) - (\kappa_d^2 \alpha M)^2. \tag{2.97}$$

Solving this equation we obtain:

$$\kappa_d^2 = -i\omega\frac{1}{D}, \tag{2.98}$$

where

$$D = \frac{k}{\eta S} \tag{2.99}$$

and

$$S = \frac{H}{(\lambda_{dr} + 2\mu_{dr})M}. \tag{2.100}$$

Quantity S is called the storage coefficient (see Jaeger *et al.*, 2007, p. 188) or sometimes also the uniaxial storage coefficient. The coefficient D is called the hydraulic diffusivity.

For plane waves propagating in the direction of positive x-values, equation (2.98) provides us with the following wave number:

$$\kappa_d = (1 - i)\sqrt{\frac{\omega}{2D}}. \tag{2.101}$$

We observe that the absolute values of the real and imaginary parts of the wave number are equal. Such a feature is typical for so-called diffusion waves (Mandelis, 2000) which are characterized by a very strong dissipation of their (initially elastic) energy. We will consider this in more detail. Frequency divided by the real part of the wave number yields the phase velocity of the corresponding wave. The imaginary part of the wave number is equal to the attenuation coefficient of the wave. Indeed, the corresponding dilatational plane wave (2.84) has the following form:

$$\mathbf{u_d} = \mathbf{u_{d0}}e^{i\omega(t-x/\sqrt{2D\omega})}e^{-x\sqrt{\omega/(2D)}},$$
$$\mathbf{w_d} = \mathbf{w_{d0}}e^{i\omega(t-x/\sqrt{2D\omega})}e^{-x\sqrt{\omega/(2D)}}. \tag{2.102}$$

Therefore, for the phase velocity c_{PII} of this second dilatational wave we obtain

$$c_{PII} = \sqrt{2D\omega} = \sqrt{4\pi D/t}, \tag{2.103}$$

where the last part of this equation yields the phase velocity of the second dilatational wave with a time period t (i.e. with an angular frequency $\omega = 2\pi/t$). The wavelength of the second dilatational wave λ_{PII} is given by

$$\lambda_{PII} = c_{PII}\frac{2\pi}{\omega} = \sqrt{\frac{8\pi^2 D}{\omega}} = \sqrt{4\pi Dt}. \tag{2.104}$$

Here again the last part of the equation yields the wavelength of the second dilatational wave with time period t (angular frequency $\omega = 2\pi/t$).

The attenuation coefficient α_{PII} of the second dilatational wave (see the second exponential factor on the right-hand side of equations (2.102)) is given by

$$\alpha_{PII} = \sqrt{\frac{\omega}{2D}} = \sqrt{\frac{\pi}{Dt}}. \tag{2.105}$$

The last part of equation (2.105) yields the attenuation coefficient of the second dilatational wave with time period t and the angular frequency $2\pi/t$, respectively.

Therefore, over a distance equal to a single wavelength the amplitude of this wave decreases in $\exp 2\pi \approx 535$ times. Over a distance of a single wavelength the energy of the second dilatation wave will be nearly completely transformed into heat. Simultaneously we observe that the wave numbers of the rotational and first dilatational waves do not have imaginary parts. Therefore, for frequencies much smaller than ω_c the attenuation of these waves is vanishingly small.

2.4.2 Particle motions in poroelastic wavefields

To clarify the reason for such contrasting behavior of the wave motions we will consider their polarizations. First we consider again the system of equations (2.87)–(2.88). Recalling in the first one that $\kappa_r^2 = \omega^2 \rho/\mu_{dr}$, and in the second one that we consider the low-frequency limit, we obtain $w_{r0} = 0$. In other words, in rotational wave motions, particles of the solid and of the fluid have the same displacements. They move together and do not move with respect to each other. Analogous consideration of equation system (2.85)–(2.86) along with equation (2.95) yields for the first dilatational wave $w_{d0} = 0$. Note that also in equation (2.86) we must consider the low-frequency limit, i.e. the last term with factor w_0 is dominant and it must vanish in this limit. Thus, also in this wave, both fluid and solid particles move with equal displacements and do not move in respect to each other. Such a polarization feature of these two wave types is responsible for their vanishing attenuation in the low-frequency range.

We consider again equation system (2.85)–(2.86). For the second dilatational wave, where the wave number satisfies relation (2.98), both equations provide the same result (again, higher-order terms in ω must be neglected):

$$Hu_{d0} = -\alpha M w_{d0}. \tag{2.106}$$

Substituting here equation (2.62) we obtain:

$$(H - \phi\alpha M)u_{d0} = -\alpha\phi M u_{d0}^f. \tag{2.107}$$

Both quantities, $H - \phi\alpha M = \lambda_{dr} + 2\mu_{dr} + \alpha(\alpha - \phi)M$ and $\alpha\phi M$, are always positive (because $\alpha \geq \phi$; see also our discussion after equation (2.51)). Thus, in this case, wave particles of the fluid and of the solid move in opposite directions. This produces a strong friction loss and leads to a strong attenuation of the wave.

Summarizing, in a homogeneous isotropic fluid-saturated poroelastic continuum there are three types of waves propagating a perturbation from a source to a point of observation. These are two types of elastic body waves, P and S (these are usual longitudinal and shear seismic waves), and a highly dissipative second dilatational (longitudinal) wave. Usually the phase velocity of the latter is very small. It tends to zero in the zero-frequency limit. Below the critical frequency the second dilatational wave has the smallest phase velocity from the three types of waves. Thus, frequently this wave is also called the slow wave.

2.4.3 Fluid flow and attenuation of seismic waves

Fluid flows can strongly contribute to the attenuation of seismic waves. In the theoretical model of a homogeneous poroelastic medium (as described earlier in this section) the dissipation of seismic waves (a very insignificant one) is caused by the global flow, i.e. a relative movement between the solid matrix and the fluid on the scale of a wavelength. In a heterogeneous medium, additional flow phenomena may cause additional (sometimes very significant) dissipation of the elastic energy. For example, seismic wavefields in a stack of layers with variable compliances cause a flow of the pore fluid (also called mesoscopic flow, local flow or interlayer flow) across interfaces from more compliant layers into stiffer layers and vice versa (White, 1983; Norris, 1993; Gurevich and Lopatnikov, 1995; Gelinsky and Shapiro, 1997; Pride *et al.*, 2004). Indeed, the fluid pressure tends to equilibrate between adjacent heterogeneities (e.g., layers) by a motion of the viscous pore fluid across the boundaries of heterogeneities. This fluid flow produces strong frictional losses, and it is a reason for the strong dissipation of seismic waves. The phenomenon of the mesoscopic fluid flow can be also described as an excitation of diffusion slow waves at interfaces of heterogeneities by scattering of the normal seismic P- (and also sometimes S-) waves.

The mesoscopic flow is especially significant at the characteristic frequencies where the scale of rock heterogeneities is comparable with the wavelength of the slow wave. Such a flow is possibly one of main causes of the attenuation and dispersion of seismic waves in the reflection seismic frequency range, 10–100 Hz (Shapiro and Hubral, 1999). The various attenuation mechanisms are approximately additive, dominated by the mesoscopic flow at low frequencies and in the seismic frequency range. Elastic scattering is significant over a broad frequency range from seismic to sonic frequencies. The global flow contributes mainly in the range of ultrasonic frequencies. One more type of fluid flow, the so-called squirt flow, seems to be mainly responsible for seismic-wave attenuation in the frequency range of several 1000 Hz. It takes place due to fluctuation of compliances of the pore space caused by its micro-scale geometry (see Gurevich *et al.*, 2010, and literature cited therein). The squirt can be also understood as a fluid flow between stiff and compliant parts of the pore space (see our discussion at the end of this chapter). From the seismic frequency range up to ultrasonic frequencies, attenuation of seismic waves due to heterogeneity is strongly enhanced compared with that in a homogeneous poroelastic medium.

2.4.4 Slow wavefields in the low-frequency range

Practical experience shows that a cloud of fluid-induced microseismic events requires hours or even days to reach a size of several hundred meters. This is definitely too slow a process to be attributed to elastic-wave propagation. Elastic waves in well consolidated rocks propagate during seconds on kilometer-scale distances. However, elastic waves are primarily responsible for the elastic stress equilibration. This indicates that the triggering of at least some microseismic events has to be related to the slow wave.

In the following discussion we will derive a wave equation for slow waves. Equation (2.98) shows that, in the range of frequencies below the critical one, the wave number of slow waves is of order $O(\omega^{1/2})$. This means that for slow wavefields time derivatives can be neglected in respect to the same- or lower-order spatial derivatives. Equivalently, inertial terms can be neglected in respect to static and dissipative terms because inertial forces are negligible in comparison to the viscous friction and elastic forces.

We simplify equations (2.71) and (2.72) according to this approximation (assuming initially an elastic and hydraulic isotropy and homogeneity of rocks):

$$(\lambda_{dr} + M\alpha^2)\frac{\partial \epsilon}{\partial x_i} - \alpha M \frac{\partial \chi}{\partial x_i} + 2\mu_{dr}\frac{\partial \epsilon_{ij}}{\partial x_j} = 0, \tag{2.108}$$

$$-\frac{\partial}{\partial x_i}(M(\chi - \alpha\epsilon)) = \frac{\eta}{k}\frac{\partial}{\partial t}w_i. \tag{2.109}$$

We recall that the slow wavefields are dilatational ones. In other words, the displacement fields $\mathbf{u} = \mathbf{u_d}$ and $\mathbf{w} = \mathbf{w_d}$ are irrotational (see equations (2.79)). Taking also into account that the third term on the left-hand side of equation (2.108) can be rewritten by using relations (1.43), (1.48) and $\nabla^2 \mathbf{u_d} = \nabla \epsilon$ we can simplify this equation further:

$$H \frac{\partial \epsilon}{\partial x_i} - \alpha M \frac{\partial \chi}{\partial x_i} = 0. \tag{2.110}$$

Taking this into account by applying the divergence operator to equation (2.109) we obtain a closed equation describing slow wavefields:

$$D \nabla^2 \chi = \frac{\partial}{\partial t} \chi. \tag{2.111}$$

This is a diffusion equation for the volume of filtrating fluid per unit volume of rocks. It is not surprising that the "wave" equation for the slow wavefields is a diffusion one. Above we have already seen that the slow waves are waves of a diffusion type.

If we consider slow wavefields in infinite media, where at infinity both perturbations, ϵ and χ, disappear, we obtain from equation (2.110):

$$\epsilon = \frac{\alpha M}{H} \chi. \tag{2.112}$$

Then, taking into account equation (2.42) we obtain that in slow wavefields

$$P_p = \frac{\chi}{S}. \tag{2.113}$$

This transforms the diffusion equation for χ into the diffusion equation for the pore pressure:

$$D \nabla^2 P_p = \frac{\partial}{\partial t} P_p. \tag{2.114}$$

Clearly, the same equation will be valid also for ϵ. Therefore, the slow wavefields describe well-known and simple physical phenomena. These are pore-pressure diffusion and fluid-mass diffusion in a porous rock.

The slow wavefields can also be expressed in terms of diffusion-like relaxation of elastic stresses. For this we must combine equation (2.44) with equation (2.112). This yields:

$$\sigma_{ii} = -\frac{4\mu_{dr} \alpha M}{H} \chi. \tag{2.115}$$

Thus, in the slow wavefield the same diffusion equation (2.114) is also valid for the quantity σ_{ii}.

Let us now consider a more general situation: elastically homogeneous and isotropic rocks, which can be hydraulically anisotropic and heterogeneous. Then, for the slow wavefields equations (2.68) and (2.69) will simplify so that equation (2.108) will be still valid and equation (2.69) will transform into

$$-\frac{k_{ij}}{\eta}\frac{\partial}{\partial x_j}(M(\chi-\alpha\epsilon)) = \frac{\partial}{\partial t}w_i. \tag{2.116}$$

Applying the divergence operator yields

$$\frac{\partial}{\partial x_i}\frac{k_{ij}}{\eta}\frac{\partial}{\partial x_j}(M(\chi-\alpha\epsilon)) = \frac{\partial}{\partial t}\chi. \tag{2.117}$$

It is clear that, if we look for a solution of equation system (2.108) and (2.117) under the separation assumption (2.79), we will obtain a closed equation system (2.110) and (2.117) describing irrotational wavefields only. Therefore, we observe that in hydraulically heterogeneous and anisotropic media dilatational slow wavefields propagate independently from other wave motions. Moreover, these two equations can be combined into the following one:

$$\frac{\partial}{\partial x_i}D_{ij}\frac{\partial}{\partial x_j}\chi = \frac{\partial}{\partial t}\chi, \tag{2.118}$$

where

$$D_{ij} = \frac{k_{ij}}{\eta S} \tag{2.119}$$

is a tensor of hydraulic diffusivity. This is a quite general equation of diffusion of the dilatational perturbation χ.

Assuming finally that we consider slow wavefields in an infinite medium with wavefields vanishing at infinity, we will still be able to apply equation (2.113) and thus:

$$\frac{\partial}{\partial x_i}D_{ij}\frac{\partial}{\partial x_j}P_p = \frac{\partial}{\partial t}P_p. \tag{2.120}$$

We observe that the dynamic equation describing slow wavefields in infinite elastically homogeneous and isotropic but hydraulically heterogeneous and anisotropic media is the pressure-diffusion equation (2.120). The diffusion equation (2.118) is more general. It does not require wavefields to be vanishing at limiting boundaries of a finite domain.

2.4.5 Elastic P- and S-waves in the low-frequency range

Let us briefly consider a possibility of the wave numbers to be of order $O(\omega)$. As we have seen above this is a feature of normal seismic waves. Then, in the limit of low frequencies in equation (2.68) we must keep all terms, because they are all of the order $O(\omega^2)$. However, equation (2.69) provides great simplifications. The leading term there (the last one on the left-hand side of the equation) is of the order $O(\omega)$. It is proportional to the filtration velocity $\mathbf{q}$. Thus, for normal seismic waves in the low-frequency range,

$$\mathbf{q} = 0. \tag{2.121}$$

This is not surprising. In the low-frequency range the influence of the global fluid flow on the propagation of normal seismic waves is vanishing. In other words, the relative motions of the fluid and of the solid in a normal seismic wave is vanishing. We have already seen this above by considering polarizations of poroelastic plane waves. However, this also means that $\chi = 0$. This leads further to the statement that in the normal seismic wavefields the following relation is valid (see equations (2.42), (2.43) and compare with (2.52) for an undrained sample):

$$P_p = -\alpha M \epsilon = -B \sigma_{kk}/3. \tag{2.122}$$

Equation (2.68) can be then simplified:

$$\frac{\partial}{\partial x_j} C^u_{ijkl} \frac{\partial u_k}{\partial x_l} = \rho \frac{\partial^2 u_i}{\partial t^2}. \tag{2.123}$$

This is the Lamé equation with parameters of the undrained rock (compare to equation (1.28)). Therefore, in the low-frequency range, normal seismic P- and S-waves propagate in the undrained regime.

2.5 The quasi-static approximation of poroelasticity

The quasi-static approximation of poroelastic dynamic equations is broadly applied for analyzing different problems of strain, stress and pore-pressure evolution in fluid-saturated rocks, including problems related to earthquake mechanics. It was developed in a series of now-classical works including Rice and Cleary (1976), Rudnicki (1986), Detournay and Cheng (1993) and others. Summarizing reviews can be found, for example, in the books of Wang (2000), Jaeger *et al.* (2007) and Segall (2010). In this approximation the following system of equations (initially containing neither sources of the fluid mass nor body forces) is considered as a starting point. This is the continuity equation (2.64) expressing the fluid mass conservation, and is Darcy's law (2.66) describing a stationary filtration of the pore fluid. Finally, this is the equilibrium equation for the poroelastic stress:

$$\frac{\partial}{\partial x_j} \sigma_{ij} = 0. \tag{2.124}$$

Taking into account the stress–strain relations (2.36) and (2.42) and substituting Darcy's law into the continuity equation we will immediately obtain:

$$(\lambda_{dr} + M\alpha^2)\frac{\partial \epsilon}{\partial x_i} - \alpha M \frac{\partial \chi}{\partial x_i} + 2\mu_{dr}\frac{\partial \epsilon_{ij}}{\partial x_j} = 0, \qquad (2.125)$$

$$\frac{\partial}{\partial x_i}\frac{k_{ij}}{\eta}\frac{\partial}{\partial x_j}(M(\chi - \alpha\epsilon)) = \frac{\partial}{\partial t}\chi. \qquad (2.126)$$

This is the system of equations of the quasi-static approximation of the poroelasticity. Note that equations (2.125) and (2.126) coincide exactly with equations (2.108) and (2.117), respectively. However, the latter pair of equations describes slow wavefields. Thus, the quasi-static approximation describes slow wavefields too.

Does the quasi-static approximation describe more phenomena than just slow wavefields? To answer this question let us, for simplicity, consider further elastically and hydraulically homogeneous and isotropic rocks. These assumptions are usually accepted in the quasi-static approximation. Equation (2.126) simplifies then to

$$\frac{k}{\eta}\nabla^2(M(\chi - \alpha\epsilon)) = \frac{\partial}{\partial t}\chi. \qquad (2.127)$$

Let us consider further the equation system (2.125) and (2.127) under undrained conditions. Note that, in this case, the latter equation reduces to $\nabla^2\epsilon = 0$, which is the rather general and well-known equation (1.50) for elastic equilibrium.

For clarity, we will denote field quantities (i.e. the pore pressure, stresses, strains, particle velocities and displacements) in undrained conditions with a superscript u. Then for the stress equilibration equation (2.125) we obtain

$$(\lambda_{dr} + M\alpha^2)\frac{\partial \epsilon^u}{\partial x_i} + 2\mu_{dr}\frac{\partial \epsilon_{ij}^u}{\partial x_j} = 0. \qquad (2.128)$$

This is the stress equilibrium equation with elastic parameters of an undrained medium. Note also that the following relations are valid under undrained conditions:

$$\nabla^2\epsilon^u = 0, \qquad (2.129)$$

$$\chi^u = 0, \qquad (2.130)$$

and from (2.46):

$$P_p^u = -B\sigma_{kk}^u/3. \qquad (2.131)$$

Let us denote solutions of equations (2.125) and (2.127) describing slow wavefields as

$$\epsilon^s, \ \epsilon_{ij}^s, \ \chi^s. \qquad (2.132)$$

Then, the following fields will also be solutions of these equations:

$$\epsilon = \epsilon^u + \epsilon^s,$$
$$\epsilon_{ij} = \epsilon_{ij}^u + \epsilon_{ij}^s, \tag{2.133}$$
$$\chi = \chi^u + \chi^s \equiv \chi^s.$$

It is clear that the stress tensor and the pore pressure will also follow this rule:

$$\sigma_{ij} = \sigma_{ij}^u + \sigma_{ij}^s,$$
$$P_p = P_p^u + P_p^s = P_p^s - B\sigma_{kk}^u/3. \tag{2.134}$$

Therefore, in addition to the dynamics of slow wavefields, the quasi-static approximation of poroelasticity describes elastic equilibrium in undrained rocks. However, a sufficient time interval is required for propagating normal shear and longitudinal elastic waves to establish elastic equilibrium. Therefore, the quasi-static approximation is not applicable for time intervals of the order of, or shorter than, several travel times of elastic waves in systems under consideration. From the analysis of plane-wave solutions we know that non-vanishing fluid flow in respect to the solid skeleton is related to slow wavefields only (see Section 2.4.2). Normal elastic waves in the quasi-static limit are characterized by vanishing relative fluid motion (i.e. vanishing χ). Solutions of the quasi-static equations can be represented in the form of a summation of dynamic slow wavefields with static stress and strain fields given by elastic equilibrium of undrained rocks. The dynamics of the slow wavefields is mainly described by diffusion equations.

Further, we will introduce some additional useful equations of the quasi-static approximation. We apply the divergence operator to equations (2.125):

$$H\nabla^2\epsilon = \alpha M\nabla^2\chi. \tag{2.135}$$

This equation has several equivalent forms following from (2.37) and (2.42)–(2.44). For example:

$$4\alpha\frac{\mu_{dr}}{\lambda_{dr} + 2\mu_{dr}}\nabla^2 P_p = -\nabla^2\sigma_{ii} \tag{2.136}$$

and

$$4\mu_{dr}\nabla^2\epsilon = -\nabla^2\sigma_{ii} \tag{2.137}$$

and

$$4\mu_{dr}\alpha M\nabla^2\chi = -H\nabla^2\sigma_{ii} \tag{2.138}$$

and

$$(\lambda_{dr} + 2\mu_{dr})\nabla^2\epsilon = \alpha\nabla^2 P_p \tag{2.139}$$

and also

$$\nabla^2 \chi = S \nabla^2 P_p. \tag{2.140}$$

Substituting (2.135) into equation (2.127) we obtain the familiar diffusion equation (coinciding with equation (2.111)):

$$D\nabla^2 \chi = \frac{\partial}{\partial t} \chi. \tag{2.141}$$

From our discussion above it is clear that this equation (equally the quantity χ generally) describes just slow wavefields. Thus it is not surprising that this equation is the same as (2.111).

Using again relations (2.37), (2.42), (2.43) and (2.44) we can rewrite the diffusion equation (2.141) in several equivalent forms. For example, using equation (2.42) we can rewrite it as follows:

$$D\nabla^2 (P_p + M\alpha\epsilon) = \frac{\partial}{\partial t}(P_p + M\alpha\epsilon). \tag{2.142}$$

Another possible form is obtained from a direct substitution of equation (2.44) into (2.141):

$$D\nabla^2 (K_u\epsilon - \frac{1}{3}\sigma_{ii}) = \frac{\partial}{\partial t}(K_u\epsilon - \frac{1}{3}\sigma_{ii}). \tag{2.143}$$

Also, equation (2.43) can be substituted into the diffusion equation (2.141) and we obtain another form derived by Rice and Cleary (1976):

$$D\nabla^2 (P_p + \frac{B}{3}\sigma_{ii}) = \frac{\partial}{\partial t}(P_p + \frac{B}{3}\sigma_{ii}). \tag{2.144}$$

Note that this equation directly describes dynamics of the pressure perturbation corrected for its undrained component (see also equations (2.134)), i.e. again, we arrive at the same diffusion equation addressing slow wavefields. All the diffusion equations derived above describe dynamics of the same slow wavefields. In the case of an undrained system they turn into the identity $0 \equiv 0$ (slow wavefields do not exist).

In addition to these forms of diffusion equations there exist different mixed forms. One can use various equivalent forms of equation (2.135). For example, using equations (2.42) and (2.140) in equation (2.141) we obtain

$$\frac{kM}{\eta}\nabla^2 P_p = \frac{\partial}{\partial t}P_p + \alpha M \frac{\partial}{\partial t}\epsilon. \tag{2.145}$$

Also, using equation (2.37) we obtain

$$\frac{kBK_{dr}}{\alpha\eta}\nabla^2 P_p = \frac{\partial}{\partial t}(P_p + \frac{B}{3}\sigma_{ii}). \tag{2.146}$$

Therefore, under rather general conditions (without assuming irrotational displacements) the pore pressure is given by solutions of diffusion equations with coupling dilatational strain (or equivalently, isotropic stress) terms. Such solutions describe a complete pore pressure resulting from an interference of slow wavefields with a pressure component given by establishing an initial elastic equilibrium in the undrained regime. These solutions also take into account elastic fields resulting from interactions of the slow wave with geometric boundaries of the medium. If the medium is completely undrained then these (mixed-form) equations turn into equation $\nabla^2 P_p = 0$. Under undrained conditions this is equivalent to the familiar equation $\nabla^2\epsilon = 0$ (compare to (1.50)). Under conditions of an irrotational displacement field and in the absence of body forces in infinite or semi-infinite domains where ϵ and P_p vanish at infinity (see also Detournay and Cheng, 1993), the undrained components of P_p will be equal to zero. Then the diffusion equations above will reduce to one describing slow wavefields only. The complete consideration will be the same as we applied in the section on slow wavefields. Equation (2.114) will then describe the diffusion of the complete pore-pressure field.

In realistic situations of point-like borehole fluid injections (including far distances from finite-size injection sources) in large-scale reservoir domains far from their boundaries, dynamical parts of the pore-pressure fields are given by slow wavefields and can be well described by diffusion equations. Such linear equations were implicitly or explicitly used in many works on hydraulically induced seismicity. Later we will also extensively use them.

2.6 Sources of fluid mass and forces in poroelastic media

In the previous sections we discussed linear differential equations of the quasi-static approximation of poromechanics. These equations describe phenomena relevant for understanding of fluid-induced seismicity. Here we will discuss several useful particular solutions to these equations. Such solutions can be found if initial and boundary conditions for the corresponding equations have been formulated (see the next section). One more possibility is to directly introduce forces and mass sources into equations and to use corresponding Green's functions. In this section we will introduce several solutions of the Green's function type. We will mainly adopt the approach of Rudnicki (1986).

We must modify the basic equations of the quasi-static approximation to introduce the forces and sources of a fluid. A fluid source can be directly introduced into continuity equation (2.64):

$$\frac{\partial \chi}{\partial t} + \frac{\partial q_i}{\partial x_i} = Q(\mathbf{r}, t), \tag{2.147}$$

where $Q(\mathbf{r}, t)$ is a source of an additional fluid volume (per unit volume of rock per unit time). Note that we work with linearized equations of poromechanics. Therefore, changes of the fluid density have been neglected here. A force can be directly introduced into the equilibrium equation for the poroelastic stress:

$$\frac{\partial}{\partial x_j}\sigma_{ij} + F_i = 0, \tag{2.148}$$

where F_j is a component of force density vector $\mathbf{F}$.

We assume further that the medium is homogeneous and isotropic. Taking into account the stress–strain relations (2.36) and (2.41) and substituting Darcy's law (2.66) into the continuity equation (2.147) we obtain the following modifications of equations (2.125) and (2.127):

$$(\lambda_{dr} + M\alpha^2)\frac{\partial \epsilon}{\partial x_i} - \alpha M \frac{\partial \chi}{\partial x_i} + 2\mu_{dr}\frac{\partial \epsilon_{ij}}{\partial x_j} + F_i = 0, \tag{2.149}$$

$$Q(\mathbf{r}, t) + \frac{k}{\eta}\nabla^2(M(\chi - \alpha\epsilon)) = \frac{\partial}{\partial t}\chi. \tag{2.150}$$

Analogously to equation (1.49), equation (2.149) can be written in the following vectorial form:

$$\nabla(H\epsilon - \alpha M\chi) - \mu_{dr}\nabla \times (\nabla \times \mathbf{u}) + \mathbf{F} = 0. \tag{2.151}$$

Applying the divergence operator to this equation yields

$$\nabla^2(H\epsilon - \alpha M\chi) + \nabla\mathbf{F} = 0. \tag{2.152}$$

Expressing from here $\nabla^2\epsilon$ and substituting it into equation (2.150) yields the known diffusion equation (2.141) along with the source terms:

$$\frac{\partial}{\partial t}\chi = D\nabla^2\chi + Q(\mathbf{r}, t) + \frac{\alpha D}{\lambda_{dr} + 2\mu_{dr}}\nabla\mathbf{F}. \tag{2.153}$$

The contribution of the body forces to the slow wavefield χ is non-trivial. Based on an earlier observation of Cleary (1977), Rudnicki (1986) proposed the following interpretation. Let us consider a fluid-mass source in the form of point-like dipoles distributed in space and time. This is a combination of fluid sources and sinks

$Q = Q_0 f(\mathbf{r}, t)/|\boldsymbol{\delta r}|$ and $Q = -Q_0 f(\mathbf{r}, t)/|\boldsymbol{\delta r}|$, respectively. Function $f(\mathbf{r}, t)$ describes the spatio-temporal distribution of them. In the case of a single dipole the source and the sink are located close to each other along vector $\boldsymbol{\delta r}$. This results in the following fluid-mass source:

$$\lim_{|\boldsymbol{\delta r}| \to 0} \frac{Q_0}{|\boldsymbol{\delta r}|} (f(\mathbf{r} - \boldsymbol{\delta r}, t) - f(\mathbf{r}, t)) = -Q_0 \hat{\delta r}_k \frac{\partial f(\mathbf{r}, t)}{\partial x_k}, \tag{2.154}$$

where $\hat{\delta r}_k$ is the k-component of the unit vector $\hat{\boldsymbol{\delta r}} \equiv \boldsymbol{\delta r}/|\boldsymbol{\delta r}|$. This source will produce exactly the same effect as the following body force:

$$\mathbf{F} = -Q_0 \hat{\boldsymbol{\delta r}} \frac{\lambda_{dr} + 2\mu_{dr}}{\alpha D} f(\mathbf{r}, t). \tag{2.155}$$

Or equivalently, the fluid-mass effect of the body-force distribution $\mathbf{F}_0 f(\mathbf{r}, t)$ is equal to the effect of the fluid-mass dipole with:

$$Q_0 \hat{\boldsymbol{\delta r}} = -\mathbf{F}_0 \frac{\alpha D}{\lambda_{dr} + 2\mu_{dr}}. \tag{2.156}$$

Thus, the following strategy can be applied to solve the system of poroelastic equations with source terms, i.e. equations (2.151) and (2.153). The case of a point-like fluid-mass source corresponds to the absence of any body forces, i.e. $\mathbf{F} = 0$ (recall that we neglect effects of gravitation forces). Such a situation approximately corresponds to a point-like fluid injection from a borehole perforation into rocks. In this case firstly a solution of equation (2.153) for the slow wavefield χ must be obtained. Owing to the symmetry of the problem such a solution of equation (2.151) for the displacement field $\mathbf{u}$ is completely irrotational. All the field quantities describe slow wavefields only. Owing to the quasistatic approximation, normal seismic waves are not described. Their contribution to any elastic equilibrium is assumed to occur infinitely quickly.

Very similar to the previous one is the case of a fluid-mass dipole only. Again, $\mathbf{F} = 0$. This is a superposition of two point-like positive and negative fluid-mass sources. Such a situation approximately corresponds to a pair of closely located boreholes with a point-like fluid injection in one of them and a point-like fluid extraction in the other (e.g. a geothermal duplet considered at distances much larger than the distance between the boreholes, i.e. in the far field). Thus, a solution for the slow wavefield χ and for other field quantities can be found as a superposition of the corresponding solutions for a single fluid-mass source. Taking into account that a dipole source is equal to two differentially closely located mass sources of opposite signs, one can show that such a superposition is proportional to a directional derivative of of the solutions for a single mass source (see Rudnicki, 1986). The field quantities describe slow wavefields only. They are completely irrotational and satisfy (2.151) with zero body forces. Normal seismic waves are not considered.

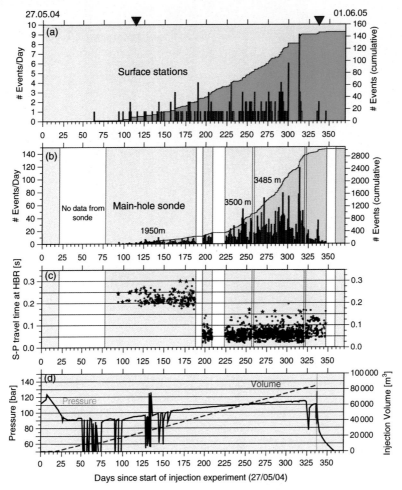

Plate 3.1 Microseismicity during the 2004–2005 KTB injection experiment.
(a) Events recorded by the surface stations. (b) Events recorded by the geophone
located in the main borehole. (c) Time differences between arrivals of S-and
P-waves. Changes in the locations of the borehole geophone are clearly seen
from the plot. (d) The injection pressure along with the cumulative volume of
the injected water. The time when the amount of previously extracted fluid was
re-injected is marked in part (a) of the figure by the inverse triangle on the top,
between the days 100 and 125. Another triangle (between the days 325 and 350)
marks the injection termination. (Modified after Shapiro *et al.*, 2006a.)

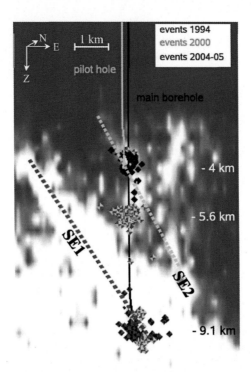

Plate 3.2 A vertical slice of a 3D depth migrated image of the KTB site (see Buske, 1999). Light tones correspond to high seismic reflection intensities and dark tones to lower ones, respectively. The SE1 reflector is clearly visible as a steeply dipping structure. Additionally, seismicity induced by the injection experiments of years 1994 (close to the injection depth of 9.1 km), 2000 (injection depth of approximately 5.6 km) and 2004–2005 (depth around 4 km) is shown. Locations of the main (black line) and pilot (light-tone line) boreholes are also plotted. (Modified after Shapiro *et al.*, 2006a.)

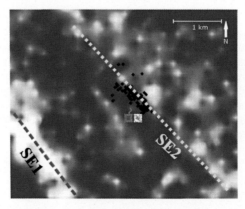

Plate 3.3 A horizontal slice at 4 km depth over the depth-migrated 3D image of the KTB site (Buske, 1999) plotted along with the slice-plane projections of seismic hypocenters induced by the injection experiment of 2004–2005. The white and gray squares are locations of the main and pilot boreholes, respectively. (After Shapiro *et al.*, 2006a.)

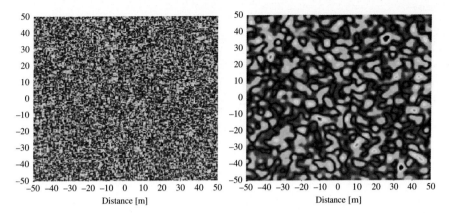

Plate 3.6 Left: realizations of the critical pore pressure C randomly distributed in space with an exponential auto-correlation function. Right: the same but with a Gaussian auto-correlation function. (Modified from Rothert and Shapiro, 2003.)

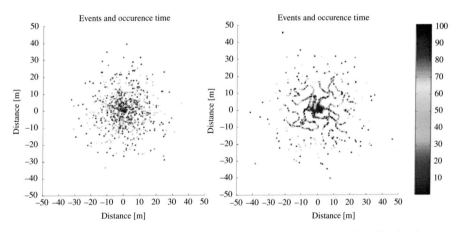

Plate 3.7 Synthetic microseismicity simulated for the two spatial distributions of the critical pore pressure shown in Figure 3.6, respectively. (Modified from Rothert and Shapiro, 2003.)

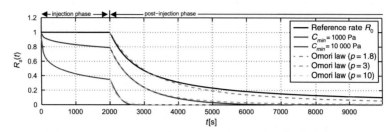

Plate 5.1 Seismicity rate in the case of stable pre-existing fractures (C_{min} is finite and significant). The rate is normalized to the reference seismicity rate ν_I defined in (5.17). Parameters of the model are $D = 1\ \mathrm{m}^2/\mathrm{s}$, $t_0 = 2000\ \mathrm{s}$, $p_0 = 1\ \mathrm{MPa}$, $a_0 = 4\ \mathrm{m}$, $C_{max} = 1\ \mathrm{MPa}$. The solid lines correspond to $C_{min} = 0\ \mathrm{Pa}$, $C_{min} = 1000\ \mathrm{Pa}$ and $C_{min} = 10\,000\ \mathrm{Pa}$ (from the upper to the lower curve). The dashed lines show the modified Omori law with $p_d = 1.8$, 3.0 and 10.0. (After Langenbruch and Shapiro, 2010.)

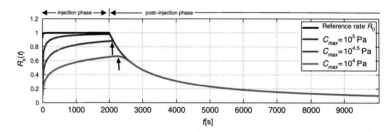

Plate 5.2 Seismicity rate in the case of unstable pre-existing fractures ($C_{min} = 0$). The rate is normalized to its maximum value, given by the reference seismicity rate ν_I. Parameters of the model are $D = 1\ \mathrm{m}^2/\mathrm{s}$, $t_0 = 2000\ \mathrm{s}$, $p_0 = 1\ \mathrm{MPa}$, $a_0 = 4\ \mathrm{m}$. The solid lines correspond to $C_{max} = 10^6\ \mathrm{Pa}$, $C_{max} = 10^5\ \mathrm{Pa}$, $C_{max} = 10^{4.5}\ \mathrm{Pa}$ and $C_{max} = 10^4\ \mathrm{Pa}$ (from the upper to the lower curve). The arrows denote the time of maximum probability to induce an event with significant magnitude. (After Langenbruch and Shapiro, 2010.)

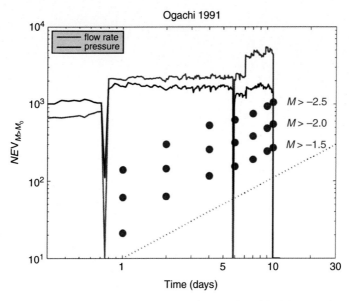

Plate 5.3 $N_{\geq M}$ as functions of injection time for the Ogachi 1991 experiment. The points are observed cumulative numbers of earthquakes with magnitudes larger than the indicated ones. The straight line has the proportionality coefficient 1, predicted by equation (5.14). The curves show the injection pressure (the lower line in the time range 1–10 days) and the injection rate (the upper line in the time range 1–10 days). (Modified from Shapiro *et al.*, 2007.)

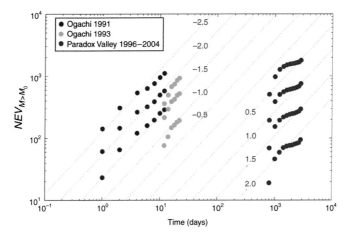

Plate 5.5 A combined plot of numbers of events with magnitudes larger then given ones as functions of injection durations at Ogachi and at Paradox Valley. Thin dashed lines correspond to equation (5.14) with $i = 0$. (Modified from Shapiro *et al.*, 2007.)

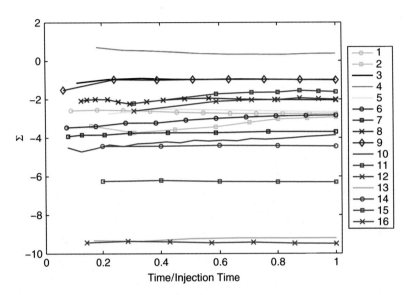

Plate 5.10 Seismogenic index computed for different locations of Enhanced Geothermal Systems, hydraulic fracturing in hydrocarbon reservoirs, and other injection locations (injection times are given in parentheses). 1: Ogachi 1991 (11 days), 2: Ogachi 1993 (16 days), 3: Cooper Basin 2003 (9 days), 4: Basel 2006 (5.5 days), 5: Paradox Valley (2500 days), 6-9: Soultz 1996 (48 hr), 1995 (11 days), 1993 (16 days) and 2000 (6 days), 10: KTB 2004/05 (194 days), 11–12: KTB 1994 (9 hr) [upper and lower bound, calculated for two b-values], 13: Barnett Shale (6 hr), 14–16: Cotton Valley Stages A (2.5 h), B (2.5 h) and C (3.5 h) (Modified from Dinske and Shapiro, 2013.)

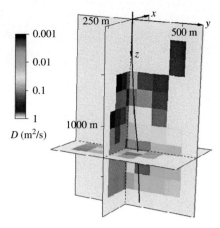

Plate 3.13 Reconstruction of the diffusivity distribution in a hydraulically heterogeneous geothermic reservoir of Soultz (corresponding to the stimulation of 1993)
using the eikonal-equation approach. The dark-tone grid cells in the upper part of
the structure have hydraulic diffusivity in the range 0.1–1 m^2/s. Below 1000 m
the structure has the diffusivity mainly in the range 0.001–0.05 m^2/s. (Modified
from Shapiro *et al.*, 2002.)

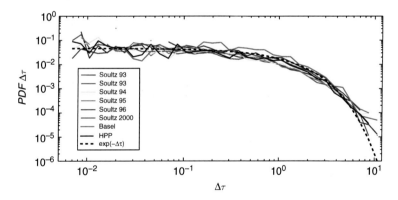

Plate 5.11 Estimated probability density functions of the normalized inter-event
time for stationary periods of injections at Soultz and Basel. (After Langenbruch
et al., 2011.)

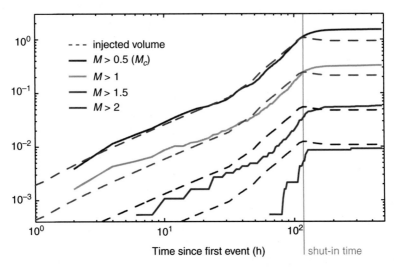

Time since first event (h)   shut-in time

Plate 5.13 Number $N_{\geq M}$ (solid lines) of induced earthquakes with magnitudes M larger than indicated values as functions of the time t elapsed from the time of the first event in the catalog (nearly the injection start) at the Basel borehole (Häring et al., 2008). The plot also shows the injected water volume $Q_c(t)$ (the upper dashed line). The quantities $Q_c(t)$ and $N_{\geq 0.5}(t)$ (the upper solid line; 0.5 is approximately a completeness magnitude) are normalized to their values at the moment of the maximum injection pressure (several hours before the injection termination). Immediately after the injection termination the curve $Q_c(t)$ starts to decrease because of an outflow of the injected water. Theoretical curves $N_{\geq M}(t)$ corresponding to (5.20) are given by lower dashed lines. They are constructed by a time-independent shifting of the curve of the injected-fluid volume. The lower solid lines shows the observed quantities $N_{\geq M}(t)$ normalized by the same value as the quantity $N_{\geq 0.5}(t)$. (Modified from Shapiro et al., 2011.)

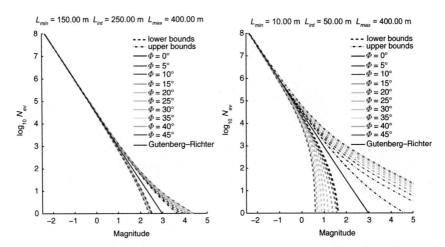

Plate 5.21 The same as Figure 5.20 but cuboidal stimulated volumes and different angles ϕ. (After Shapiro et al., 2013.)

The corresponding elastic part of the equilibrium is assumed to establish itself instantaneously.

In both situations discussed above, slow wavefields fully represent the dynamics of systems and, thus, corresponding solutions satisfy equations (2.112)–(2.115).

The case of an applied body force ($\mathbf{F} \neq 0$) and a vanishing source of the fluid mass ($Q = 0$) is different. A solution for the slow wavefield χ can be first found for a fluid-mass dipole (2.156). This solution corresponds to an irrotational part of the displacement field satisfying (2.151) with zero body forces. This solution describes the slow wavefield. The complete solution will be the sum of the slow-wavefield solution and the solution of equation (2.151) for $\chi = 0$ and the force $\mathbf{F}$. This latter solution corresponds to the zero slow wavefield. It describes elastic equilibrium established in the undrained rock by normal seismic waves (again, these waves are not considered in the quasi-static approximation; therefore, this approximation assumes that the undrained elastic equilibrium is established momentarily).

Other solutions of quasi-static poroelastic equations with source terms can be obtained by superposition of the solutions discussed above for the cases with loading of the form $Q \neq 0$ and $\mathbf{F} = 0$ and with loading $Q = 0$ and $\mathbf{F} \neq 0$. A rather general form of the superposition principle (1.37) for a linear equation system (including the linear equation system of poroelasticity considered above) uses Green's functions. A review of poroelastic Green's functions valid in the total frequency range was given by Karpfinger *et al.* (2009). Numeric inverse Fourier transforms are required for their computations. In this book we are interested in the low-frequency (quasi-static) range. In this range several explicit Green-function-like solutions are known (see Rudnicki, 1986). However, quasi-static poroelastic equations do not describe dynamic effects on small time scales (time intervals on the order of normal seismic travel times). Thus, in this approximation it is convenient to work with source terms having Heaviside-type temporal distributions $h(t)$ ($h(t) = 0$ if $t < 0$ and $h(t) = 1$ if $t \geq 0$).

2.6.1 Fluid injection at a point of a poroelastic continuum

Let us consider a point-like source of an instantaneous fluid injection of a volume Q_0 at the time moment $t = 0$ located at the point $\mathbf{r} = 0$. The source function has the following form:

$$Q(\mathbf{r}, t) = Q_0\delta(\mathbf{r})\delta(t). \tag{2.157}$$

In this case $\mathbf{F} = 0$ and, as discussed above, all field quantities represent a slow wavefield. Thus we must find a solution for the quantity χ. Therefore, we consider diffusion equation (2.153) in an infinite poroelastic 3D medium with $\chi = 0$ at $t < 0$:

$$\frac{\partial}{\partial t}\chi = D\nabla^2\chi + Q_0\delta(\mathbf{r})\delta(t).\tag{2.158}$$

Solution χ_G of this equation is Green's function of the diffusion equation (2.153). This classical solution can be found by reformulating the problem into an initial-value one. A spatial Fourier transform is used to turn the homogeneous diffusion equation into an ordinary one-dimensional (1D) differential equation (see the derivation in paragraph 51, chapter V, of Landau and Lifshitz, 1991):

$$\chi_G(\mathbf{r}, t) = \frac{Q_0}{(4\pi Dt)^{3/2}} \exp\left(-\frac{r^2}{4Dt}\right).\tag{2.159}$$

We see that, because of the spherical symmetry of the problem, χ_G is a function of the distance $r = |\mathbf{r}|$ and time t only (no angular dependencies). Owing to the diffusional dynamics the normalized distance

$$r_0 = \frac{r}{\sqrt{4Dt}} = \sqrt{\pi}\frac{r}{\lambda_{PII}}\tag{2.160}$$

is a convenient variable to describe spatio-temporal behavior of the field quantities. The right-hand part of this equation uses the wavelength of the slow wave with the time period t (see equation (2.104)).

In reality an instantaneous fluid injection is a mathematical abstraction. The following situation is more relevant in practice. We consider further a point-like source of a continuous fluid injection of a constant volume rate Q_I located at $\mathbf{r}_0 = 0$ and starting at $t = 0$ (i.e. the injection rate is the following function of time: $Q_I h(t)$). The source function is given by the time integration of this rate:

$$Q(\mathbf{r}, t) = Q_I t\delta(\mathbf{r}).\tag{2.161}$$

Also in this case $\mathbf{F} = 0$ and, as discussed above, all field quantities represent a slow wavefield. Owing to the symmetry of the problem the displacement vector $\mathbf{u}$ will have a radial direction. Scalar-dilatational and pressure-like field quantities ϵ, χ, P_p and σ_{ii} will be functions of the distance r and time t only (no angular dependencies).

The solution of diffusion equation (2.153) in this case is very well known (it is given by an integration over time of Green's function (2.159); see, for example, Carslaw and Jaeger, 1973):

$$\chi(\mathbf{r}, t) = \frac{Q_I}{4\pi Dr} \mathrm{erfc}(r_0),\tag{2.162}$$

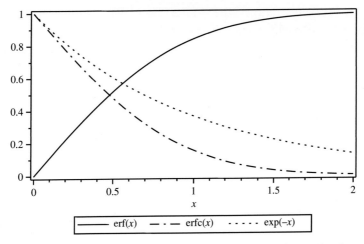

Figure 2.6 An error function, a complimentary error function and a decreasing exponential function.

where $\mathrm{erfc}(x)$ stands for the complementary error function, $\mathrm{erfc}(x) = 1 - \mathrm{erf}(x)$, and the error function $\mathrm{erf}(x)$ is defined as follows (see also Figure 2.6):

$$\mathrm{erf}(x) \equiv \frac{2}{\sqrt{\pi}} \int_0^x \exp(-z^2)\,dz. \tag{2.163}$$

Consequently, the pore pressure is given by equation (2.113):

$$P_p(\mathbf{r}, t) = \frac{Q_I \eta}{4\pi k r}\, \mathrm{erfc}(r_0). \tag{2.164}$$

Using equations (2.162) and (2.151) and taking into account radial symmetry, irrotationality and physical dimensionality of the displacement, Rudnicki (1986) derives the following result:

$$u_i(\mathbf{r}, t) = \frac{Q_I \eta x_i}{16\pi k r}\frac{n_s}{\mu_{dr}} s_{rr}(r_0), \tag{2.165}$$

where the function $s_{rr}(r_0)$ is defined as follows:

$$s_{rr}(r_0) \equiv 2\,\mathrm{erfc}(r_0) + \frac{\mathrm{erf}(r_0)}{r_0^2} - 2\frac{\exp(-r_0^2)}{r_0\sqrt{\pi}}, \tag{2.166}$$

and we have introduced the poroelastic stress coefficient n_s given by

$$n_s \equiv \frac{\alpha(1 - 2\nu^{dr})}{2(1 - \nu^{dr})} = \frac{\alpha\mu_{dr}}{\lambda_{dr} + 2\mu_{dr}} = \alpha\mu_{dr}\frac{SM}{H}. \tag{2.167}$$

Here v^{dr} is Poisson's ratio of the drained rock. Substituting result (2.165) into Hooke's law (2.36) yields the stress tensor (Rudnicki, 1986):

$$\sigma_{ij}(\mathbf{r}, t) = -n_s \frac{Q_I \eta}{4\pi k r} \left[\delta_{ij} s_{tt}(r_0) + \frac{x_i x_j}{r^2} s_{dif}(r_0) \right], \tag{2.168}$$

where the functions $s_{tt}(r_0)$ and $s_{dif}(r_0)$ are defined as follows:

$$s_{tt}(r_0) \equiv \text{erfc}(r_0) - \frac{\text{erf}(r_0)}{2r_0^2} + \frac{\exp(-r_0^2)}{r_0 \sqrt{\pi}}, \tag{2.169}$$

$$s_{dif}(r_0) \equiv s_{rr} - s_{tt} = \text{erfc}(r_0) + 3\frac{\text{erf}(r_0)}{2r_0^2} - 3\frac{\exp(-r_0^2)}{r_0 \sqrt{\pi}}. \tag{2.170}$$

Stress tensor (2.168) corresponds to the spherical slow wavefield radiated by the point-like mass source considered above.

One can write these results in terms of the radial and tangential components of the corresponding field quantities. For the radial component of the displacement (its tangential components vanish) we have

$$u_r(\mathbf{r}, t) = \frac{Q_I \eta}{16\pi k} \frac{n_s}{\mu_{dr}} s_{rr}(r_0). \tag{2.171}$$

For the radial component of the stress σ_{rr} (it is equal, e.g. σ_{11} at $\mathbf{r} = (x_1, 0, 0)$) we obtain

$$\sigma_{rr}(\mathbf{r}, t) = -n_s \frac{Q_I \eta}{4\pi k r} s_{rr}(r_0). \tag{2.172}$$

For the tangential components $\sigma_{\theta\theta} = \sigma_{\phi\phi}$ (they are equal, e.g. σ_{33} at $\mathbf{r} = (x_1, 0, 0)$) we obtain

$$\sigma_{\theta\theta}(\mathbf{r}, t) = -n_s \frac{Q_I \eta}{4\pi k r} s_{tt}(r_0). \tag{2.173}$$

Rudnicki (1986) also found explicit analytical expressions of Green's functions for source terms like a fluid-mass dipole, a point force, a line source of a fluid mass, a line of fluid-mass dipoles, and a line force.

2.7 Boundary loading of poroelastic media

Let us consider a loading of a poroelastic medium over its boundaries. Detournay and Cheng (1993) proposed decomposing the problem into two loading modes. Mode I denotes the following load distribution: to a given surface a normal stress σ_n (we recall that this is a normal confining, i.e. a normal total stress) is applied so that $\sigma_n(t) = -h(t)p_0$ and the applied pore pressure is absent, $P_p = 0$. Mode II denotes the following loading: $\sigma_n = 0$ and $P_p = h(t)p_0$. A corresponding superposition of solutions of quasi-static poroelastic equations for these two modes will provide solutions for arbitrary distributions of pore pressure and normal stress at the boundary.

2.7.1 Loading of a poroelastic half-space by a fluid reservoir

Let us use this strategy to compute poroelastic field perturbations excited by a sudden application of fluid pressure p_0 to the surface of a poroelastic half-space. Clearly this is an example of a 1D problem. All quantities are functions of one variable (e.g. of depth z) only. The boundary conditions at the surface of the half-space $z = 0$ are $\sigma_{zz} = -h(t)p_0$ and $P_p = h(t)p_0$. It is clear also that there is only one non-vanishing strain component ϵ_{zz} (because no displacements along horizontal directions are possible). Thus, this problem is also frequently called a uniaxial-strain problem.

For such a uniaxial-strain problem the stress–strain relation (2.36) takes the following simplified forms:

$$\sigma_{zz} = (\lambda_{dr} + 2\mu_{dr})\epsilon_{zz} - \alpha P_p \tag{2.174}$$

and

$$\sigma_{xx} = \sigma_{yy} = \lambda_{dr}\epsilon_{zz} - \alpha P_p. \tag{2.175}$$

The pore-pressure relation (2.42) takes the following form:

$$P_p = M(\chi - \alpha\epsilon_{zz}). \tag{2.176}$$

Also we will need the equilibrium equation for the poroelastic stress (2.148). Its correspondingly simplified form is

$$\frac{\partial}{\partial z}\sigma_{zz} = 0. \tag{2.177}$$

Further, it is convenient to use and simplify the diffusion equation for the slow wavefield in the mixed form (2.145):

$$\frac{kM}{\eta}\frac{\partial^2}{\partial z^2}P_p = \frac{\partial}{\partial t}P_p + \alpha M\frac{\partial}{\partial t}\epsilon_{zz}. \tag{2.178}$$

Using the stress–strain relation (2.174) we obtain:

$$D\frac{\partial^2}{\partial z^2}P_p = \frac{\partial}{\partial t}P_p + \frac{\alpha M}{H}\frac{\partial}{\partial t}\sigma_{zz}. \tag{2.179}$$

This equation coincides with equation (141) of Detournay and Cheng (1993).

Further, we will denote field quantities resulting from the loading modes I and II by the superscripts I and II, respectively. In loading mode I the stress applied to the surface of the half-space is $\sigma_{zz}^I(0, t) = -h(t)p_0$. Simultaneously, $P_p^I(0, t) = 0$.

We must now find field quantities satisfying these boundary conditions and equations (2.174)–(2.179).

From equation (2.177) it follows that the total stress is independent of z. Thus,

$$\sigma_{zz}^{\mathrm{I}}(z, t) = -h(t)p_0. \tag{2.180}$$

Therefore, for times $t > 0$ this stress is constant in time too. Thus, equation (2.179) takes a form of a 1D diffusion equation for the pore pressure:

$$D\frac{\partial^2}{\partial z^2}P_p^{\mathrm{I}} = \frac{\partial}{\partial t}P_p^{\mathrm{I}}. \tag{2.181}$$

It is clear that in the first instant after the load application the system was in the undrained state (i.e. at this moment the fluid flow was vanishing: $\chi^{\mathrm{I}}(z, 0) = 0$). Correspondingly with equation (2.176) the pore pressure under undrained conditions is directly related to the uniaxial strain:

$$P_p^{\mathrm{I}}(z, 0) = -M\alpha\epsilon_{zz}^{\mathrm{I}}. \tag{2.182}$$

Under undrained conditions using this relation in equation (2.174) and using definition (2.167) we can further directly relate the pore pressure to the loading stress:

$$P_p^{\mathrm{I}}(z, 0) = -\frac{n_s}{S\mu_{dr}}\sigma_{zz}^{\mathrm{I}}(z, 0), \tag{2.183}$$

Therefore, in the first instant the load should lead to a step-like increase of the pore pressure equal to $n_s p_0 h(t)/(S\mu_{dr})$ everywhere in the medium (also at infinity). On the other hand, in the loading mode I the boundary condition for the pore pressure is $P_p^{\mathrm{I}}(0, t) = 0$. The solution of equation (2.181) satisfying these two facts (we have already seen that the erfc-function is a solution of a homogeneous 1D diffusion equation) is:

$$P_p^{\mathrm{I}}(z, t) = \frac{n_s p_0}{S\mu_{dr}}(h(t) - \mathrm{erfc}(\frac{z}{\sqrt{4Dt}})). \tag{2.184}$$

In the loading mode II $\sigma_{zz}^{\mathrm{II}}(z, t) = 0$ and $P_p^{\mathrm{II}}(0, t) = p_0 h(t)$. The corresponding solution of equation (2.181) is

$$P_p^{\mathrm{II}}(z, t) = p_0\,\mathrm{erfc}(\frac{z}{\sqrt{4Dt}}). \tag{2.185}$$

Note that this expression will not change after multiplication by $h(t)$. Therefore, the complete solution of the boundary-value problem is the sum of the solutions for the modes I and II. It has the following form:

$$P_p(z, t) = p_0 h(t)\left[\frac{n_s}{S\mu_{dr}} + (1 - \frac{n_s}{S\mu_{dr}})\,\mathrm{erfc}(\frac{z}{\sqrt{4Dt}})\right]. \tag{2.186}$$

This form of the pore-pressure distribution for a poroelastic half-space loaded by a water reservoir was first obtained by Roeloffs (1988). In this classical paper Roeloffs (1988) also obtained a semi-analytical semi-numerical solution for a poroelastic half-space loaded by a finite-width fluid reservoir (in the plane-strain geometry) located on the half-space surface. She applied her solution to analyze reservoir-induced seismicity. These poroelastic solutions and several related approximations were used in an extensive series of interpretation and modeling studies of induced seismicity at the Koyna–Warna artificial water reservoirs (see Talwani, 1997; Gupta, 2002; Kalpna and Chander, 2000; Gavrilenko et al., 2010; Guha, 2000; and further references therein).

Using (2.186) and (2.174)–(2.175) and taking into account that $\sigma_{zz}(z,t) = \sigma_{zz}^I(z,t)$ is given by (2.180) we obtain:

$$\epsilon_{zz}(z,t) = -\frac{p_0 h(t)}{\lambda_{dr} + 2\mu_{dr}}\left[1 - \alpha\left[\frac{n_s}{S\mu_{dr}} + \left(1 - \frac{n_s}{S\mu_{dr}}\right)\operatorname{erfc}\left(\frac{z}{\sqrt{4Dt}}\right)\right]\right]$$
(2.187)

and the lateral components of the stress:

$$\sigma_{xx} = \sigma_{yy} = -p_0 h(t)\left[\frac{\lambda_{dr}}{\lambda_{dr} + 2\mu_{dr}} + 2n_s\left[\frac{n_s}{S\mu_{dr}} + \left(1 - \frac{n_s}{S\mu_{dr}}\right)\operatorname{erfc}\left(\frac{z}{\sqrt{4Dt}}\right)\right]\right].$$
(2.188)

Finally, the fluid-filtration caused change of the pore-space volume per unit rock volume, $\chi(z,t)$, can be computed by using equation (2.176):

$$\chi(z,t) = p_0 h(t) S\left(1 - \frac{n_s}{S\mu_{dr}}\right)\operatorname{erfc}\left(\frac{z}{\sqrt{4Dt}}\right).$$
(2.189)

2.7.2 Fluid injection into a spherical cavity of a poroelastic continuum

In this book we are interested in fluid injections in rocks over surfaces of cavities. For example, a fluid injection into an open short borehole section is used for development of enhanced geothermal systems. Very roughly, one can consider such an injection as an application of a fluid pressure to the surface of a spherical cavity.

Let us consider a spherical cavity of radius a_0 located at the origin of a fluid-saturated porous elastic infinite continuum with initially absent stress and pressure perturbations. We assume a sudden application of a fluid pressure $h(t)p_0$ to the inner surface of the cavity. This loading is equal to a direct superposition of modes I and II described above. For $r > a_0$ we must solve the system of poroelastic equations, i.e. equations (2.151) and (2.153) without source terms.

First we consider the loading mode I. In this mode no sources of fluids are present in the medium. The normal stress applied to the surface of the cavity is $\sigma_{rr}^I(a_0, t) = -h(t)p_0$. A pore-pressure perturbation vanishes at both limits, at the cavity surface as well as at infinity: $P_p^I(a_0, t) = P_p^I(\infty, t) = 0$. Thus, no slow wavefields are being excited. Therefore, we must consider elastic deformations of the undrained rock. The solution of such a boundary-value problem is well known (Landau and Lifshitz, 1987, paragraph 7, problem 2). There is a radial displacement component only:

$$u_i^I = h(t) \frac{p_0}{2\mu_{dr}} \frac{a_0^3 x_i}{r^3}. \tag{2.190}$$

This can be reformulated for the radial displacement directly:

$$u_r^I = h(t) \frac{p_0}{2\mu_{dr}} \frac{a_0^3}{r^2}. \tag{2.191}$$

This also yields $\epsilon^I = \nabla \mathbf{u} = 0$ and $\sigma_{kk}^I = 0$. Therefore, no pore-pressure perturbations have been induced by this undrained deformation. For the radial and tangential strains one obtains:

$$\epsilon_{rr}^I = -h(t) \frac{p_0}{\mu_{dr}} \frac{a_0^3}{r^3}, \tag{2.192}$$

$$\epsilon_{\theta\theta}^I = \epsilon_{\phi\phi}^I = h(t) \frac{p_0}{2\mu_{dr}} \frac{a_0^3}{r^3}. \tag{2.193}$$

The radial and tangential stresses are:

$$\sigma_{rr}^I = -h(t) p_0 \frac{a_0^3}{r^3}, \tag{2.194}$$

$$\sigma_{\theta\theta}^I = \sigma_{\phi\phi}^I = h(t) p_0 \frac{a_0^3}{2r^3}. \tag{2.195}$$

Let us further consider loading mode II. The normal stress applied to the surface of the cavity is vanishing, $\sigma_{rr}^{II}(a_0, t) = 0$. A pore-pressure perturbation vanishes at infinity. However, at the cavity surface we have $P_p^{II}(a_0, t) = p_0 h(t)$. Owing to the symmetry of the problem it is clear that only irrotational wavefields can be excited. Because of the nature of the boundary problem it is convenient to work directly with the diffusion equation for the pore pressure instead of equation (2.153). The corresponding initial- and boundary-value problem for the diffusion equation (2.114) is very well known. For example, it can be solved in the same way as the problem 1 of paragraph 52, chapter V of Landau and Lifshitz (1991):

$$P_p^{II}(\mathbf{r}, t) = \frac{a_0 p_0}{r} \operatorname{erfc}(r_a), \tag{2.196}$$

where

$$r_a = \frac{r - a_0}{\sqrt{4Dt}} = \sqrt{\pi} \frac{r - a_0}{\lambda_{PII}} \tag{2.197}$$

is the normalized distance r_0 counted from the cavity surface. By this loading mode only slow wavefields can be radiated (i.e. slow wavefields define completely the field quantities). Thus, we use equation (2.113) to obtain the field χ:

$$\chi^{II}(\mathbf{r}, t) = \frac{a_0 p_0 S}{r} \text{erfc}(r_a). \tag{2.198}$$

Results (2.196) and (2.198) will coincide with equations (2.164) and (2.162), respectively, if we make the following substitutions:

$$r_a \rightarrow r_0 \tag{2.199}$$

$$a_0 p_0 \rightarrow \frac{Q_I \eta}{4\pi k}. \tag{2.200}$$

Furthermore, because only slow wavefields are excited, we can directly apply equations (2.112) and (2.115) to compute the corresponding dilatational field quantities, respectively:

$$\epsilon^{II}(\mathbf{r}, t) = \frac{a_0 p_0 n_s}{\mu_{dr} r} \text{erfc}(r_a), \tag{2.201}$$

$$\sigma_{ii}^{II}(\mathbf{r}, t) = -\frac{4 a_0 p_0 n_s}{r} \text{erfc}(r_a). \tag{2.202}$$

On the other hand, the dilatation ϵ^{II} is given by the divergence of the displacement vector field. Owing to the spherical symmetry of the problem the displacement can have a radial component only. Therefore, in the spherical coordinate system with the origin at the cavity's center, we have

$$\epsilon^{II}(\mathbf{r}, t) = \nabla \mathbf{u}^{II} = \frac{1}{r^2} \frac{\partial r^2 u_r^{II}}{\partial r}. \tag{2.203}$$

Using this equation and equation (2.201) we obtain the displacement by the integration of the dilatation over r:

$$u_r^{II}(\mathbf{r}, t) = \frac{(r - a_0)^2 a_0 p_0 n_s}{\mu_{dr} r^2} \left[\frac{1}{4} s_{rr}(r_a) + \frac{a_0}{r - a_0} s'(r_a) \right] + \frac{C_{II}}{r^2}, \tag{2.204}$$

where we have introduced the following new notation

$$s'(r_a) \equiv \text{erfc}(r_a) + \frac{1}{r_a \sqrt{\pi}} (1 - \exp(-r_a^2)), \tag{2.205}$$

and C_{II} is an integration constant. This constant must be found from the boundary conditions. Later we will see that $C_{II} = 0$.

Because of the radial symmetry the components of the displacement vector in a Cartesian coordinate system with the origin at the cavity's center are

$$u_i^{II}(\mathbf{r}, t) = u_r^{II}(\mathbf{r}, t) \frac{x_i}{r}. \tag{2.206}$$

From here we compute the strain ϵ_{rr}^{II}, which is equal to ϵ_{xx}^{II} at points $(x, 0, 0)$:

$$\epsilon_{rr}^{II} = \frac{a_0 p_0 n_s}{\mu_{dr}} \left[\frac{\mathrm{erfc}(r_a)}{r} - 2\frac{(r - a_0)^2}{r^3}(\frac{1}{4}s_{rr}(r_a) + \frac{a_0}{r - a_0}s'(r_a)) \right] - 2\frac{C_{II}}{r^3}.$$
(2.207)

To compute the radial-stress component it is convenient to directly apply equation (2.36). The stress σ_{rr}^{II} is equal to σ_{xx}^{II} at points $(x, 0, 0)$:

$$\sigma_{rr}^{II} = \lambda_{dr}\epsilon^{II} + 2\mu_{dr}\epsilon_{rr}^{II} - \alpha P_p.$$
(2.208)

Using here results (2.201) and (2.207) and taking into account the boundary condition of loading mode II that $\sigma_{rr}^{II}(a, t) = 0$ (this yields $C_{II} = 0$) we obtain:

$$\sigma_{rr}^{II}(\mathbf{r}, t) = -\frac{(r - a_0)^2 a_0 p_0 n_s}{r^3} \left[s_{rr}(r_a) + \frac{4a_0}{r - a_0}s'(r_a) \right].$$
(2.209)

Further, the twofold tangential-stress component can be obtained by extraction of this result from equation (2.202). Because $\sigma_{\theta\theta}^{II} = \sigma_{\phi\phi}^{II}$ we obtain

$$\sigma_{\phi\phi}^{II} = \sigma_{\theta\theta}^{II} = -\frac{a_0 p_0 n_s}{r} \left[2\,\mathrm{erfc}(r_a) - \frac{(r - a_0)^2}{2r^2}s_{rr}(r_a) - \frac{2a_0(r - a_0)}{r^2}s'(r_a) \right].$$
(2.210)

The solution of the complete boundary-value problem for the corresponding field quantities is given by the following sums:

$$\mathbf{u} = \mathbf{u}^I + \mathbf{u}^{II},$$
(2.211)

where (2.191) and (2.204) must be substituted (note again that, in (2.204), $C_{II} = 0$). Further,

$$\boldsymbol{\sigma} = \boldsymbol{\sigma}^I + \boldsymbol{\sigma}^{II},$$
(2.212)

where (2.194), (2.195), (2.209) and (2.210) must be substituted, respectively. Finally, the following two quantities are given directly by

$$\chi = \chi^{II}, \qquad P_p = P_p^{II},$$
(2.213)

where (2.198) and (2.196) must be substituted, respectively.

However, at distances $r \gg a_0$ (i.e. in the far field of the injection cavity) one can easily see that the contributions $\mathbf{u}^I$ and $\boldsymbol{\sigma}^I$ become negligible compared to contributions $\mathbf{u}^{II}$ and $\boldsymbol{\sigma}^{II}$, respectively. We note also that, at large distances, r_a can be replaced by r_0. Therefore, the solution for the boundary-value problem considered here nearly coincides with the solution for the mass source from Section 2.6.1. In the far field their only difference is a constant factor describing the strength of the injection source. In the case of a fluid-mass source this factor is $Q_1\eta/(4\pi k)$. In the case of a pressure applied to the surface of an injection cavity this factor is $a_0 p_0$.

Using this fact along with the symmetry considerations we conclude that in far field of the cavity the corresponding solution of equation (2.151) for the displacement and stress fields can be obtained by the direct modification of equations (2.165) and (2.168), respectively:

$$u_i(\mathbf{r}, t) = \frac{a_0 p_0 x_i}{4r} \frac{n_s}{\mu_{dr}} S_{rr}(r_0), \tag{2.214}$$

and

$$\sigma_{ij}(\mathbf{r}, t) = -n_s \frac{a_0 p_0}{r} \left[\delta_{ij} S_{tt}(r_0) + \frac{x_i x_j}{r^2} S_{dif}(r_0) \right]. \tag{2.215}$$

Again, we write these results in terms of the radial and tangential components of the corresponding field quantities. For the radial component of the displacement (its tangential components are vanishing) we have

$$u_r(\mathbf{r}, t) = \frac{a_0 p_0}{4} \frac{n_s}{\mu_{dr}} S_{rr}(r_0). \tag{2.216}$$

For the radial component of the stress σ_{rr} we obtain

$$\sigma_{rr}(\mathbf{r}, t) = -n_s \frac{a_0 p_0}{r} S_{rr}(r_0). \tag{2.217}$$

For the tangential components we obtain

$$\sigma_{\phi\phi}(\mathbf{r}, t) = \sigma_{\theta\theta}(\mathbf{r}, t) = -n_s \frac{a_0 p_0}{r} S_{tt}(r_0). \tag{2.218}$$

Concluding this section, we note that the direct applicability of the homogeneous diffusion equation (2.114) to the pore pressure in the boundary-value problem considered here is a consequence of the following circumstances. First, the symmetry of the problem allows for an irrotational displacement field only. This corresponds to the exclusive radiation of the slow wavefield in loading mode II. Second, and more importantly, we have considered here propagation of pore-pressure perturbations in an infinite spatial domain. In spatially bounded bodies, even under conditions of irrotational displacements, the homogeneous diffusion equation for the pore pressure (2.114) is not directly applicable. The physical reason of this is the fact that slow wavefields interact with boundaries (surfaces or interfaces). As a result of such interactions at the boundaries the slow wavefields would be reflected into secondary slow wavefields and transformed into normal elastic waves (converted P-waves). Pore-pressure perturbations corresponding to the secondary slow wavefields will still satisfy equation (2.114) directly. However, pressure-perturbation components corresponding to the normal elastic waves (propagating infinitely quickly in the quasi-static poroelastic approximation) will not satisfy this equation.

The well-known (in the quasi-static approximation) Mandel–Cryer effect (see Detournay and Cheng, 1993; Jaeger *et al.*, 2007; and references therein) can be also understood in terms of conversion (reflection) of a normal P-wave into the slow wave at the boundaries of finite poroelastic domains. Indeed, the Mandel–Cryer effect is clearly observed in the following situations. An instantaneous pressure pulse $p_0 \delta(t)$ or a constant normal stress $p_0 h(t)$ is instantly applied to the surface of a fluid-saturated porous elastic spherical or cylindrical sample with radius a_0 and a vanishing initial pore pressure. In the second situation, the sample is subjected to loading mode I discussed above. The pore pressure in the center of the sample will instantaneously increase up to the level corresponding to the undrained sample (note that in the quasi-static approximation this jump in the pore pressure occurs instantaneously, because normal elastic waves propagate in this approximation with infinite velocity). Then, the pore pressure will start to decrease slowly (due to the fluid filtration). However, in contrast to the expectation of a simple diffusion approximation for the pore pressure (neglecting the poroelastic coupling of the pore pressure and the stress in a finite spatial domain), the temporal evolution of the pore pressure is not monotonic. First the pore pressure in the center of the sample will rise slightly above the undrained level instantaneously reached at the time $t = 0$ and only after some time will it start to slowly decrease.

Detournay and Cheng (1993) show an example of numerical simulation of the loading of an infinitely long cylinder sample with the undrained and drained Poisson's ratios $\nu_u = 0.15$ and $\nu_{dr} = 0.31$ (see their figure 3). The pore pressure in the center of the sample exceeds the undrained level by about 10% at most. According to the standard explanation this occurs because the filtration of the fluid from the surface vicinity of the sample leads to softening of this vicinity. This vicinity of the sample surface can be compressed even more. This produces an additional load to the center of the sample and leads to an additional increase of the pore pressure there.

Alternatively one can look at this effect from the point of view of wave propagation. Instantaneous loading of the surface of the sample produces normal P-waves propagating in the sample. These waves converge in the center of the sample and spread further towards the sample's surface. The normal P-waves quickly establish the undrained pore-pressure level. On the other hand, they interact again with the sample surface and a part of their energy becomes converted into slow P-waves. These waves correspond to the fluid filtration in the sample and from the sample outwards. On the surface of the sample the slow waves are also converted into the normal P-waves responsible for adjusting the elastic equilibrium. Slow waves require more time for propagation in the sample. This time delay leads to a slow rise of the pore pressure in the sample's center above the undrained level. Interestingly, from figure 3 from Detournay and Cheng (1993), one can estimate that the time of the maximum pore pressure in the center corresponds approximately to the propagation time required by the slow wave of the sample-radius length to reach the center.

Detournay and Cheng (1993) give analytical solutions of several poroelastic boundary-value problems in finite spatial domains (e.g. a poroelastic cylinder whose surface has been suddenly exposed to a fluid pressure). They show that direct application of equation (2.114) does not provide the exact solution of the problem. Under such circumstances the homogeneous diffusion equation must first be solved for the quantity χ.

Descriptions of many boundary-value problems of linear quasi-static poroelasticity and their analytical solutions can be also found in the book by Wang (2000).

2.8 Stress and pressure coupling for radially symmetric fluid sources

Poroelastic coupling can affect seismicity induced by borehole-fluid injections and extractions. In this section we consider effects of poroelastic coupling onto a change ΔFCS of the failure criterion (1.73). For this we use the exact solutions by Rudnicki (1986) for a point-like fluid-mass source in a poroelastic medium presented in the previous section, (2.164), (2.172) and (2.173), and far-field limits of solutions for a pressure-cavity source, (2.196), (2.217) and (2.218).

Poroelastic contributions to the tectonic stress as functions of time t and distance r are given by the pore-pressure perturbation P_p and by radial and tangential stress perturbations. The pore-pressure perturbation is counted relative to the natural level of the *in situ* pore pressure in rocks (which is commonly, but not always, close to the hydrostatic one). In spherical coordinates the radial component of the stress perturbation is equal to the radial component of the poroelastic traction perturbation acting on a plane with a normal parallel to the radius vector of a given location. The tangential stress perturbation is equal to the normal component of the additional (caused by poroelastic coupling) traction acting on a plane that includes the radius vector of a given location. The radial poroelastic-perturbation-stress component is σ_{rr}. Because of the spherical symmetry the tangential component is equal to $\sigma_{\phi\phi} = \sigma_{\theta\theta}$. Figure 2.7 schematically shows the poroelastic contributions at two points of a fluid-saturated tectonically compressed medium in the case of fluid injection.

In a spherical coordinate system with the origin coinciding with the injection source, the poroelastic contributions (see also definitions (2.166) and (2.169)) are:

$$\sigma_{\theta\theta} = -n_s C_f \frac{S_{tt}(r_0)}{r}, \quad \sigma_{rr} = -n_s C_f \frac{S_{rr}(r_0)}{r}, \quad P_p = C_f \frac{\mathrm{erfc}(r_0)}{r}. \quad (2.219)$$

The source term C_f is given by

$$C_f = \frac{Q_I \eta}{4\pi k}, \quad (2.220)$$

where Q_I is a constant injection volume rate (mass rate divided by an initial density of the injection fluid, which is assumed be the same as the fluid saturating the medium). For an injection with a constant pressure (p_0) from a spherical cavity of the radius a_0 the source term will be

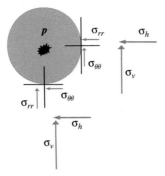

Figure 2.7 A sketch of poroelastic-coupling-related tractions in two points of a
tectonic domain in the case of fluid injection. The injection source is symbolized
by a dark irregular spot indicated with p. We consider two points on an arbitrary
spherical surface in the medium. One of the points is at the bottom of the sphere
and another one is on the right. The crossed black lines indicate horizontal and
vertical planes at these points. Small arrows indicate tractions normal to these
planes. These are poroelastic-coupling-caused tractions σ_{rr} and $\sigma_{\theta\theta}$. Note that
their directions depend on the point location on the sphere. However, directions
of the tectonic tractions σ_v and σ_h are independent of the location on the sphere
(we assume a uniformly distributed tectonic stress tensor). The complete stress
tensor is equal to the sum of the tectonic and poroelastic parts. At the bottom of
the sphere the resulting stress components will be $\sigma_v + \sigma_{rr}$ and $\sigma_h + \sigma_{\theta\theta}$. At the
right the stress components are $\sigma_v + \sigma_{\theta\theta}$ and $\sigma_h + \sigma_{rr}$. Therefore, even at equal
distances from the injection source poroelastic coupling leads to a heterogeneous
distribution of the resulting stress tensor.

$$C_f = p_0 a_0. \tag{2.221}$$

Figure 2.8 shows source-normalized poroelastic stresses and pore pressure corrected for geometrical spreading (or spherical divergence) given by the functions
$s_{tt}(r_0)$, $s_{rr}(r_0)$ and $\mathrm{erfc}(r_0)$, respectively. Let us first assume for simplicity that the
maximum compressive tectonic stress is vertical. This is a normal-faulting (exten-
sional) tectonic regime. We will then denote the maximum compressive stress as
σ_v. The corresponding minimum compressive stress is a horizontal one, σ_h. Note
that the values of these stress components are negative. An extension environment
and the fact that the maximum compressive stress is vertical mean that (recall the
notations introduced in Section 1.2):

$$\sigma_1 = -\sigma_v > \sigma_3 = -\sigma_h. \tag{2.222}$$

Let us consider poroelastic-coupling contributions to the stress tensor along the
directions of the maximum compressive stress and the minimum compressive stress
i.e. along the directions corresponding to σ_1 and σ_3, respectively.

First we consider points located exactly vertically above (or below) the injection source (Figure 2.7). These are locations whose radius vectors **r** are parallel to the axis $\mathbf{x_v}$ of the maximum compression stress (with the absolute value σ_1, in this case). Thus, the principal stresses in this location will be modified by the poroelastic coupling as follows:

$$\sigma_{mod_h} = \sigma_h + \sigma_{\theta\theta}, \qquad \sigma_{mod_v} = \sigma_v + \sigma_{rr}, \tag{2.223}$$

where we denote by σ_{mod_v} and σ_{mod_h} the poroelastic-coupling-modified stresses σ_v and σ_h, respectively. In turn, the values of principal stresses σ_1 and σ_3 will also be modified. We denote them as σ_{mod_1} and σ_{mod_3}, respectively. Note that we assume that absolute values of the tectonic stresses are significantly larger than absolute values of the poroelastic contributions, and that poroelastic effects do not change the order of these stresses in condition (1.52). Then (see also equation (2.222)):

$$\sigma_{mod_3} = \sigma_3 - \sigma_{\theta\theta}, \qquad \sigma_{mod_1} = \sigma_1 - \sigma_{rr}. \tag{2.224}$$

Therefore, the differential and mean stresses at such locations will be also modified, respectively:

$$\sigma_{mod_1} - \sigma_{mod_3} = \sigma_1 - \sigma_3 + \sigma_{\theta\theta} - \sigma_{rr}, \tag{2.225}$$

$$(\sigma_{mod_1} + \sigma_{mod_3})/2 = (\sigma_1 + \sigma_3)/2 - (\sigma_{\theta\theta} + \sigma_{rr})/2. \tag{2.226}$$

In these two equations the second terms on their right-hand sides give the poroelastic-coupling contributions (i.e. changes of the differential and mean stresses, $\Delta\tau_v$ and $\Delta\sigma_v$, respectively). Using equations (2.219) we obtain for these coupling contributions, respectively (see also (2.170)):

$$\Delta\tau_v = \sigma_{\theta\theta} - \sigma_{rr} = n_s C_f s_{dif}(r_0)/r, \tag{2.227}$$

$$\Delta\sigma_v = -(\sigma_{\theta\theta} + \sigma_{rr})/2 = n_s C_f s_m(r_0)/r, \tag{2.228}$$

where we have introduced

$$s_m(r_0) \equiv (s_{tt}(r_0) + s_{rr}(r_0))/2. \tag{2.229}$$

Analogously, at points located exactly horizontally to the left and right from the injection source (see Figure 2.7), i.e. points with radius vectors parallel to axis $\mathbf{x_h}$ of the minimum-compression *in situ* stress:

$$\sigma_{mod_h} = \sigma_h + \sigma_{rr}, \qquad \sigma_{mod_v} = \sigma_v + \sigma_{\theta\theta}. \tag{2.230}$$

In terms of the quantities σ_1 and σ_3, this means:

$$\sigma_{mod_3} = \sigma_3 - \sigma_{rr}, \qquad \sigma_{mod_1} = \sigma_1 - \sigma_{\theta\theta}. \tag{2.231}$$

Correspondingly, the differential and mean stresses at such locations will also be modified:

$$\sigma_{mod_1} - \sigma_{mod_3} = \sigma_1 - \sigma_3 + \sigma_{rr} - \sigma_{\theta\theta}, \tag{2.232}$$

$$(\sigma_{mod_1} + \sigma_{mod_3})/2 = (\sigma_1 + \sigma_3)/2 - (\sigma_{\theta\theta} + \sigma_{rr})/2. \tag{2.233}$$

Again, in these two equations the second terms on their right-hand sides give the poroelastic-coupling contributions. Using equations (2.219) we obtain these coupling contributions:

$$\Delta\tau_h = -(\sigma_{\theta\theta} - \sigma_{rr}) = -n_s C_f s_{dif}(r_0)/r, \tag{2.234}$$

$$\Delta\sigma_h = -(\sigma_{\theta\theta} + \sigma_{rr})/2 = n_s C_f s_m(r_0)/r. \tag{2.235}$$

At all other points σ_{mod_1} and σ_{mod_3} will have intermediate values between those given by (2.224) and (2.231). The values of principal stress σ_2 will be modified in a similar manner. It will be $\sigma_2 - \sigma_{rr}$ in points located along the acting direction of this stress, and $\sigma_2 - \sigma_{\theta\theta}$ in points located on the plane normal to this direction. Therefore, if the difference $|\sigma_{rr} - \sigma_{\theta\theta}|$ is greater than $\sigma_2 - \sigma_3$ or $\sigma_1 - \sigma_2$ then the poroelastic coupling can change the stress regime. In practice this would require nearly vanishing differences between the principal stresses at least in one of principal-stress planes.

Figure 2.8 shows that normalized poroelastic stress contributions s_{dif} and s_m are positive. Also the source term C_f is positive for the case of fluid injection. For example, it is equal to $p_0 a_0$ for a constant pressure p_0 switched on at time $t = 0$ on a surface of a spherical cavity of radius a_0. The radius of Mohr's circle increases with increasing $\Delta\tau$. Thus, one can see from (2.227) and (2.234) that during an injection the maximum increase of the radius of the Mohr's circle will be at the top and bottom locations. Its maximum decrease will be along the horizontal axis.

For fluid extractions the source term C_f is negative. Thus, from (2.227) and (2.234) we conclude that during fluid production the maximum increase of Mohr's circle will be along the horizontal direction, and its maximum decrease will be along the vertical direction.

A change of the differential stress (and, therefore, radius of Mohr's circle) is only one of possible contributions to changes of the failure criterion stress FCS. Because of fluid injection/extraction, the following changes of failure criterion stress (1.73)

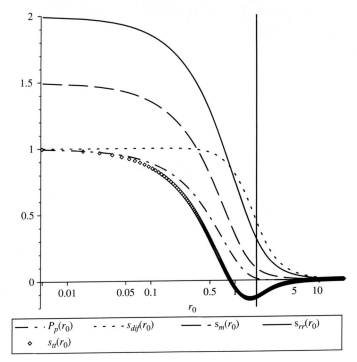

Figure 2.8 Dimensionless functions (from top to bottom at the vicinity of $r_0 = 0.01$) $s_{rr}(r_0)$, $s_m(r_0)$, $s_{dif}(r_0)$, $\text{erfc}(r_0)$ (it corresponds to P_p in equation 2.219), and $s_{tt}(r_0)$ of the dimensionless distance $r_0 = r/\sqrt{4Dt}$. These functions show the pore pressure and poroelastic-stress perturbations without influences of the source, of the geometrical spreading and of the poroelastic stress coefficient. The vertical line shows $r_0 = \sqrt{\pi}$ being an approximation of the microseismic triggering front (see the next Chapter and equation 3.6) used in this book for the case of a linear-diffusion induced seismicity.

are caused along the maximum compressive-stress direction (i.e. the vertical axis for the normal-stress regime):

$$\Delta FCS_v = C_f \left[\frac{1}{2} n_s s_{dif}(r_0) - (n_s s_m(r_0) - \text{erfc}(r_0)) \sin \phi_f \right] / r. \qquad (2.236)$$

Along the minimum compressive-stress direction (i.e. the horizontal axis for the normal-stress regime) these changes are

$$\Delta FCS_h = C_f \left[-\frac{1}{2} n_s s_{dif}(r_0) - (n_s s_m(r_0) - \text{erfc}(r_0)) \sin \phi_f \right] / r. \qquad (2.237)$$

Equations (2.236) and (2.237) show that during fluid injection (a positive C_f) the maximum change of FCS is $\Delta FCS_{max} = \Delta FCS_v$ and its minimum change is $\Delta FCS_{min} = \Delta FCS_h$. During fluid extraction (a negative C_f) $\Delta FCS_{max} = \Delta FCS_h$, and its minimum change is $\Delta FCS_{min} = \Delta FCS_v$.

The poroelastic coupling is given in (2.236) and (2.237) by terms with the factor n_s. Therefore, in addition to the radially symmetric destabilizing effect of the increasing pore pressure, the poroelastic-coupling effect of fluid injection enhances this destabilization below and above the injection source, i.e. along the direction of the maximum *in situ* compressive stress. The coupling effect becomes stabilizing along the minimum *in situ* compressive stress (i.e. it patially compensates the destabilizing effect of the increasing pore pressure in such locations). In the case of fluid extraction, and under conditions of a sufficiently strong poroelastic coupling (i.e. sufficiently high values of parameter n_s), destabilization of rocks is also possible. This effect reaches its maximum in the horizontal direction. In other words, the destabilizing effect of fluid extraction is strongest along the minimum *in situ* compressive stress direction.

These conclusions can be reformulated in the following way. In the case of an extensional tectonic stress regime, the maximum compressive stress is vertical. This stress regime leads to normal faulting. As a result of poroelastic coupling, fluid injection will stimulate normal faulting, especially strongly above and below the injection source. In contrast, fluid extraction will stimulate normal faulting, especially strongly on a horizontal plane on flanks of the fluid-sink position.

In the case of compressional tectonic-stress regime, the maximum compressive stress is horizontal. This leads to reverse faulting. Fluid injection will stimulate reverse (thrust) faulting especially strongly on a horizontal plane on flanks of the fluid-source position. In contrast, fluid extraction will stimulate reverse faulting, especially strongly above and below the fluid-sink position.

Similar observations and conclusions were made by Segall (1989) and Segall and Fitzgerald (1998), who computed strains and stresses for a finite poroelastic reservoir imbedded into an elastic continuum.

The difference in poroelastic-coupling contributions (2.236) and (2.237) in the vertical and horizontal directions, respectively, can cause anisotropy of the shape of microseismic clouds. In the case of fluid injection, the clouds of induced seismic events should have a tendency to grow faster (i.e. to get longer) along the maximum principal stress direction. Schoenball *et al.* (2010) and Altmann *et al.* (2010) observed this behavior on numerically simulated clouds of microseismic events.

Depending on the parameter n_s the effects of poroelastic coupling can be strong or weak. One of the main aims of the next section is to better understand the nature of this parameter, and to estimate a realistic range of its values.

2.9 Several non-linear effects of poroelastic deformations

A non-linear effect of poroelastic deformations relevant for understanding the fluid-induced seismicity is a stress and pore-pressure dependence of rock properties. Especially interesting for applications are changes of elastic parameters and of permeability as functions of applied loads. Understanding stress dependence of elastic properties of rocks (including seismic velocities) is important, for example, for interpretation of reflection seismic data, for analysis of 4D-seismic monitoring of hydrocarbon reservoirs and for overpressure prediction based on seismic data.

Practical experience collected in exploration seismology shows that elastic non-linearity of rocks is usually rather weak. This is purely a consequence of the very small amplitude of seismic waves at the depth of typical hydrocarbon and geothermal reservoirs (and hence, is a result of geometrical spreading). Seismic-wave-propagation phenomena and other small deformation (incremental) effects can be considered to be linear to a good approximation. On the other hand, laboratory experiments on drained and fluid-saturated rock samples, as well as seismic imaging and interpretation in depleting reservoirs and their overburden, show significant changes in linear elastic parameters induced by variations in confining stress and/or pore pressure.

Usually, in well-consolidated rocks, elastic strains due to moderate stress changes (less than or of the order of 10^2 MPa) are on the order of 10^{-2} to 10^{-3} or less. This is also the order of rotations and size changes in corresponding rock samples. Geometric non-linearity effects are even smaller; these are second-order terms in respect to displacement gradients, and thus they are of the order of 10^{-4}–10^{-6}. However, relative variations in elastic velocities can realistically reach the order of 10^{-1} in typical laboratory experiments. On the other hand, elastic properties of minerals (i.e. elastic moduli of the grain material) depend only weakly on stresses in the stress range up to 10^8 Pa.

Natural cracks, fractures and grain-contact domains are most sensitive to deformations. They are the most compliant parts of the pore space. Therefore, the stress dependence of elastic moduli of drained rocks must be mainly controlled by deformations of the pore space. This is the main reason of the physical (elastic) non-linearity of rocks. Many authors related changes of contact geometries (including spherical contact models) and crack geometries (including crack contact models) with changes of the elastic properties; see, for example, Mindlin (1949), Duffy and Mindlin (1957), Merkel *et al.* (2001), Gangi and Carlson (1996), and others. We have seen in the previous sections that the theory of poroelasticity provides a general approach for describing deformations of the pore space. Let us consider this effect in more detail.

2.9.1 Deformation of the pore and fracture space

Let us assume that in an initial stress state the internal surface Ψ_p of the rock sample was deformed. The geometry of the pore space can be characterized by an initial location of the surface Ψ_p plus its displacement due to the load. The corresponding strain of the pore space averaged over the volume of the sample is given by the quantity ξ_{ij} as defined in equation (2.22). If the rock is statistically homogeneous then the quantity ξ_{ij} is independent of the size of the sample. Thus, it is reasonable to assume that this quantity (along with the initial configuration of Ψ_p and the grain parameters) will control the elastic moduli of the drained rock.

Let us consider changes of the average strain ξ_{ij} due to changes of the load state $(\delta\sigma_{ij}^e; \delta P_p)$. These changes of the strain will describe average changes of geometry of the pore space. Changes of quantity ξ_{ij} satisfy the following rule:

$$\delta\xi_{ij} \equiv \delta(-\frac{\zeta_{ij}}{V}) = -\frac{\delta\zeta_{ij}}{V} - \xi_{ij}\frac{\delta V}{V}. \tag{2.238}$$

The quantity ζ_{ij} is defined by equation (2.5). The quantity ξ_{ii} is closely related to the porosity. Indeed, for ξ_{ii} equation (2.238) yields:

$$\delta\xi_{ii} \equiv \delta(-\frac{\zeta_{ii}}{V}) = -\frac{\delta\zeta_{ii}}{V} - \xi_{ii}\frac{\delta V}{V}. \tag{2.239}$$

On the other hand, if V_{p0} is the volume of the pore space at state with $\zeta_{ij} = 0$, then the complete change of the porosity due to a change in the applied load will be

$$\delta\phi = \delta(\frac{V_p}{V}) = \delta(\frac{V_{p0} - \zeta_{ii}}{V}) = -\frac{\delta\zeta_{ii}}{V} - \frac{V_{p0} - \zeta_{ii}}{V}\frac{\delta V}{V} = -\frac{\delta\zeta_{ii}}{V} - \phi\frac{\delta V}{V}. \tag{2.240}$$

Thus, the computing rule for a change of ϕ nearly coincides with the rule for a change of ξ_{ii}.

Let us consider Hooke's laws permitting us to write these rules in forms of stress–strain relations. We us assume that the grain material is homogeneous (in other words we assume the validity of equation (2.11); a more-general consideration can be found in Shapiro and Kaselow, 2005). Then, we obtain from equation (2.21):

$$-\frac{\delta\zeta_{ij}}{V} = (S_{klij}^{dr} - S_{klij}^{gr})\delta\sigma_{kl} + (S_{ijkk}^{dr} - S_{ijkk}^{gr} - \phi S_{ijkk}^{gr})\delta P_p. \tag{2.241}$$

From equation (2.20) we obtain

$$\frac{\delta V}{V} = S_{klmm}^{dr}\delta\sigma_{kl} + (C^{dr} - C^{gr})\delta P_p. \tag{2.242}$$

Substituting these two equations into equations (2.238) and (2.240) yields

$$\delta\xi_{ij} = (S_{klij}^{dr} - S_{klij}^{gr} - \xi_{ij}S_{klmm}^{dr})\delta\sigma_{kl}^e - (\phi S_{ijmm}^{gr} - \xi_{ij}C^{gr})\delta P_p, \tag{2.243}$$

and, for the porosity,

$$\delta\phi = (S_{klmm}^{dr} - S_{klmm}^{gr} - \phi S_{klmm}^{dr})\delta\sigma_{kl}^e. \tag{2.244}$$

We observe that load-caused changes of the porosity depend on a single combination of the confining stress and pore pressure. This combination is the effective stress, σ_{kl}^e. Again, this is a consequence of the condition $C^{gr} = C^p$. This condition follows from equation (2.11) and is a consequence of the assumption of statistical homogeneity of the grain material. It is also consistent with the Gassmann equation (see also Brown and Korringa, 1975). In relation to this fact it is commonly said that the effective-stress coefficient for the porosity (i.e. the coefficient in front of the pore pressure in the effective stress) is equal to 1.

Finally, under a completely hydrostatic load (i.e. $\sigma_{kl}^e = -\delta_{kl} P_d$, where P_d is called the differential pressure), these equations are reduced to

$$\delta \xi_{ij} = -(S_{llij}^{dr} - S_{llij}^{gr} - \xi_{ij} C^{dr}) \delta P_d - (\phi S_{ijkk}^{gr} - \xi_{ij} C^{gr}) \delta P_p, \tag{2.245}$$

$$\delta \phi = -(C^{dr} - C^{gr} - \phi C^{dr}) \delta P_d. \tag{2.246}$$

Alternative derivations of equation (2.246) were presented by different authors (see, for example, Zimmerman *et al.*, 1986, Detournay and Cheng, 1993, and literature cited therein; see also Goulty, 1998, and Gurevich, 2004). This equation can be also understood as a differential equation for the porosity as a function of the differential pressure:

$$\frac{d\phi}{d P_d} = C^{gr} - (1 - \phi) C^{dr}. \tag{2.247}$$

Since the solid grain materials (common minerals) are nearly linear elastic media, their compliances S^{gr} and the compressibility C^{gr} are practically independent of loads. Thus, in equation (2.247), only two quantities are significantly stress dependent: ϕ and C^{dr}. In order to quantify the stress dependence of these quantities, at least one more equation relating them is required. This equation cannot be formulated exactly because C^{dr} is defined by the complete geometry of the pore space rather than by ϕ only. Thus, further analysis requires involving empirical observations and heuristic assumptions. Below we concentrate our consideration on the stress dependence of the porosity of an isotropic rock in the case of a hydrostatic load (i.e. a change of a pore pressure and/or of a confining pressure) and follow mainly the derivation proposed in Shapiro (2003). Analogous consideration for anisotropic rocks and non-hydrostatic loads must involve a treatment of strain ξ_{ij}. More details can be found in Shapiro and Kaselow (2005).

2.9.2 Stiff and compliant porosities

We separate total porosity ϕ into two parts

$$\phi = \phi_c + [\phi_{s0} + \phi_s], \tag{2.248}$$

where the first part, ϕ_c, is a compliant porosity supported by thin cracks and grain-contact vicinities. According to laboratory observations, the compliant porosity usually closes up by the differential pressure of a few tens of MPa. This corresponds to the porosity with aspect ratio γ (a relationship between the minimal and maximal dimensions of a pore) less than 0.01 (see Zimmerman *et al.*, 1986). The second part, $[\phi_{s0} + \phi_s]$, is a stiff porosity supported by more-or-less isomeric pores (i.e. equidimensional or equant pores; see also Hudson *et al.*, 2001, and Thomsen, 1995) and worm-like pores (having their stiff cross-section of an isomeric shape). The aspect ratio of such pores (or of their stiff cross-section, respectively) is typically larger than 0.1. Such a subdivision of the porosity into a compliant and stiff parts is frequently done in seismic rock physics. For example, it is very similar to the definitions of the stiff and soft porosity of Mavko and Jizba (1991).

In turn, we separate the stiff porosity into a part ϕ_{s0} that is equal to the stiff porosity in the case of $P_d = 0$, and to a part ϕ_s that is a change of the stiff porosity due to a deviation of the differential pressure from zero. We assume that the relative changes of the stiff porosity, ϕ_s/ϕ_{s0}, are small. In contrast, the relative changes of the compliant porosity $(\phi_c - \phi_{c0})/\phi_{c0}$ can be very large, i.e. of the order of 1 (ϕ_{c0} denotes the compliant porosity in the case of $P_d = 0$). However, both ϕ_c and ϕ_{c0} are usually very small quantities. As a rule, they are much smaller than ϕ_{s0}. For example, in porous sandstones typical orders of magnitude of these quantities are $\phi_{s0} = 0.1$, $|\phi_s| = 0.01$ and ϕ_c can be less than, or of the order of, 0.01.

Under such circumstances, it is logical to assume the first, linear approximations of the compressibility C^{dr} as a function of the both parts of the porosity:

$$C^{dr}(\phi_{s0} + \phi_s, \phi_c) = C^{drs}[1 + \theta_s\phi_s + \theta_c\phi_c],   (2.249)$$

where C^{drs} is the drained compressibility of a hypothetical state of the rock with a closed compliant porosity (i.e. $\phi_c = 0$) and the stiff porosity equal to ϕ_{s0}. The acronym "*drs*" stands for "drained and stiff." Becker *et al.* (2007) proposed to call such a state of a rock a "swiss cheese model." Furthermore,

$$\theta_s = \frac{1}{C^{drs}}\frac{\partial C^{dr}}{\partial \phi_s}, \qquad \theta_c = \frac{1}{C^{drs}}\frac{\partial C^{dr}}{\partial \phi_c},   (2.250)$$

where the derivatives are taken at $\phi_s = 0$ and $\phi_c = 0$.

Approximation (2.249) implies that both quantities $\theta_s\phi_s$ and $\theta_c\phi_c$ are significantly smaller than 1. Numerous laboratory experiments and practical experience show that the drained compressibility depends strongly on changes in the compliant porosity, and it depends much more weakly on changes in the stiff porosity. We will express this empirical observation by the assumption

$$\theta_s\phi_s \ll \theta_c\phi_c \ll 1.   (2.251)$$

The quantity θ_s can be approximately estimated by using the upper bound of Hashin and Strikman (1963) for K_{dr}. Such an estimate corresponds to a small concentration of spherical voids in a solid matrix (see also Mavko *et al.*, 1998):

$$\theta_s \approx 1 + \frac{3}{4}\frac{K_{gr}}{\mu_{gr}}. \tag{2.252}$$

This shows that θ_s has values of the order of 1. Thus, $\theta_s\phi_s$ is a quantity of the order of 0.01.

An idea about the order of magnitude of θ_c can be obtained from results of various effective-medium theories for penny-shaped cracks, which usually coincide in the limit of $\phi_c \longrightarrow 0$ (see Mavko *et al.*, 1998, for effective bulk-modulus formulations in self-consistent or differential effective-medium approximations):

$$\theta_c \approx \frac{K_{gr}(3K_{gr} + 4\mu_{gr})}{\pi\gamma\mu_{gr}(3K_{gr} + \mu_{gr})}. \tag{2.253}$$

This estimate shows that θ_c is usually of the order of $1/\gamma$. Therefore, θ_c can be of the order of 10^2 or larger. Thus, in spite of a very small porosity ϕ_c the quantity $\theta_c\phi_c$ can reach the order of 0.1 or become even larger. If so, the approximation (2.249) is further simplified as follows:

$$C^{dr}(\phi_{s0} + \phi_s, \phi_c) = C^{drs}[1 + \theta_c\phi_c]. \tag{2.254}$$

Using this approximation, and neglecting ϕ in comparison with 1, we obtain the following relationship instead of equation (2.247):

$$\frac{d\phi_s}{dP_d} + \frac{d\phi_c}{dP_d} = C^{gr} - C^{drs} - \theta_c\phi_c C^{drs}. \tag{2.255}$$

We further assume that stiff-porosity changes with stress are independent of the changes of the compliant porosity. This also means that changes of the stiff porosity are independent of whether the compliant porosity is closed or not. If the compliant porosity is closed then $\phi_c = 0$ and we obtain from (2.255)

$$\frac{d\phi_s}{dP_d} = C^{gr} - C^{drs}. \tag{2.256}$$

Furthermore, if the assumption above is valid then this relationship will be valid also for an arbitrary (however, because of other assumptions, small) ϕ_c. Therefore,

$$\frac{d\phi_c}{dP_d} = -\theta_c\phi_c C^{drs}. \tag{2.257}$$

Solution of equations (2.256) and (2.257) provides us with the following approximations of the stress dependence of the stiff and compliant porosities:

$$\phi_s = P_d(C^{gr} - C^{drs}), \tag{2.258}$$

$$\phi_c = \phi_{c0} \exp\left(-\theta_c P_d C^{drs}\right). \tag{2.259}$$

Note that equation (2.258) is not valid for very large P_d. This is due to the fact that in equation (2.254) we neglected the stiff-porosity dependence of the compressibility C^{drs}, which becomes equal to C^{gr} if $P_d \longrightarrow \infty$. The validity of such a simplification and the validity of equation (2.258) are restricted by the condition (2.251). For very high stresses, the stiff porosity will also obey an exponentially saturating decreasing behavior.

2.9.3 Stress dependence of elastic properties

Usually, the dependence of a seismic velocity on a load is phenomenologically described by the following simple relationship (Zimmerman *et al.*, 1986; Eberhart-Phillips *et al.*, 1989; Freund, 1992; Jones, 1995; Prasad and Manghnani, 1997; Khaksar *et al.*, 1999; Carcione and Tinivella, 2001; Kirstetter and MacBeth, 2001):

$$c(P_d) = A_c + K_c P_d - B_c \exp\left(-P_d D_c\right), \tag{2.260}$$

The coefficients A_c, K_c, B_c and D_c of equation (2.260) are fitting parameters for a given set of measurements. It is often observed that equation (2.260) provides very good approximations for velocities in dry (see, for example, Figure 2.9) as well as in saturated rocks. Below we will derive this equation and clarify the nature of the fitting parameters.

Substituting equations (2.258) and (2.259) into equation (2.249), we obtain

$$C^{dr}(P_d) = C^{drs}\left[1 - \theta_s(C^{drs} - C^{gr})P_d + \theta_c\phi_{c0}\exp\left(-\theta_c P_d C^{drs}\right)\right]. \tag{2.261}$$

For the bulk modulus this gives approximately

$$K_{dr}(P_d) = K_{drs}\left[1 + \theta_s(C^{drs} - C^{gr})P_d - \theta_c\phi_{c0}\exp\left(-\theta_c P_d C^{drs}\right)\right]. \tag{2.262}$$

Using for the skeleton shear modulus μ_{dr}, an expansion similar to (2.249) yields

$$\mu_{dr}(P_d) = \mu_{drs}\left[1 + \theta_{s\mu}(C^{drs} - C^{gr})P_d - \theta_{c\mu}\phi_{c0}\exp\left(-\theta_c P_d C^{drs}\right)\right], \tag{2.263}$$

where μ_{drs} is the shear modulus of the dry rock in the case of a closed compliant porosity and stiff porosity equal to ϕ_{s0} (i.e. the case of a "swiss cheese model") and

$$\theta_{s\mu} = -\frac{1}{\mu_{drs}}\frac{\partial\mu_{dr}}{\partial\phi_s}, \qquad \theta_{c\mu} = -\frac{1}{\mu_{drs}}\frac{\partial\mu_{dr}}{\partial\phi_c}. \tag{2.264}$$

By analogy with the quantity θ_s the quantity $\theta_{s\mu}$ can be estimated using the upper Hashin–Strikman bound of μ_{dr}:

$$\theta_{s\mu} \approx 1 + \frac{6K_{gr} + 2\mu_{gr}}{9K_{gr} + 8\mu_{gr}}. \tag{2.265}$$

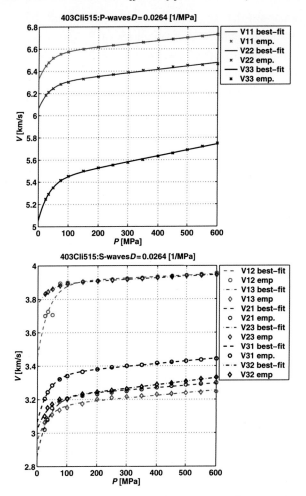

Figure 2.9 Measurements (dots) of quasi P-wave velocities (top) and quasi S-wave velocities (bottom) of a dry, metamorphic rock sample from the German KTB deep borehole. Velocities were measured in three orthogonal directions in laboratory experiments by Kern and Schmidt (1990) and Kern *et al.* (1994). Best-fit lines corresponding to equation (2.260) with different coefficients A_c, K_c, B_c reveal nearly the same value of the parameter, $D_c = 0.026$ MPa^{-1}. (After Shapiro and Kaselow, 2005.)

We see that the quantity $\theta_{s\mu}$ has the order of magnitude of 1. Again, by using effective-media theories for small concentrations of penny-shaped cracks, we can estimate the order of magnitude of $\theta_{c\mu}$ (see Mavko *et al.*, 1998, for effective shear modulus formulations in self-consistent or differential effective-medium approximations):

$$\theta_{c\mu} \approx \frac{1}{5}\left[1 + \frac{4(3K_{gr} + 4\mu_{gr})(9K_{gr} + 4\mu_{gr})}{3\pi\gamma(3K_{gr} + \mu_{gr})(3K_{gr} + 2\mu_{gr})}\right]. \qquad (2.266)$$

This estimate shows that for small γ, the quantity $\theta_{c\mu}$ is usually of the order of $1/\gamma$.

Approximations similar to (2.262) and (2.263) can be obtained for the undrained bulk modulus and for the velocities of P- and S-waves (for details, see Shapiro, 2003). A comparison of results for elastic moduli (2.261)–(2.263) with equation (2.260) shows that they all have the same functional form. This also leads to the same form of the stress dependencies of velocities. For example, for the P-wave and S-wave velocities in drained rocks, c_{Pdr} and c_{Sdr}, the corresponding relations will have the following forms:

$$c_{Pdr}(P_d) = A_P + K_P P_d - B_P \exp(-P_d D_P),\qquad (2.267)$$

$$c_{Sdr}(P_d) = A_S + K_S P_d - B_S \exp(-P_d D_S).\qquad (2.268)$$

Using results (2.265) and (2.266) we can estimate the order of terms in equations (2.262) and (2.263). If the differential pressure is smaller than, or of the order of, 10^2 MPa and C_{drs} is of the order of 10^{-5}–10^{-4} MPa^{-1}, the terms $\theta_{s\mu}(C^{drs} - C^{dr})P_d$ and $\theta_s(C^{drs} - C^{dr})P_d$ in equations (2.262) and (2.263) will be of the order of 0.001–0.01. These terms can often be neglected because the other two terms in equations (2.262) and (2.263) are of the order of 0.1 and 1. Therefore, if P_d is measured in MPa, both terms with coefficients K_P and K_S in equations (2.267) and (2.268) are of the order of 1 and, thus, can be neglected in comparison with other, much larger terms. This fact is in agreement with observations of Zimmerman *et al.*, (1986), Khaksar *et al.*, (1999), and Kirstetter and MacBeth (2001).

The modified equations have similar forms for the both, P- and S-wave velocities:

$$c_{Pdr}(P_d) = A_P - B_P \exp(-P_d D_P),\qquad (2.269)$$

$$c_{Sdr}(P_d) = A_S - B_S \exp(-P_d D_S).\qquad (2.270)$$

These equations also keep their forms for saturated rocks. Moreover, we should expect that, for dry as well as for saturated isotropic rocks, in the first approximation all coefficients D_c are identical:

$$D_P = D_S = \theta_c C^{drs}.\qquad (2.271)$$

This is also in good agreement with laboratory estimates of these quantities for porous sandstones (see Figure 2.10). Moreover, an approximate universality of coefficients D_c has been observed for different velocities in anisotropic rocks (see Ciz and Shapiro (2009) and Figure 2.9) and even for elastic stiffnesses and electrical resistivity (Kaselow and Shapiro, 2004).

Summarizing, to the first approximation the seismic velocities as well as the porosity depend on the differential pressure, i.e. the difference between the confining pressure and the pore pressure. The stress dependence of the geometry of

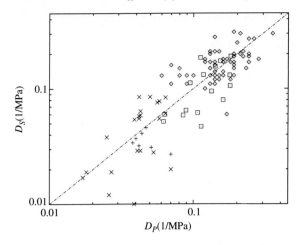

Figure 2.10 Coefficient D_S versus D_P for several sandstones (saturated as well as dry). The straight line is given by (2.271). Data have been collected from different literature sources: Eberhart-Phillips *et al.* (1989); Freund (1992); Jones (1995); Khaksar *et al.* (1999). (After Shapiro, 2003.)

the pore space controls the elastic moduli and velocity changes with stress. Here, the compliant porosity, which can be just a very small part of the total porosity, plays the most important role. Closing compliant porosity with increasing differential pressure explains the experimentally observed exponentially saturating increase of seismic velocities. Coefficients of this relationship are defined by the compliant-porosity dependence of the drained bulk modulus. The coefficients of equations (2.267) and (2.268) can be used to compute such characteristics of rocks as ϕ_{c0}, $\theta_{c\mu}$ and θ_c.

The dimensionless quantity θ_c defines the sensitivity of the elastic moduli to the differential pressure. Shapiro (2003) proposed calling it the elastic piezo-sensitivity. The piezo-sensitivity is a physical property of a rock. From the derivation above it is clear that it is controlled by the compliant porosity. Moreover, it is approximately proportional to an averaged (effective) reciprocal aspect ratio of the compliant porosity. Taking into account the data of Figure 2.10 we obtain that the realistic range of the magnitude orders of the piezo-sensitivity for sandstones is 10^2–10^4. For low-permeable tight rocks like the crystalline rocks of the German KTB borehole, basalts, granites and shale (including organic reach shale), the piezo-sensitivity is of the order of 10^2 or even smaller. This can be estimated from values of D_c obtained in different works (Ciz and Shapiro, 2009; Kaselow *et al.*, 2006; Becker *et al.*, 2007; Pervukhina *et al.*, 2010) and also from frequently reported observation that elastic properties of some shales are nearly stress independent in the range of loads below 30 MPa.

2.9.4 Non-linear nature of the Biot–Willis coefficient α

The piezo-sensitivity can be related to the poroelastic coefficient α. Indeed, substituting (2.254) into (2.35) we obtain:

$$\alpha = 1 - \frac{(1 - \theta_c \phi_c)}{C^{drs} K_{gr}}. \tag{2.272}$$

We can obtain upper and lower bounds for α. Following Zimmerman *et al.* (1986) and Norris (1989), we use the Hashin–Strikman bounds for the drained compressibility C^{drs}. Its upper bound is infinitely large and yields $\alpha = 1$. Its lower bound corresponds to spherical void inclusions distributed in the grain material so that the porosity of the resulting composite is ϕ_{s0}. As a result we obtain the following bounds for α:

$$1 \geq \alpha \geq 1 - (1 - \theta_c \phi_c) \frac{(1 - \phi_{s0})}{1 + \frac{3}{4} \phi_{s0} K_{gr}/\mu_{gr}} \geq \frac{2\theta_c \phi_c + 3\phi_{s0}}{2 + \phi_{s0}} \approx 1.5 \phi_{s0} + \theta_c \phi_c, \tag{2.273}$$

where we assumed the positiveness of Poisson's ratio for the rock-grain material (i.e. $K_{gr}/\mu_{gr} \geq 2/3$) and have neglected terms with products of porosities. If $\phi_c = 0$ we obtain

$$1 \geq \alpha \geq \frac{3\phi_{s0}}{2 + \phi_{s0}} \approx 1.5 \phi_{s0}. \tag{2.274}$$

These are the bounds of α derived by Zimmerman *et al.* (1986). However, the values of α usually observed (see, for example, table 4 of Detournay and Cheng, 1993) are much larger than just 1.5ϕ or even 2ϕ (the latter corresponds to situations with the grain material Poisson's ratio close to 0.25). Equation (2.273) clearly explains the reason of such observations. The compliant porosity increases the lower bound of α. The greater is the piezo-sensitivity and/or compliant porosity of rocks the greater is the lower bound and, correspondingly, the greater the quantity α tends to be. Obviously, the smaller is the effective aspect ratio of the compliant porosity the higher is the lower bound of α. The upper bound of α is always 1, of course.

On the other hand, the piezo-sensitivity is related to the non-linear elastic moduli of rocks. Indeed, for small values of P_d we obtain from equation (2.262)

$$K_{dr}(P_d) = [1 + \beta_K P_d]/C^{drs}, \tag{2.275}$$

where the coefficient

$$\beta_K = \theta_c^2 \phi_{c0} C^{drs} \tag{2.276}$$

is equal to a rational function of the bulk modulus and of the non-linear elastic moduli (see Zarembo and Krasilnikov, 1966, pp. 299–309). It is clear that the larger the piezo-sensitivity and the compliant porosity, the larger the elastic non-linearity of rocks. Taking into account the values of the piezo-sensitivity following from Figure 2.10, we can expect that for well-consolidated sedimentary rocks the non-linearity is 10^2–10^4 times larger than for such materials as metals, where the coefficient β_K is roughly of the order of $1/K$.

2.9.5 Magnitude of the poroelastic-stress coupling

Equations (2.276), (2.272) and (2.273) show that coefficient α is closely related to the piezo-sensitivity and to the coefficient β_K and, thus, to the non-linear elastic moduli of drained rock. This fact can be used to estimate the strength of the poroelastic-stress coupling. Indeed, in Sections 2.6–2.8 we have seen that the stress effects of the poroelastic coupling are controlled by parameter n_s, which is propotional to the Biot–Willis coefficient (see also (2.167)):

$$n_s = \alpha n',$$ (2.277)

where we used notation

$$n' \equiv \frac{1 - 2v^{dr}}{2(1 - v^{dr})}.$$ (2.278)

Figure 2.11 shows n' as a function of the drained Poisson's ratio. We conclude that generally for realistic rocks with positive Poisson's ratio, $n_s < 0.5$.

In the following chapters of this book we will be often interested in low-permeability crystalline and sedimentary tight rocks, where stimulation of the fluid mobility is necessary. The stiff porosity of such rocks belongs effectively to the solid material composing their drained skeleton. Thus, the product $K_{gr}C_{drs}$ is approximately equal to 1. Then from (2.272) and (2.259) we obtain:

$$\alpha \approx \theta_c\phi_c = \theta_c\phi_{c0}exp(-\theta_cC_{drs}P_d).$$ (2.279)

Generally, $1 \gg \theta_c\phi_{c0}$. The product $\theta_c\phi_{c0}$ is equal to the largest correction of the bulk compressibility of drained rock due to the presence of compliant porosity. For low-permeability tight rocks this correction is usually lower than 0.3. For such rocks Poisson's ratio is rather close to 0.25 (see Figure 2.11, top) and, therefore, usually $n_s < 0.1$.

Moreover, equation (2.279) implies that α decreases with increasing depth. Usually for low-permeability tight rocks one can expect that ϕ_{c0} is of the order of 0.001 or smaller. Even for relatively permeable rocks like sandstones the compliant porosity can also be of the order of 0.001 (see, for example, Pervukhina *et al.*, 2010). Realistic values of θ_c for tight rocks are of the order of 10–1000 (see, for example, estimates obtained for different rocks including shales: Pervukhina

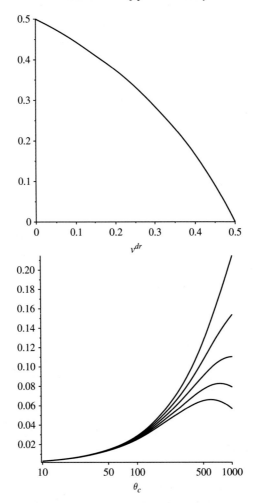

Figure 2.11 Poroelastic-stress coefficient and its components. Top: the function $n' \equiv (0.5 - \nu^{dr})/(1 - \nu^{dr})$ of the drained Poisson's ratio. Bottom: the poroelastic-stress coefficient n_s as a function of the piezo-sensitivity, θ_c. The parameters of the curves are: $n' = 0.3$, $\phi_{c0} = 0.001$ and $K_{drs} = 30$ GPa, and P_d is equal to 10, 20, 30, 40 and 50 MPa from the upper curve to the lower curve, respectively. In the case of hydrostatic conditions these differential pressures roughly correspond to depths of 0.5, 1, 1.5, 2 and 2.5 km, respectively.

et al., 2010; Ciz and Shapiro, 2009; Kaselow *et al.*, 2006). Figure 2.11 (bottom) shows some estimates of the poroelastic-stress coefficient. For a realistically chosen C_{drs} and for the depth greater than 1.5 km, the poroelastic-stress coefficient is significantly smaller than 0.1 (nearly the maximum value). It is quickly vanishing with depth. Correspondingly, the role of the poroelastic-stress coupling will quickly vanish with depth.

In the previous sections we have seen that, in the case of fluid injection, a strong poroelastic coupling will lead to stronger growth of a cloud of fluid-induced seismicity along a direction of the maximum tectonic stress (maximum absolute value). In the case of fluid extraction this will lead to a possible rock failure, firstly along the minimum tectonic stress (minimum absolute value). However, the magnitudes of stress perturbations caused by poroelastic coupling are controlled by the poroelastic-stress coefficient. Our analysis in this section shows that, for low-permeability tight rocks at depths of 1.5 km and more, this coefficient is smaller than 0.1. Under such conditions the pore-pressure perturbation will make a dominant contribution to poroelastic modifications of a failure-criterion stress. Thus, in the following we will usually neglect the poroelastic coupling effects (see also Section 3.4).

2.9.6 Effective-stress coefficients

Confining stress and pore pressure are applied to different surfaces of a porous sample. Their impacts on the rock can be to some extent mutually compensating or enhancing. In the previous sections we have seen that different quantities characterizing poroelastic media depend on the applied confining stress and on the pore pressure in different manners. Generally this dependence can be expressed as a function of the following combination of the loads:

$$L_{ij}^q \equiv -\sigma_{ij} - \rho_{ij}^q P_p. \tag{2.280}$$

Berryman (1992, 1993) considered different forms of load combinations L_{ij}^q as effective-stress rules for the corresponding property q of a rock. The quantity ρ_{ij}^q is called an effective-stress coefficient for the physical quantity q. It is clear that, generally, this coefficient can be a tensor. For example, for the strain tensor ϵ_{kl} of a porous fluid-saturated rock sample (i.e. a strain of the skeleton) equation (2.31) gives $\rho_{ij}^q = \alpha_{ij}$.

Further, we will follow Berryman (1992, 1993) and consider a simplified situation of a hydrostatic load defined by a confining pressure P_c and by a pore pressure P_p. We will assume also a statistically homogeneous grain material. We will also consider load-dependent quantities such as scalars. In this case, effective stress (2.280) for a given physical quantity q reduces to the following quantity:

$$L_q = P_c - \rho_q P_p. \tag{2.281}$$

For example, equation (2.37) shows that for the sample dilatation ϵ (i.e. skeleton dilatation) the effective-stress coefficient is equal to $\rho_\epsilon = \alpha$. Equation (2.43) shows that, for the quantity χ, which is a change of the pore-space volume in a unit rock volume caused by a fluid-mass migration, the effective-stress coefficient is equal to $\rho_\chi = 1/B$, where B is the Skempton coefficient. Finally, equation (2.246) shows that the effective-stress coefficient for the connected porosity, ρ_ϕ is 1. Moreover, equations (2.241)–(2.246) show that, in the case of small porosity and small strains, the effective-stress coefficient for the geometry of the pore (and fracture) space is close to 1.

The effective-stress coefficient for the rock failure is not directly related to poroelastic coefficients ρ_q. However, the failure of a rock must be related to deformations of its pore (and fracture) space. Thus, it must be related to ρ_q describing load dependence of the pore-space geometry. Terzaghi (1936) concluded from heuristic reasoning that the effective-stress coefficient for a load combination controlling the failure of saturated rocks (we will denote this combination as L^f) is equal to 1. This corresponds to $\rho_{ij}^f = \delta_{ij}$ in equation (2.280) and to $\rho_f = 1$ in equation (2.281). Experimental observations and the fact that the stress concentrations at microcracks are proportional to the differential pressure also support the assumtion that $\rho_f = 1$ (see Jaeger *et al.*, 2007).

2.9.7 Stress dependence of permeability

Deformations of the pore space are also related to the stress dependence of the permeability. To understand the effective-stress law for the permeability, Berryman (1992) proposed the following argument. The physical dimension of the permeability is (length)^2. Thus, uniform rescaling of the size of an isotropic porous sample (leading also to the corresponding rescaling of sizes of pores) increases or decreases permeability proportionally to (volume)$^{2/3}$. Note that such a geometric transformation of the pore sample does not change its porosity. In other words, this rescaling does not change specific geometric relations between the solid and the pore space. However, additionally to the sample scale, permeability must depend on a non-dimensional measure of mutual geometric relations between the solid and the pore space (i.e. the relative positioning of grains). A power-law function of the porosity frequently seems to be such an adequate measure. For example, it is in agreement with a Kozeny–Carman-type of relation between the porosity and permeability for laminar flow in tube-like pores and other pore channels of simple shapes (see Coussy, 2004). Taking this into account we obtain

$$k \propto \phi^n V^{2/3}, \tag{2.282}$$

where n is a nearly empiric exponent close to 4 or larger.

Equation (2.282) implies that the permeability is not just a simple function of the differential pressure. This is clear from the fact that the permeability depends not only on the porosity but also on a characteristic length. In equation (2.282), the characteristic length is related to the sample volume. On the other hand, our discussion above indicates that the effective-stress law for the volume includes α as the effective-stress coefficient. Therefore, the permeability depends on a combination of the load components P_p and P_c that is more complex than the differential pressure.

Using relation (2.282), Berryman (1992) shows that the effective-stress coefficient for permeability ρ_k must be smaller than 1 and it is a function of α, ϕ and n. However, real rocks can show ρ_k lower, equal to or larger than 1 (Berryman, 1992; Al-Wardy and Zimmerman, 2004; Li *et al.*, 2009). The reason for this is the fact that equation (2.282) has been derived from too-general dimensionality considerations. By such considerations the main dimensional parameters are the volume of the sample and the volume of its pore space. Thus, the length scale in (2.282) can be substituted by a pore diameter or a grain diameter instead of a scale derived from the macroscopic volume. However, two length scales (the pore diameter and the sample size) are not sufficient to describe the pressure-dependent permeability. In other words, equation (2.282) treats the porosity in a too-general way. It does not specify which porosity component mainly controls the permeability. The total connected porosity can be substituted in this equation by any other porosity function.

Later we will be especially interested in situations where the main changing load component is an increasing pore pressure (i.e. a decreasing differential pressure) leading to an expansion of the pore space. Equation (2.282) can be taken as a basis for modeling pressure-dependent permeability if we use additional information on the pore-space geometry.

For example, in the previous sections we argued that the stress dependence of drained elastic properties is mainly controlled by the compliant porosity. The stiff porosity is frequently insignificant. Roles of compliant and stiff parts of the pore space can also be very different in the pressure dependence of permeability. To model such a pressure dependence we combine equation (2.282) with equations (2.248), (2.259) and (2.258). To account for possibly different contributions of the stiff and compliant porosities in the permeability we take these porosity components with different weighting factors (Φ_s and Φ_c, respectively) introduced *ad hoc*:

$$k \propto \left[\Phi_s (\phi_{s0} - P_d(C^{drs} - C^{gr})) + \Phi_c \phi_{c0} \exp\left(-\theta_c P_d C^{drs}\right) \right]^n V^{2/3}. \quad (2.283)$$

Note that it is sufficient to work with a single relative weighting factor only, e.g. Φ_c / Φ_s. However, we keep both factors for convenience.

From relation (2.283) we can derive several simple models. By moderate loads the volume V will not change a lot. Thus, we can include it approximately into a proportionality factor. Thus, we do approximate the effective stress by the differential pressure. Let us further assume a rock where the permeability is controlled by a stiff porosity only (i.e. $\Phi_c = 0$). For example, this can be the case for high-porosity sand, where compliant pores are mainly confined to grain contacts and do not significantly contribute to permeability (which is controlled by stiff pore throats). For $\Phi_c = 0$ we obtain

$$k \propto (1 - \frac{C^{drs} - C^{gr}}{\phi_{s0}} P_d)^n. \tag{2.284}$$

This function resembles theoretical models derived by Gangi (1978) and Walsh (1981).

Let us further assume that the permeability is controlled by stiff porosity, but the initial (connected) stiff porosity is vanishingly small (i.e. $\phi_{s0} = 0$). Then the increasing pore pressure will open new connected stiff pores. Under such conditions $P_d < 0$. Equation (2.284) is still applicable. We obtain a power-law model of the permeability:

$$k \propto (-P_d)^n. \tag{2.285}$$

Conversely, assume that the contribution of the stiff porosity in the connected porosity is vanishingly small. This will lead to $\Phi_s = 0$. Then the permeability is controlled by the compliant porosity. This corresponds, for instance, to fractured carbonates, where compliant voids (fractures) provide the only conduits between isolated vugs. This yields

$$k \propto \exp(-\kappa P_d). \tag{2.286}$$

Here κ is the so-called permeability compliance. Equation (2.283) implies that

$$\kappa = \theta_c C^{drs} n. \tag{2.287}$$

Both types of pressure dependence (2.285) and (2.286) can be considered as simple modeling alternatives. Then quantities n and b_p are fitting constants. For example, function (2.286) has been used in experimental studies (Li *et al.*, 2009).

2.10 Appendix. Reciprocity-based relationship between compliances of porous media

In this appendix we will derive equation (2.13). We apply the reciprocity theorem (Amenzade, 1976; Brown and Korringa, 1975):

$$I = \int_{\Omega_g} [\delta u_i(\hat{x}) \delta' \tau_i(\hat{x}) - \delta' u_i(\hat{x}) \delta \tau_i(\hat{x})] d^2 \hat{x} = 0, \tag{2.288}$$

where the total surface of the linear grain material is introduced: $\Omega_g = \Sigma_c + \Psi_p$. Note that we assume the existence of a vanishing thin film sealing pores on the external surface of the rock specimen. Any influence of this film on the drained compressibilities is neglected. Such an enveloping thin film is assumed to allow an application of independent forces to Σ_c and Ψ_p (i.e. a jacketed sample as shown in Figure 2.2) Further, $\delta\tau_i$ and $\delta'\tau_i$ are components of two systems of traction increments acting on the surface Ω_g in two different hypothetical experiments. These small traction increments cause small displacements of the surface Ω_g. Quantities δu_i and $\delta'u_i$ are components of these displacements, respectively.

Let us consider the following pair of traction increments uniformly distributed on Σ_c or Ψ_p:

$$\tau_i = \delta\sigma_{ij}n_j \text{ on } \Sigma_c \quad \text{and} \quad \tau_i = 0 \text{ on } \Psi_p, \tag{2.289}$$

and

$$\tau_i' = 0 \text{ on } \Sigma_c \quad \text{and} \quad \tau_i' = \delta\sigma_{ij}^f n_j \text{ on } \Psi_p. \tag{2.290}$$

The reciprocity theorem then gives:

$$\delta\sigma_{ij}^f \int_{\Psi_p} \delta u_i(\hat{x})n_j d^2\hat{x} = \delta\sigma_{kl} \int_{\Sigma_c} \delta u_k'(\hat{x})n_l d^2\hat{x}. \tag{2.291}$$

Assuming that the traction increments and displacements are small, and using the notations from equations (2.3) and (2.5), equation (2.291) can be written as

$$\delta\sigma_{ij}^f \delta\zeta_{ij} = \delta\sigma_{kl}\delta\eta_{kl}. \tag{2.292}$$

Since this holds for any $\delta\sigma_{ij}^f$ and $\delta\sigma_{kl}$ (i.e. we can select several components being non-vanishing only) we have:

$$\left(\frac{\partial\zeta_{ij}}{\partial\sigma_{kl}}\right)_{\sigma^f} = \left(\frac{\partial\eta_{kl}}{\partial\sigma_{ij}^f}\right)_{\sigma}. \tag{2.293}$$

By changing independent variables from σ_{ij}^f and σ_{ij} to σ_{ij}^f and σ_{ij}^e (see definition (2.6)), we obtain

$$\left(\frac{\partial\zeta_{ij}}{\partial\sigma_{kl}}\right)_{\sigma^f} = \left(\frac{\partial\zeta_{ij}}{\partial\sigma_{kl}^e}\right)_{\sigma^f} = \left(\frac{\partial\eta_{kl}}{\partial\sigma_{ij}^f}\right)_{\sigma^e} - \left(\frac{\partial\eta_{kl}}{\partial\sigma_{ij}^e}\right)_{\sigma^f}. \tag{2.294}$$

Dividing this by the volume, V, we obtain (2.13).

3

Seismicity and linear diffusion of pore pressure

In this chapter we consider situations of fluid injections with the injection pressure (i.e. the bottom hole pressure or the pressure at the perforation borehole interval) smaller than the absolute value of the minimum principal compressive stress, σ_3. Then, usually, permeability enhancement on the reservoir scale is significantly smaller than one order of magnitude. In such a situation the behavior of the seismicity triggering in space and in time is mainly controlled by a linear process of relaxation of stress and pore pressure.

We will start this chapter by considering fluid-injection experiments in crystalline rocks at the German Continental Deep Drilling site KTB (Kontinentale Tiefbohrung). Results of these experiments are helpful for identifying factors controlling fluid-induced seismicity. Then we consider kinematic features of induced seismicity such as its triggering front and its back front. They provide envelopes of clouds of microseismic events in the spatio-temporal domain. We further consider features of the triggering front in a hydraulically anisotropic medium.

We introduce several approaches for the quantitative interpretation of fluid-induced seismicity. These approaches can be applied to characterize hydraulic properties of rocks on the reservoir scale. Thus, they can be useful for constructing models or constraints for reservoir simulations. These approaches are based on the assumption of a linear fluid–rock interaction. Some of these approaches have a heuristic nature. These are the triggering-front-based estimation of hydraulic properties of rocks (in the next chapter we will observe that in some cases of a nonlinear pressure diffusion the triggering front can be introduced exactly) and the related eikonal-equation-based approach for characterization of spatially heterogeneous hydraulic diffusivity. We will also consider a dynamic property of induced seismicity: its spatial density. A related issue of statistical properties of the strength of pre-existing defects will be a subject of our consideration too.

3.1 Case study: KTB

The German KTB site (see a series of papers with an introductory overview of Emmermann and Lauterjung, 1997) is located in Windischeschenbach (Southeastern Germany, Bavaria). It was placed near the western margin of the Bohemian Massif at the contact zone of the Saxo-Thuringian and the Moldanubian tectonic units. The corresponding crystalline crustal segment is mainly composed of metabasites and gneisses. The site includes two boreholes, the pilot and the main one, drilled between 1987 and 1994. The pilot hole reached a depth of 4 km, whereas the main hole penetrated the crust down to 9.1 km.

Three large-scale series of borehole-fluid injection-extraction experiments were performed at the site with the aim of investigating fluid-transport processes and crustal stresses.

The first fluid-injection experiment was carried out in 1994 (Zoback and Harjes, 1997). The duration of the injection was approximately 24 h. About 200 m^3 of KBr/KCl brine were injected into a 70 m open section of the main borehole in the depth range 9030–9100 m (with *in situ* temperature of 260 °C). During the active injection period the flow rates varied approximately from 50 to 550 l/min and the corresponding well-head pressure varied approximately from 10 to 50 MPa. The downhole injection pressure was approximately 130 MPa larger due to the static pressure of the borehole-fluid column.

The seismic observation system included 73 surface stations and one three-component borehole seismometer. Approximately 400 microearthquakes were recorded. All events were considered to be induced by the injected fluids: 94 of these earthquakes could be localized with respect to master events with a relative location accuracy of several tens of meters (Zoback and Harjes, 1997; Jost *et al.*, 1998). Later, the data were precisely relocalized by using various hypocenter location improvements (Baisch *et al.*, 2002). The seismically active zone comprised a volume of approximately 0.35 km^3 around the bottom of the borehole. By analyzing modeling results, Zoback and Harjes (1997) estimated that a small increase in pore pressure was sufficient to trigger the earthquakes. A perturbation of the pore pressure of 1 MPa corresponds roughly to the distance of 30 m from the injection interval. However, there were many significantly more-distant events. Surprisingly, only events above a depth of 9 km were observed. Hypothetical reasons for this were proposed by Zoback and Harjes (1997). One of the reasons is the possible proximity of the brittle–ductile transition zone at this depth and, as a consequence, the presence of an impermeable lower half-space. On the other hand, it is possible that, at greater depth, the level of the differential stress is significantly smaller than the rocks' frictional strength. Thus significantly larger pore-pressure perturbations would be required for inducing earthquakes.

The second fluid-injection experiment at the KTB was performed in the year 2000 (Baisch *et al.*, 2002). This was a two-month-long injection. It was designed to enable fluid migration and to perturb the pore pressure farther away from the injection interval than in the previous experiment. About 4000 m^3 of water were injected at the well head of the main borehole at an injection rate of 30–70 l/min and a well-head pressure between 20 and 30 MPa. It was assumed that the main borehole was hydraulically isolated at least down to 6 km depth. However, due to several leakages in the borehole casing, about 80% of the fluid was already injected between 5.35 and 5.4 km depth.

A temporary seismic network consisting of a borehole seismometer in the pilot hole at 3827 m depth and 39 surface stations was installed (Baisch *et al.*, 2002). Nearly 2800 microseismic events were detected, of which 237 were localized with an accuracy better than 100 m on average. Seismic events concentrated at two depth levels: 5.0–6.0 km (81% of total seismicity) and 8.8–9.2 km (11% of total seismicity).

In June 2002 the third series of long-term fluid experiments (2002–2005) was started (Kümpel. *et al.*, 2006; Erzinger and Stober, 2005; Shapiro *et al.*, 2006a). Features of particular interest of the research program were two dominant fault systems encountered at 7.2 km and 4.0 km depths. These two fault systems were identified as seismic reflectors at the KTB site and denoted as SE1 and SE2 reflectors, respectively (Hirschmann and Lapp, 1995; Simon *et al.*, 1996; Harjes *et al.*, 1997; Buske, 1999). The first major experiment was a one-year-long fluid production test in the KTB pilot hole (June 2002–June 2003). A total volume of 22 300 m^3 of saline crustal fluids (with temperature of 119 °C *in situ*) was produced from the open hole section (3850–4000 m, approximately). The final draw down was 605 m below the surface. The corresponding fluid yield was 58 l/min. The KTB main hole was equipped with a seismometer installed at 4000 m depth and two water-level sensors. The fluid level in the main hole – at 200 m distance from the pilot hole – steadily fell from zero to 50 m below the surface, indicating a hydraulic connection between the pilot and the main hole. During the fluid-production phase of the experiment, induced seismicity was absent. Hydraulic permeability was estimated to be around 2×10^{-15} m^2 (Stober and Bucher, 2005; Gräsle *et al.*, 2006).

After 12 months of hydraulic recovery, a fluid injection test in the pilot hole was started in June 2004. Over ten months, 84 600 m^3 of water were injected into the open hole section at 4 km depth of the pilot hole, where the SE2 reflector intersected the borehole. The injection rate was 200 l/min on average, at about 10 MPa well-head injection pressure. The fluid level in the main hole clearly responded to the injection in the pilot hole. In October 2004, the main hole became artesian and produced some 1 m^3 of water per day. Significant induced seismicity started in September 2004 and increased slowly (see Figure 3.1).

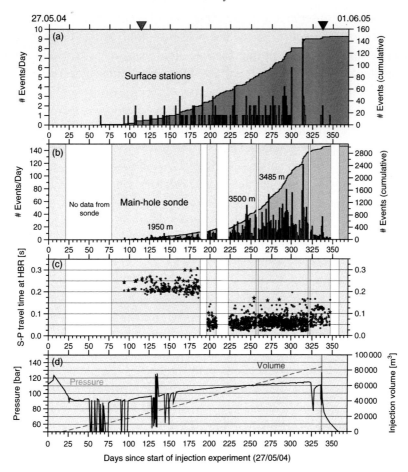

Figure 3.1 Microseismicity during the 2004–2005 KTB injection experiment. (a) Events recorded by the surface stations. (b) Events recorded by the geophone located in the main borehole. (c) Time differences between arrivals of S- and P-waves. Changes in the locations of the borehole geophone are clearly seen from the plot. (d) The injection pressure along with the cumulative volume of the injected water. The time when the amount of previously extracted fluid was re-injected is marked in part (a) of the figure by the inverse triangle on the top, between the days 100 and 125. Another triangle (between the days 325 and 350) marks the injection termination. (Modified after Shapiro *et al.*, 2006a.) A black and white version of this figure will appear in some formats. For the color version, please refer to the plate section.

During the injection experiment, about 3000 microseismic events were detected by the borehole sonde and about 140 events were localized by using the borehole sonde and a specially deployed monitoring array of three-component seismometers (five sensors in 25–50 m shallow boreholes and eight surface stations).

One important observation of the 2002–2005 KTB hydraulic experiments was the absence of seismicity during the phase of fluid extraction. Moreover, the induced seismicity started during the injection phase approximately at the time when the fluid volume previously extracted had been compensated by the injection. It was approximately 110 days after the start of the injection (see Figure 3.1).

Horizontal projections of the located events compose a NW–SE elongated zone. This zone is nearly parallel to the Franconian Lineament. Before and during the drilling phase at the KTB site, intensive seismic studies were carried out. From a 2D seismic survey (KTB8502), a sharp northeast-dipping seismic reflector (SE1) was identified in seismic profiles (Simon *et al.*, 1996; Harjes *et al.*, 1997; Buske, 1999). This reflector is regarded as the continuation of the Franconian Lineament through the upper crust. A pre-stack Kirchhoff depth migration of the KTB8502 profile as well as of a 3D seismic reflection survey (ISO89-3D) was presented by Buske (1999). During this survey, an area of about 21 km × 21 km, with the main borehole located at the center, was investigated. Figure 3.2 shows the relevant parts of this data set after migration. Above the SE1 reflector and quasi-parallel to it, a slightly reflecting linear structure intersecting the boreholes approximately at the depth of 4 km can be also seen. This is the SE2 reflector. The induced seismicity of the 2004–2005 injection experiment is directly related to this reflector (Figure 3.2). This is even more evident in Figure 3.3, where a horizontal slice of the reflectivity at the depth of 4 km is shown. From Figures 3.2 and 3.3 we can conclude that the seismicity seems to be guided by the SE2 fault structure. A possible explanation is that the SE2 fault system is characterized by an enhanced permeability due to the presence of natural fractures. Such permeability is anisotropic. Pore-pressure fluctuations due to the fluid injection then propagate mainly along the direction of the largest principal component of the permeability tensor. This component is directed along the fractures, i.e. along the fault. This leads up to the phenomenon of fault-guided induced seismicity. Moreover, Figures 3.2 and 3.3 indicate that the permeability along the SE2 fault into the horizontal (strike) direction is larger than the permeability along its up-dip direction. This is possibly related to the geometry of fractures composing the SE2 fault system.

In spite of only a small number of located events, the fluid-induced seismicity at the KTB site shows several important features of events triggering by pore-pressure perturbations. First, it was observed that seismicity can be triggered by pressure perturbations as low as 0.01–0.1 MPa or possibly even less. This is seen from the following. Most of the microseismic events occurred several hundred meters away from the injection borehole (see Figures 3.2 and 3.3). Simultaneously the water level in the main hole at the start of seismic activity had increased by a few tens of

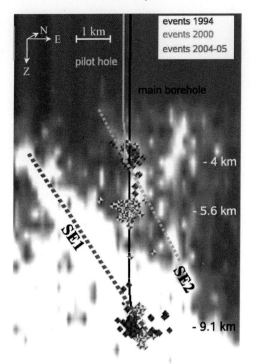

Figure 3.2   A vertical slice of a 3D depth migrated image of the KTB site (see Buske, 1999). Light tones correspond to high seismic reflection intensities and dark tones to lower ones, respectively. The SE1 reflector is clearly visible as a steeply dipping structure. Additionally, seismicity induced by the injection experiments of years 1994 (close to the injection depth of 9.1 km), 2000 (injection depth of approximately 5.6 km) and 2004–2005 (depth around 4 km) is shown. Locations of the main (black line) and pilot (light-tone line) boreholes are also plotted. (Modified after Shapiro *et al.*, 2006a.) A black and white version of this figure will appear in some formats. For the color version, please refer to the plate section.

meters. Therefore pore-pressure perturbations less than few 0.1 MPa at hypocenters are required to trigger seismic events at the KTB site. Second, seismicity triggering is much more likely in the case of positive pore-pressure perturbation (i.e. in the case of an injection). For example, the probability of event triggering by the fluid extraction at KTB was less than 1/3000 of the probability during fluid injection. No events occurred during the fluid extraction and more than 3000 occurred during the injection. Moreover, the onset of seismicity roughly coincides with the time of compensation of the extracted fluid volume by the following injection. This confirms that pressure diffusion is the dominant mechanism of seismicity triggering by fluid injections. Finally, it was shown that heterogeneity and anisotropy of hydraulic properties of rocks can influence spatial evolution of microseismic

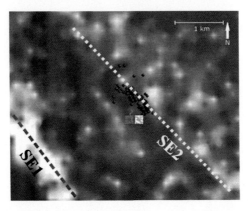

Figure 3.3 A horizontal slice at 4 km depth over the depth-migrated 3D image of the KTB site (Buske, 1999) plotted along with the slice-plane projections of seismic hypocenters induced by the injection experiment of 2004–2005. The white and gray squares are locations of the main and pilot boreholes, respectively. (After Shapiro *et al.*, 2006a.) A black and white version of this figure will appear in some formats. For the color version, please refer to the plate section.

clouds. For example, the seismicity induced by the fluid injection directly into the SE2 structure remained guided by this crustal fault.

3.2 Linear relaxation of pore pressure as a triggering mechanism

A heterogeneous distribution of elastic stresses is a consequence of a heterogeneity of rock elastic properties (Langenbruch and Shapiro, 2014). At some locations in the Earth's crust, the tectonic stress is close to the critical stress necessary for a brittle failure of rocks. Formally this means that the failure criterion stress FCS given by equation (1.73) is still negative. However, it has small absolute values. Increasing fluid pressure in a reservoir also leads to an increase of the pore pressure at critical locations (i.e. at rock defects such as pre-existing cracks). Such an increase in the pore pressure causes a decrease in the effective normal stress (poroelastic-coupling stress contributions considered in Section 2.8 are neglected in the first approximation; we will discuss their role again in Section 3.4). At some locations the decrease in the effective normal stress produces a sufficiently high positive change of ΔFCS. This leads to sliding along pre-existing favorably oriented cracks (sub-critical before the pore-pressure rise). This mechanism of event triggering has been considered by, for example, Nur and Booker (1972), Fletcher and Sykes (1977), Ohtake (1974), Talwani and Acree (1985), Ferreira *et al.* (1995) and Zoback and Harjes (1997).

The linear relaxation of stress and pore pressure is described by the system of Frenkel–Biot equations for small linear deformations of poroelastic systems

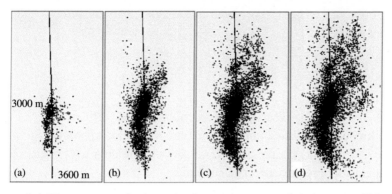

Figure 3.4 Hypocenters of microseismic events occurred during (a) 100 h, (b) 200 h, (c) 300 h and (d) 400 h after starting a water injection in a borehole of the geothermal system at Soultz, France (after Shapiro *et al.*, 1999; see also Section 3.8.2 for a description of the case study).

(Biot, 1962). We discussed this system of equation in the previous chapter. This equation system shows that, in a homogeneous isotropic fluid-saturated poroelastic medium, there are three waves propagating a strain perturbation from a source to a point of observation. These are two elastic body waves, P and S (the longitudinal and shear seismic body waves, respectively), and a highly dissipative longitudinal slow wave.

Clouds of fluid-induced microseismic events grow slowly. They need hours or days to become several hundred meters long (see Figure 3.4). This process cannot be explained by the elastic stress transfer produced mainly by quick seismic body waves. Thus, the growth of microseismic clouds has to be related to the slow wave. The pore-pressure perturbation in the slow wave in the limit of frequencies extremely low in comparison with the global-flow critical frequency (which is usually of the order of 0.1–100 MHz for realistic geologic materials) is described by a linear partial-differential equation of diffusion (2.114). It is exactly the same diffusion equation that can be obtained by uncoupling the pore pressure from the complete Frenkel–Biot equation system in the low-frequency range. In addition, the uncoupling of the pore-pressure diffusion equation requires an assumption of an irrotational displacement field in the solid skeleton (Detournay and Cheng, 1993). In weakly heterogeneous and weakly elastically anisotropic rocks far from high-contrast poroelastic boundaries, this assumption is approximately valid. Then the diffusion equation describes well enough the linear relaxation of pore-pressure perturbations in a poroelastic fluid saturated medium.

Shapiro *et al.* (1999, 2002, 2003) proposed using the diffusion equation (2.120) to describe the spatio-temporal evolution of clouds of hydraulically induced

seismic events in terms of pore-pressure relaxation in anisotropic and heterogeneous porous media:

$$\frac{\partial P_p}{\partial t} = \frac{\partial}{\partial x_i} \left[D_{ij} \frac{\partial}{\partial x_j} P_p \right]. \tag{3.1}$$

Here, D_{ij} are the components of the tensor of hydraulic diffusivity, x_j ($j = 1, 2, 3$)
are the components of the radius vector from the injection point to an observation
point in the medium, P_p is the pore-pressure perturbation with respect to the pre-
injection *in situ* pore pressure, and t is time.

Under the assumptions mentioned above, by only allowing additionally for
heterogeneity and anisotropy of the hydraulic permeability, (3.1) can still be uncou-
pled from the complete system of the Frenkel–Biot equations in the low-frequency
range (see Section 2.4.4). The coefficient **D** here is the same coefficient of hydraulic
diffusivity derived by Rice and Cleary (1976) and Van der Kamp and Gale (1983)
in the notation of Biot (1956). It has the following relation to the permeability
tensor (see equation (2.119)):

$$\mathbf{D} = \frac{\mathbf{k}}{\eta S}, \tag{3.2}$$

where **k** is the tensor of hydraulic permeability, η is the pore-fluid dynamic
viscosity and S is the storage coefficient (2.100):

$$S = \frac{K_{dr} + \frac{4}{3}\mu_{dr} + \alpha^2 M}{M(K_{dr} + \frac{4}{3}\mu_{dr})}, \tag{3.3}$$

where $M = (\phi/K_f + (\alpha - \phi)/K_{gr})^{-1}$ and $\alpha = 1 - K_{dr}/K_{gr}$. Here $K_{f,dr,gr}$ are
bulk moduli of the fluid, dry frame and grain material, respectively; μ_{dr} is the shear
modulus of the frame and ϕ is the porosity.

Note, again, that here we neglected the elastic anisotropy in comparison with
the anisotropy of the permeability. It is clear that, in reality, anisotropy and heterogeneity of the permeability are related to anisotropy and heterogeneity of
elastic properties of rocks, at least on the micro-scale. However, usually the elastic
anisotropy has the order of 5%–30%. On the other hand, the permeability has an
anisotropy of the order of several hundred percent or more. Thus, in equation (3.2),
elastic anisotropy can be neglected for simplicity.

For the case of highly porous rocks, an approximation

$$S \approx \frac{\phi}{K_f} \tag{3.4}$$

can be used to roughly estimate this quantity. For the case of low-porosity crystalline rocks an order-of-magnitude approximation of S is:

$$S \approx \frac{\phi}{K_f} + \frac{\alpha}{K_{gr}}. \tag{3.5}$$

In the following discussion we use the diffusion equation in order to construct an approximate model of the process of microseismicity triggering by borehole fluid injections. Of course, it would be possible to introduce the diffusion equation without any relation to the slow wave. However, understanding of this relation provides us with several advantages. First, we are able to derive a relationship between the hydraulic diffusivity and other poroelastic parameters in terms of physical quantities used in exploration seismology. Second, we understand that the diffusion equation describes the phenomenon in the low-frequency approximation only. For example, this explains an unphysical result of this equation: an infinitely quick propagation of high-frequency components of pore-pressure perturbations. In reality, the slow wave always has a finite (or zero) propagation velocity. Finally, we understand that the diffusion equation in its classical simple form is valid under conditions of pore-pressure uncoupling from the elastic stress only. These conditions are, in turn, related to the low-frequency range under consideration. In the previous chapter we discussed the poroelastic coupling effects. In Section 3.4 we will return once more to this subject.

3.3 Triggering front of seismicity

The spatio-temporal features of the hydraulically induced seismicity can be identified in a very natural way from the concept of triggering fronts (Shapiro *et al.*, 1997, 2002). For the sake of simplicity we approximate a real configuration of a fluid injection in a borehole by a point source of pore-pressure perturbation in an infinite, hydraulically homogeneous and isotropic poroelastic fluid-saturated medium. The time evolution of the pore pressure at the injection point is taken to be a step function switched on at time 0. It is natural to assume that the probability of the triggering of microseismic events is an increasing function of the magnitude of the pore-pressure perturbation. Thus, at a given time t it is probable that events will occur at distances that are smaller than, or equal to, the size of the relaxation zone (i.e. a spatial domain of significant changes) of the pore pressure. The events are characterized by a significantly lower occurrence probability for larger distances. The spatial surface that separates these two spatial domains will be called the "triggering front."

Because the size of the relaxation zone is a rather heuristic parameter, we introduce a more formal quantity, which is directly proportional to this size. At time t,

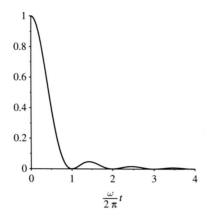

Figure 3.5 The power spectrum of a boxcar-function signal.

the boxcar-function signal of the duration t produces pore-pressure perturbations at any point of the medium. The frequency $2\pi/t$ is an upper bound of the dominant frequencies of this boxcar signal (see Figure 3.5). We define the size of the relaxation zone (i.e. the relaxation radius) to be equal to a distance traveled in a time t by the phase front of a harmonic pore-pressure diffusion wave of the frequency $2\pi/t$. Note that lower-frequency pressure-diffusion waves propagate with lower velocities.

In a homogeneous and isotropic medium this definition results in the following equation of the triggering front (we use equation (2.103) for the phase velocity of the slow wave):

$$R_t = \sqrt{4\pi D t}, \qquad (3.6)$$

where t is the time from the injection start, D is a scalar hydraulic diffusivity and R_t is the radius of the triggering front (which is a sphere in a homogeneous isotropic medium). Note that the numeric coefficient 4π in (3.6) is still heuristic. Defined in this way, the relaxation radius is equal to the wavelength of the slow wave with the period equal to t (see equation (2.104)). Geological media are usually hydraulically heterogeneous. Equation (3.6) is then an equation for the triggering front in an effective (replacing) isotropic homogeneous poroelastic medium with the scalar hydraulic diffusivity D.

Because a seismic event is much more probable in the relaxation zone than at larger distances, equation (3.6) corresponds approximately to the upper bound of the cloud of events in a plot of their spatio-temporal distribution (i.e. the plot of r versus t called in the following discussion the r–t plot). We will first demonstrate this by using the following simple numerical experiment. A point source of a fluid injection is embedded into an infinite homogeneous porous continuum.

As a result of a fluid injection and the consequent process of pressure relaxation, pore pressure will change throughout the pore space. A step-function pressure perturbation is used as an input injection signal. We solve the diffusion equation for the pore-pressure perturbation P_p at any point of the model at any time. For a hydraulically heterogeneous model one can use a finite element algorithm (see Rothert and Shapiro, 2003, for more-complex model examples). For a homogeneous isotropic medium solutions (2.196) and (2.164) can be used directly. After time t_0 the injection is stopped. The medium is divided into small cells. We assume that a critical value C of the pore-pressure perturbation is randomly distributed in the medium.

If at a given point $\mathbf{r}$ of the medium at a given time t the pore pressure $P_p(t, \mathbf{r})$ exceeds $C(\mathbf{r})$ then this point will be considered as a hypocenter of an earthquake that occurred at the time t. For simplicity, we assume that no earthquake will be possible at this point again. We address the quantity C as criticality.

Figures 3.6–3.8 show the distribution of the criticality C and resulting synthetic clouds of events generated by such numerical experiments on rock stimulations by fluid injections in two hydraulically identical homogeneous models with the hydraulic diffusivity $D = 1 \text{ m}^2/\text{s}$. These models differ only in the spatial distribution of the critical pore pressure C. Figure 3.8 shows distances r from the injection points $(0, 0)$ located at the centers of the models to points with simulated seismic events versus event-occurrence times t. In our terminology, this figure shows the r–t plots. It is evident that a vast majority of points corresponding to the seismic events is located below approximately parabolic envelopes. Curves (3.6) with the

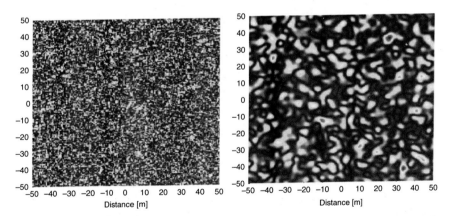

Figure 3.6 Left: realizations of the critical pore pressure C randomly distributed in space with an exponential auto-correlation function. Right: the same but with a Gaussian auto-correlation function. (Modified from Rothert and Shapiro, 2003.) A black and white version of this figure will appear in some formats. For the color version, please refer to the plate section.

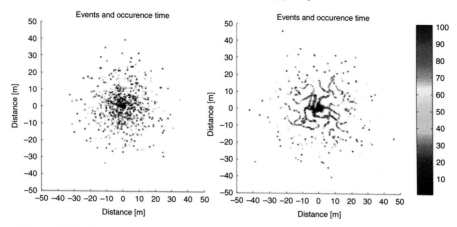

Figure 3.7   Synthetic microseismicity simulated for the two spatial distributions of the critical pore pressure shown in Figure 3.6, respectively. (Modified from Rothert and Shapiro, 2003.) A black and white version of this figure will appear in some formats. For the color version, please refer to the plate section.

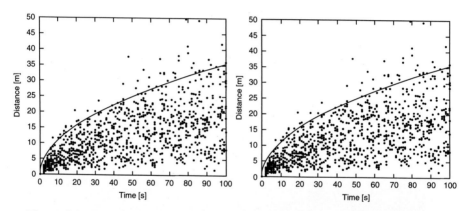

Figure 3.8   Plots of distance versus times (r–t plots) for the synthetic microseismic clouds shown in Figure 3.7, respectively. The dots are events. The lines correspond to $D = 1 \, \text{m}^2/\text{s}$. (Modified from Rothert and Shapiro, 2003.)

hydraulic diffusivity $D = 1 \, \text{m}^2/\text{s}$ are also shown in Figure 3.8. It is obvious that the spatio-temporal distribution of events agrees well with the behavior described by equation (3.6). Let us assume that we do not know the hydraulic diffusivity of the medium. In this case equation (3.6) can be used to estimate the diffusivity. Thus, a parabolic-like envelope of the cloud of events in an r–t plot is an important signature of the hydraulically triggered seismicity.

Such a spatio-temporal distribution of the microseismicity is often observed for microseismic data in reality. Two examples of fluid-injection-induced microseismicity in crystalline rocks are shown in Figure 3.9. We see that r–t plots provide us with estimates of hydraulic diffusivity. We will address them later in this chapter.

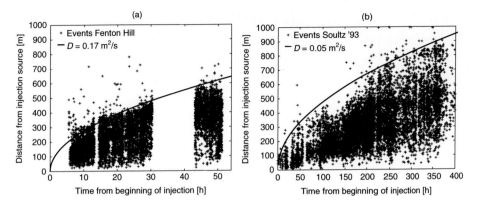

Figure 3.9 The r–t plots for microseismic clouds induced by two water injection experiments of two different Enhanced Geothermal System projects. One is at Fenton Hill, USA (a) and another is at Soultz, France (b). More details on the case studies are given in Sections 3.8.2 and 3.8.3. (Modified from Shapiro *et al.*, 2002.)

3.4 Seismicity fronts and poroelastic coupling

Let us consider further the concept of the triggering front from the point of view of a perturbation of the poroelastic failure criterion stress. Such perturbations are expressed by equations (2.236) and (2.237) of the previous chapter. They show that poroelastic contributions to the failure criterion stress depend on the dimensionless functions $\mathrm{erfc}(r_0)$, $s_{dif}(r_0)$ and $s_m(r_0)$ of the normalized distance r_0 defined in (2.160). They depend also on the coupling parameter n_s and on the friction angle ϕ_f. In such a criterion, different stress functions contribute with different magnitudes.

To compare these contributions let us consider Figure 3.10. This figure provides plots of changes of the failure criterion stress (2.236)–(2.237) corrected for geometrical spreading and normalized to the source term:

$$\delta FCS_h(r_0) = \frac{r\Delta FCS_h}{C_f}, \tag{3.7}$$

$$\delta FCS_v(r_0) = \frac{r\Delta FCS_v}{C_f}. \tag{3.8}$$

They are denoted in Figure 3.10 by the dashed lines and dotted lines, respectively. For these plots the following parameters have been used: the friction angle $\phi_f = 30°$ and the poroelastic stress coefficient $n_s = 0.3$ for the left-hand part of Figure 3.10 and $n_s = 0.05$ for the right-hand part of Figure 3.10. The solid curves in Figure 3.10 show functions

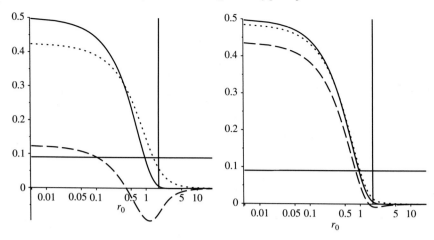

Figure 3.10 Normalized perturbations of failure criterion stress $\delta FCS(r_0)$ caused by a constant-rate fluid injection and taking into account poroelastic coupling effects. The left-hand part shows a very strong poroelastic coupling with $n_s = 0.3$. The right-hand part shows a weak poroelastic coupling with $n_s = 0.05$. By fluid stimulations of extremely low-permeable tight rocks, even weaker coupling is expected (see Figure 2.11). Dotted curves correspond to locations along the compressive stress with the maximum absolute value. These are functions $\delta FCS_v(r_0)$. Dashed curves correspond to locations along the compressive stress with the minimum absolute value. These are functions $\delta FCS_h(r_0)$. Solid curves correspond to the effect of the pore-pressure perturbation only, $\delta FCS_p(r_0)$. The values of $\delta FCS(r_0)$ relevant for fitting domains of seismicity on r–t plots in practical situations are close to or above 0.1 (the horizontal solid lines given by equation (3.11)). For example, this corresponds to ratios $r/a_0 > 30$ (a_0 is the radius of the injection source) and $C_s/p_0 > 0.003$ (see Figure 3.9). The vertical solid lines correspond to the triggering front defined by (3.6)

$$\delta FCS_p(r_0) = \frac{r \Delta FCS_p(r_0)}{C_f} = \mathrm{erfc}(r_0) \sin \phi_f. \tag{3.9}$$

These curves describe a pure pore-pressure effect onto the failure criterion stress.

We consider the case of a constant-rate (or a constant-pressure) fluid injection. This means that C_f is positive (see equations (2.220) and (2.221)). The dotted curves in Figure 3.10 correspond to the maximum change of the FCS. One could define a triggering front as a solution of the following equation:

$$C_s = \Delta FCS_{max}(r, t), \tag{3.10}$$

where C_s is a finite stress perturbation (close to the minimum criticality C). In the case of a fluid-injection, equation (3.10) simplifies to the following one:

$$\delta FCS_v(r_0) = \frac{r C_s}{C_f}. \tag{3.11}$$

A solution of this equation will provide r as a function of t. This function will be a proxy of the triggering front defined by a constant effective-stress perturbation assumed to be necessary for inducing seismic events. We will denote such a front estimate as $R_c(t)$. Usually very early microseismic events, often located very close to the injection source, are not considered for further interpretation. Such a restriction to the distance from the injection source to events is frequently necessary due to a high probability of event identification and location errors in the close vicinity of the injection source. Let us consider a range of distances above $r = 30$ m. This is a reasonable restriction for nearly all case studies from our book. At these distance ranges, $\Delta FCS_{max}(r, t)$ corresponds in Figure 3.10 to the dotted lines in the domains close to and above the solid horizontal lines. These lines are given by the right-hand side of equation (3.11) for parameters $C_s = 0.03$ MPa, $p_0 = 10$ MPa and $a_0 = 1$ m. The function $R_c(t)$ is given by the values of the argument r_0 of the dotted curves in the domains above the horizontal lines. The vertical lines correspond to the triggering front (3.6). This vertical line constrains quite well the zone of significantly perturbed pore pressure (solid curves). It still provides quite reasonable constraints even for the failure criterion stress perturbations taking the maximum and minimum poroelastic coupling contribution into account (the dotted and dashed curves). The triggering-front equation (3.6) describes the size of a spatial domain of significant perturbations of the pore pressure around the injection source. This size was assumed to be equal to the wavelength of the pore-pressure diffusional wave (corresponding to Biot's slow P-wave) of period t. Such a definition of the triggering front does not use any information on the hypothetic quantity C_s.

Note that equation (3.6) can be written as $r_0 = \sqrt{\pi}$ (see again equation (2.160)). For any monotonic in time (non-decreasing) finite-stress perturbation proportional to a function $C_f f(r_0)/r$ this equation will give a time instant at which this perturbation reaches the $f(\sqrt{\pi})/f(0)$-part of its maximally possible value $C_f f(0)/r$ at any given distance r from the source. Therefore, if the quantity C_s is not a constant but rather it is heterogeneously distributed in rocks (for example, it takes values from a broad range between nearly zero and several MPa) then a time dependence of the triggering front will be given by an equation of the type of (3.6). This is similar to a front of a spherical seismic wave (also having a $1/r$-type of geometric spreading), which is given not by a constant amplitude but rather by a constant argument of the phase function. In other words the $\sqrt{t}$-dependence will be a feature of the triggering front independently of the nature of the stress perturbation: pure pore pressure or poroelastic stress coupled with the pore pressure. Such a temporal dependence is a universal feature of a linear diffusion. It is applicable under the assumption that no significant changes of the hydraulic diffusivity D occur in the medium during the fluid injection. In the next chapter we will see that stress perturbations caused by a non-linear diffusion can deviate from the $\sqrt{t}$-dependence.

The quantity $R_c(t)$ defined by (3.11) and a constant C_s does not have a universal $\sqrt{t}$-dependence. Moreover, even in the case of a complete absence of poroelastic effects (i.e. pure pore-pressure-caused FCS perturbations) $R_c(t)$ does not have a universal $\sqrt{t}$-dependence. The smaller the value of C_s takes, the larger will be the difference between R_c and R_t. For small C_s or large r_0 one can use a series expansion of (3.11) and find an approximate solution:

$$R_c(t) = \left(\frac{C_f D n_s (3 - \sin(\phi_f))t}{C_s} \right)^{1/3}. \tag{3.12}$$

Therefore, R_c depends not only on the diffusivity of the medium D but also on the poroelastic coupling parameter n_s. Moreover, the parameter combination Dn_s/C, if kept constant, will provide the same function $R_c(t)$ for an infinite number of different poroelastic media.

In respect to the triggering front we conclude the following. The poroelastic coupling can influence the shape of the triggering front of induced seismicity (see also Section 2.8). This influence depends strongly on the coupling parameter n_s. If n_s becomes vanishingly small, all curves of Figure 3.10 will coincide. If in a data-fitting domain these curves are close, then poroelastic coupling contributions are not very significant and the pore-pressure contribution can be used as a good approximation. In such a case, the triggering front R_t (vertical line in Figure 3.10) and equation (3.6) provide a reasonable estimate of D. Such an estimate is independent of any choice of the hypothetic quantity C_s.

A triggering front defined by an isobar C_s of the pore-pressure perturbation after injection termination can also be fitted to microseismic data (see Edelman and Shapiro, 2004, and Langenbruch and Shapiro, 2010).

3.5 Seismicity and hydraulic anisotropy

Hydraulic properties of rocks are often characterized by a strong anisotropy. Let us now assume that components of the tensor of hydraulic diffusivity D_{ij} are homogeneously distributed in the medium. Under such an assumption we replace a heterogeneous, seismically active rock volume by an effectively homogeneous, anisotropic, poroelastic, fluid-saturated medium with an upscaled permeability tensor. By using equation (3.1) for homogeneous anisotropic media, Shapiro *et al.* (1999) showed how to generalize equation (3.6).

Indeed, assuming a homogeneous anisotropic medium, the equation of diffusion can be written in the principal coordinate system of the tensor of hydraulic diffusivity:

$$\frac{\partial P_p}{\partial t} = D_{11} \frac{\partial^2}{\partial x_1^2} P_p + D_{22} \frac{\partial^2}{\partial x_2^2} P_p + D_{33} \frac{\partial^2}{\partial x_3^2} P_p, \tag{3.13}$$

where D_{11}, D_{22} and D_{33} are the principal components of the diffusivity tensor and x_i are the spatial coordinates of points in the corresponding Cartesian system.

Let us consider the following new system of coordinates, X_i (note summation over repeating indices):

$$X_1 = x_1 \sqrt{\frac{D_{ii}}{3 D_{11}}}, \quad X_2 = x_2 \sqrt{\frac{D_{ii}}{3 D_{22}}}, \quad X_3 = x_3 \sqrt{\frac{D_{ii}}{3 D_{33}}}. \tag{3.14}$$

In this system, equation (3.13) becomes

$$\frac{\partial P_p}{\partial t} = \frac{D_{ii}}{3} \left(\frac{\partial^2}{\partial X_1^2} + \frac{\partial^2}{\partial X_2^2} + \frac{\partial^2}{\partial X_3^2} \right) P_p. \tag{3.15}$$

This equation is equivalent to equation (2.114); that is, it describes isotropic relaxation of the pore pressure with the isotropic hydraulic diffusivity $D_{ii}/3$. Thus, in the new coordinate system X_i, equation (3.6) keeps its form. The radius of the triggering front is given by equation:

$$\sqrt{X_i X_i} = \sqrt{4\pi \frac{D_{ii}}{3} t}. \tag{3.16}$$

The left-hand side of this equation is a radius in the new coordinate system. Taking the square of the radius and returning to the principal coordinate system x_i, we obtain

$$\frac{x_1^2}{D_{11}} + \frac{x_2^2}{D_{22}} + \frac{x_3^2}{D_{33}} = 4\pi t. \tag{3.17}$$

On the other hand it is easy to see that in the principal coordinate system of the diffusivity tensor (the matrix D_{ij} is diagonal) the following equivalence is valid:

$$\frac{x_1^2}{D_{11}} + \frac{x_2^2}{D_{22}} + \frac{x_3^2}{D_{33}} = x_i [D^{-1}]_{ik} x_k, \tag{3.18}$$

where $[D^{-1}]_{ik}$ are components of the tensor inverse to the hydraulic-diffusivity tensor in the principal coordinate system.

Thus, the right-hand sides of equations (3.17) and (3.18) must be equivalent. They can be written in a coordinate-independent form:

$$\mathbf{r}^{\mathrm{T}} \mathbf{D}^{-1} \mathbf{r} = 4\pi t, \tag{3.19}$$

where the superscript $\mathbf{T}$ denotes that the matrix (vector) is transposed, $\mathbf{r}$ is a radius vector of a point on the triggering front and $\mathbf{D}^{-1}$ is the tensor inverse of $\mathbf{D}$.

Equation (3.19) replaces equation (3.6) in the more general case of a hydrauli-cally homogeneous anisotropic medium. This equation holds its form in an arbitrary rotated (i.e. not necessary principal) coordinate system. Written in a sim-pler form, characteristic for the principal coordinate system (see equation (3.17)), it shows that the triggering front is an ellipsoidal surface. If we scale the principal coordinate system as

$$x_{sj} = x_j / \sqrt{4\pi t}, \tag{3.20}$$

then the equation of the triggering front will become an equation of an ellipsoid with the half-axes equal to the square roots of the principal diffusivities. Therefore, if we scale the coordinates of all events by the square root of their occurrence time, by analogy with equation (3.6) this ellipsoid will be an envelope of the cloud of events, but now in a 3D space, normalized accordingly with (3.20). Such an ellip-soidal envelope (see Figure 3.11) can be used to estimate spatial orientations and values of principal components of the hydraulic diffusivity. Shapiro *et al.* (2003) demonstrated applications of this approach to different case histories. For example, in the case of the Fenton Hill experiment (Figure 3.11 and Section 3.8.3) they esti-mated anisotropic tensor of hydraulic diffusivity with an approximate relation of the principal components of 1:4:9. In the case of the Soultz experiment (see Sec-tion 3.8.2 for more details) a tensor of hydraulic diffusivity with an approximate relation of the principal components of 1:2.5:7.5 was obtained. Corresponding estimates of maximal principal components of the diffusivity are 0.17 m^2/s and 0.05 m^2/s, respectively (see Figure 3.9).

Shapiro *et al.* (1999) described a slightly different approach for estimating the diffusivity tensor. They proposed dividing the entire space into directional sectors centered at the injection source. Then, they evaluated r–t plots in these sectors independently. Parabolic envelopes of seismicity at the r–t plots represent corre-sponding sections of an ellipsoidal triggering front growing with time in space. A set of diffusivity estimates obtained in the sectors can be further used to reconstruct the complete 3×3 diffusivity matrix.

The diffusivity estimates shown above also provide us with a possibility to assess the hydraulic permeability of rock. For this, equation (3.2) can be used:

$$\mathbf{k} = \mathbf{D}S\eta. \tag{3.21}$$

For the Fenton Hill experiment, we accept the following estimates used in the liter-ature for crystalline rocks at the depth of 3500 m: $\phi = 0.003$, $\eta = 1.9 \cdot 10^{-4}$ Pa·s. (dynamic viscosity of salt water at 150 °C), $K_d = 49$ GPa, $K_g = 75$ GPa and $K_f = 2.2$ GPa. From these values we obtain $S^{-1} \approx 1.68 \cdot 10^{11}$ Pa, and the permeability tensor in the principal coordinate system is

$$\mathbf{k} = \text{diag} \, (0.2 \, ; \, 0.8 \, ; \, 1.8) \cdot 10^{-16} \, \text{m}^2. \tag{3.22}$$

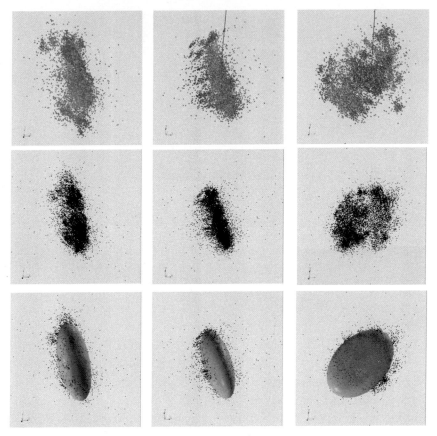

Figure 3.11 Top: a microseismic cloud (Fenton Hill, see Section 3.8.3 for more details about the case study) shown in three different projections in conventional coordinates. Middle: the same cloud projections in coordinates normalized correspondingly to equation (3.20). Bottom: the same as before, but now with a fitted ellipsoid. (Modified from Shapiro *et al.*, 2003.)

Approximate estimates of the permeability at Fenton Hill by other authors give 10^{-17}–10^{-15} m^2 (see Pearson, 1981).

The microseismic experiment in Soultz mentioned above provided the following estimates (Shapiro *et al.*, 1999):

$$\mathbf{k} = \mathrm{diag}\,(0.7;\ 1.9;\ 5.2) \cdot 10^{-17}\, \mathrm{m}^2. \tag{3.23}$$

Determinations of the apparent permeability from independent hydraulic data (see Jung *et al.*, 1996) provided the estimate of $2.5 \cdot 10^{-17}$ m^2.

3.6 Seismicity in hydraulically heterogeneous media

The equations given above are able to provide scalar or tensorial estimates of the hydraulic diffusivity in an effective homogeneous medium only. Shapiro (2000)

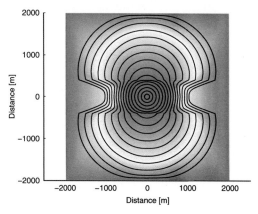

Figure 3.12   A numerical example of pore-pressure distribution (both the gray scale and the isolines) from a point source of an injection located in a layer with hydraulic diffusivity smaller by a factor of five than that of the surrounding medium. Note that the isolines of the pore pressure are similar to fronts of propagating waves. Figure courtesy of Elmar Rothert. (From Shapiro, 2008, EAGE Publications bv.)

and Shapiro *et al.*, (2002) proposed an approach to map 3D heterogeneously distributed hydraulic diffusivity. They heuristically used an analogy between pressure diffusion and wave propagation.

Figure 3.12 shows an example of the pore-pressure distribution created at a given time by a point source of a constant strength acting in a 2D numerical model. The source of pressure perturbation is located in a layer of low hydraulic diffusivity. This layer is embedded into a medium with a five-times-larger hydraulic diffusivity. Figure 3.12 clearly illustrates similarities of the pressure isolines with fronts of propagating waves.

By using the analogy between diffusion and wave propagation it is possible to derive a differential equation that approximately describes the kinematics of the triggering front in the case of quasi-harmonic pore-pressure perturbation. This approximation is similar to the geometrical-optics approach for seismic waves. Propagation of the triggering front is considered in a limited frequency range, which can be called an intermediate asymptotic one in the following sense. The dominant frequency of the injection-induced pressure perturbations must be much smaller than the critical Biot frequency. On the other hand it is assumed that the dominant wavelength of the slow wave (i.e. the pore-pressure diffusion wave) is smaller than the characteristic size of the heterogeneity of the hydraulic diffusivity. Under these conditions one can show that arrival times of the triggering front are described by an eikonal equation.

3.6.1 Eikonal-equation approach

Let us consider the relaxation of a time-harmonic pore-pressure perturbation. First we recall the form of the solution of equation (3.1) in the case of a homogeneous isotropic poroelastic medium (see, for example, chapter V of Landau and Lifshitz, 1991). If a time-harmonic perturbation $p_0 \exp(i\omega t)$ of the pore pressure is given on a spherical surface of the radius a_0 with its center at the injection point, then the solution is

$$P_p(\mathbf{r}, t) = p_0 e^{i\omega t} \frac{a_0}{r} \exp\left[-(i+1)(r - a_0)\sqrt{\frac{\omega}{2D}}\right], \tag{3.24}$$

where ω is the angular frequency and r is the distance from the injection point (i.e. the center of the sphere) to the point where the solution is sought. Comparing this result with equations (2.102) we note that equation (3.24) describes a spherical slow compressional wave with attenuation coefficient equal to $\sqrt{\omega/(2D)}$ and slowness equal to $1/\sqrt{2D\omega}$.

By analogy to solution (3.24) we look for solutions of equation (3.1) of the following form:

$$P_p(\mathbf{r}, t) = p_0(\mathbf{r}) e^{i\omega t} \exp\left[-\tau_p(\mathbf{r})\sqrt{\omega}\right]. \tag{3.25}$$

We assume that $p_0(\mathbf{r})$, $\tau_p(\mathbf{r})$, and $D_{ij}(\mathbf{r})$ are frequency-independent functions slowly changing with $\mathbf{r}$ (later in this section we will quantify this condition). Substituting equation (3.25) into equation (3.1), accepting ω as a large parameter, and keeping only terms with the largest exponents (these are terms on the order of $O(\omega)$; other terms, on the orders of $O(\omega^0)$ and $O(\sqrt{\omega})$, are neglected), we obtain

$$i = D_{ij}(\mathbf{r}) \frac{\partial \tau_p(\mathbf{r})}{\partial x_i} \frac{\partial \tau_p(\mathbf{r})}{\partial x_j}. \tag{3.26}$$

On the other hand, by analogy to equations (2.102) and (3.24) the frequency-independent quantity τ_p is related to the frequency-dependent phase travel time T_p (see the comment after equation (3.30)) as follows:

$$\tau_p = (i+1)T_p\sqrt{\omega}. \tag{3.27}$$

Substituting this into equation (3.26), we obtain

$$1 = 2\omega D_{ij}(\mathbf{r}) \frac{\partial T_p(\mathbf{r})}{\partial x_i} \frac{\partial T_p(\mathbf{r})}{\partial x_j}. \tag{3.28}$$

In the case of an isotropic poroelastic medium, $D_{ij}(\mathbf{r}) = \delta_{ij}D(\mathbf{r})$ and equation (3.28) reduces to

$$|\nabla T_p(\mathbf{r})|^2 = \frac{1}{2\omega D(\mathbf{r})}. \tag{3.29}$$

This is a so-called eikonal equation. The right-hand side of this equation is the squared phase slowness of a wave (the slow wave in this particular case). One can show (Cerveny, 2005) that eikonal equations are equivalent to Fermat's principle, which ensures minimum-time (stationary-time) signal propagation between two points of a medium. Because of equation (3.27), the minimum travel time corresponds to the minimum signal attenuation of the slow wave. In this sense, equations (3.28) and (3.29) describe the minimum-time, maximum-energy front configuration.

Let us further consider a more realistic situation, where the pressure perturbation at a point source can be roughly approximated by a step function. Equations (3.28) and (3.29) describe the phase travel time T_p of a harmonic pressure perturbation. We can attempt to use these eikonal equations further to derive an equation for the triggering time t of a step-function pressure perturbation. For this we can express T_p for an arbitrary frequency ω in terms of the triggering time t.

From our earlier discussion we know that the triggering time t roughly corresponds to the frequency

$$\omega_0 = \frac{2\pi}{t}. \tag{3.30}$$

Thus, for the particular frequency $\omega = \omega_0$ we can roughly approximate the time T_p of the corresponding harmonic pressure perturbation by t. On the other hand, we know that generally $T_p(\omega) \propto 1/\sqrt{\omega}$. Therefore, if the phase travel time T_{p0} is known for a particular frequency ω_0, then for another frequency ω (from a narrow frequency range around ω_0) the corresponding time T_p will be given by $T_p = T_{p0}\sqrt{\omega_0/\omega}$. Thus, for an arbitrary frequency ω, we obtain

$$T_p = \sqrt{\frac{2\pi t}{\omega}}. \tag{3.31}$$

Substituting this equation into equations (3.28) and (3.29) we obtain the following results. In the general case of an anisotropic heterogeneous poroelastic medium,

$$t = \pi D_{ij}(\mathbf{r}) \frac{\partial t(\mathbf{r})}{\partial x_i} \frac{\partial t(\mathbf{r})}{\partial x_j}. \tag{3.32}$$

In the case of an isotropic poroelastic medium, this equation reduces to

$$D(\mathbf{r}) = \frac{t}{\pi |\nabla t|^2}. \tag{3.33}$$

In the case of a homogeneous medium this equation is in agreement with the definition of the triggering front (3.6). Expressing the triggering time as a function of the distance by using (3.6) and substituting the triggering time into (3.33) leads to the identity $D = D$.

It follows from the previous discussions that equations (3.32) and (3.33) are limited to low-frequency diffusion-type Biot slow waves. However, they can be helpful for the characterization of hydraulic properties of rocks. In the case of an isotropic poroelastic medium, equation (3.33) can be used directly to reconstruct a 3D heterogeneous field of hydraulic diffusivity. In turn, by using equation (3.32) for an anisotropic medium, it is impossible to reconstruct a 3D distribution of the diffusivity tensor (because this is a single equation with six unknowns). Here additional information is required. For example, if one assumes that the orientation and the principal-component proportions are constant, then the tensor of hydraulic diffusivity can be expressed as $D_{ij}(\mathbf{r}) = D(\mathbf{r})d_{ij}$. Here d_{ij} is a non-dimensional constant tensor of the same orientation and principal-component proportion as the diffusivity tensor and D is the heterogeneously distributed magnitude of this tensor. If the tensor d_{ij} is known (for example, it can be estimated from the symmetry of the microseismic cloud, as discussed in the previous section), then D can be computed directly as

$$D(\mathbf{r}) = t \left[\pi d_{ij} \frac{\partial t(\mathbf{r})}{\partial x_i} \frac{\partial t(\mathbf{r})}{\partial x_j} \right]^{-1}. \tag{3.34}$$

If d_{ij} is normalized so that $d_{ii} = 3$, then in the case of an isotropic medium it is equal to the unit matrix.

Figure 3.13 shows an example of a reconstructed spatial distribution of the hydraulic diffusivity obtained for the Soultz data set according to equation (3.33). Assuming that d_{ij} has the same orientation and principal-component proportion as permeability tensor (3.23), equation (3.34) can be applied to obtain the diffusivity tensor magnitude. Without showing this in detail, it is interesting to note that there is no significant difference between results of the isotropic and anisotropic variants of the method. Both show larger diffusivity in the upper part of the stimulated volume than in its lower part. In addition, a highly permeable channel leading to the upper right-hand part of the structure is visible in the reconstructed hydraulic diffusivity. This agrees with the observation of a number of early events in the upper right-hand corner of the rock volume.

3.6.2 Validity domain of the eikonal-equation approach

The main limitations of the validity range of equation (3.26) can be formulated roughly from the following consideration of the right-hand part of equation (3.1) in a 1D medium:

$$\frac{\partial}{\partial x}\left[D \frac{\partial}{\partial x} P_p \right] = \frac{\partial D}{\partial x} \frac{\partial P_p}{\partial x} + D \frac{\partial^2 P_p}{\partial x^2}. \tag{3.35}$$

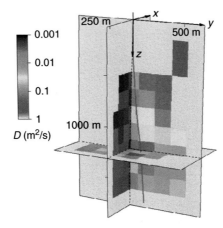

Figure 3.13 Reconstruction of the diffusivity distribution in a hydraulically heterogeneous geothermic reservoir of Soultz (corresponding to the stimulation of 1993) using the eikonal-equation approach. The dark-tone grid cells in the upper part of the structure have hydraulic diffusivity in the range 0.1–1 m^2/s. Below 1000 m the structure has the diffusivity mainly in the range 0.001–0.05 m^2/s. (Modified from Shapiro *et al.*, 2002.) A black and white version of this figure will appear in some formats. For the color version, please refer to the plate section.

Our approach is expected to be valid if the local heterogeneity of the medium can be neglected. Thus, the following inequality must be satisfied:

$$\left| \frac{\partial D}{\partial x} \frac{\partial P_p}{\partial x} \right| / \left| D \frac{\partial^2 P_p}{\partial x^2} \right| \ll 1 \tag{3.36}$$

This can be roughly reduced to

$$\left| \frac{\partial D}{\partial x} \right| / |D\kappa_d| \ll 1, \tag{3.37}$$

where κ_d is the wavenumber of the slow wave. Taking into account that $|\kappa_d|^2 \approx \omega/D$ (see equation (2.98)), we obtain the following simplified condition:

$$\left| \frac{\partial D}{\partial x} \right|^2 D^{-1} \ll \omega. \tag{3.38}$$

This inequality relates a spatial gradient of the hydraulic diffusivity and the frequency of the pressure perturbation. It is rather typical for the geometrical-optics approximation. It shows that if the frequency is high enough and the medium heterogeneity is smooth, the eikonal equation can be applied.

   In the case of a step-function pressure perturbation, the frequency corresponding to the triggering front is roughly given by (3.30). Using equation (3.6) of the triggering front in homogeneous poroelastic media, we can approximate the occurrence time of early events as $t \approx x^2/(4D)$. Note that x denotes the distance from the injection source. Thus, inequality (3.38) can be approximately reduced to

$$\left| \frac{x}{D} \frac{\partial D}{\partial x} \right| < 3. \tag{3.39}$$

This condition is rather restrictive. In addition, it shows that the smaller the distance x, the higher the resolution of the method. In spite of the restrictive character of the inequalities above, the geometrical-optics approximation is still applicable to propagating microseismicity triggering fronts under rather common conditions. This is based on the causal nature of the triggering-front definition. When considering the triggering front, we are interested in the quickest possible configuration of the phase traveltime surface for a given frequency. Thus, we are interested in kinematic aspects of the front propagation only. Corresponding description is given by the eikonal equation. However, conditions (3.38) and (3.39) necessarily take into account not only kinematic aspects of the front propagation but also dynamic aspects, i.e., amplitudes of the pressure perturbation. In other words, the eikonal equation is usually valid in a much broader domain of frequencies than those given by inequality (3.38). Therefore, the method will give useful results, at least semiquantitatively.

Shapiro *et al.* (2002) and Rothert and Shapiro (2003) performed some numerical tests of applicability of the eikonal equations derived above to a kinematic description of the diffusion process and the evolution of the triggering front in a heterogeneous media. They shown that the approach provides reasonable approximate results for models with heterogeneous distributions of the hydraulic diffusivity in rocks.

3.6.3 Effective-medium approach

Let us assume that a medium is heterogeneous on a small scale below the level of our resolution. For example, we estimate the diffusivity in a heterogeneous medium using just a microseismic triggering front at its final observed position (sometimes, at the moment of the injection termination). In this case we attempt to replace the real medium by an effective homogeneous one. Thus, we obtain an effective (upscaled) diffusivity estimate. Below we derive some rules of computing such an effective diffusivity in heterogeneous spherically symmetric d-dimensional structures and discuss their applicability for interpretation of microseismic data.

We consider the system of continuity and Darcy equations (2.64) and (2.66), respectively. The medium is assumed to be isotropic and heterogeneous. A point fluid source is located at the origin of the coordinate system. Fluid filtration from such a source corresponds to radiation of a slow wavefield (see Section 2.6.1). Under such conditions, according to equation (2.113), the quantity χ, the volumetric deformation of the pore space due to additional fluid-mass filtration, is proportional to the pore-pressure perturbation P_p. Then, in a spherically symmetric

d-dimensional medium, diffusion equation (3.1) is obtained from the following equation system:

$$S(r)\frac{\partial P_p}{\partial t} = -\frac{1}{r^{d-1}}\frac{\partial r^{d-1}q_r}{\partial r}, \tag{3.40}$$

$$q_r = -D(r)S(r)\frac{\partial P_p}{\partial r}, \tag{3.41}$$

where r is a radial distance, q_r is the radial filtration velocity, and $S(r)$ and $D(r)$ are the radial-distance-dependent storage coefficient and hydraulic diffusivity, respectively. It is clear that, by substituting the filtration velocity from the second equation into the first one, we obtain the diffusion equation

$$\frac{\partial r^{d-1} P_p}{\partial t} = \frac{1}{S}\frac{\partial}{\partial r}SDr^{d-1}\frac{\partial}{\partial r}P_p. \tag{3.42}$$

Let us first consider steady-state filtration. From (3.40) we obtain

$$\frac{\partial r^{d-1}q_r}{\partial r} = 0. \tag{3.43}$$

This corresponds to a system having an infinitely long time to equilibrate. From equation (3.43) we see that the quantity $r^{d-1}q_r$ is a constant independent of the radial distance. Note that effective permeability of a heterogeneous medium should be independent of fluid properties. Thus, usually for simplicity, it is derived under the assumption of an incompressible fluid. In this case $r^{d-1}q_r$ is just proportional to the mass flux of the fluid. Using this constant we rewrite Darcy's law in the following form:

$$\frac{\partial}{\partial r}P_p = -\left[\frac{r^{1-d}}{D(r)S(r)}\right]r^{d-1}q_r. \tag{3.44}$$

For an arbitrary d a direct spatial averaging of (3.44) will not necessarily provide any distance-independent hydraulic diffusivity. To find parameters of a homogeneous medium replacing the heterogeneous one we could apply spatial averaging over a length $2L$ of the equation above and compare such an averaged equation with a correspondingly averaged equation for a homogeneous medium. The spatial averaging of a quantity $X(r)$ around a point r_x is given by the integral

$$\langle X(r_x, L)\rangle = (2L)^{-1}\int_{r_x-L}^{r_x+L} X(r)dr. \tag{3.45}$$

Taking into account that the fluid-mass flux is constant we obtain from equation (3.44):

$$\left\langle \frac{\partial}{\partial r}P_p\right\rangle = -\left\langle \frac{r^{1-d}}{D(r)S(r)}\right\rangle r^{d-1}q_r. \tag{3.46}$$

For a replacing homogeneous medium with the effective averaged parameters D_{eff} and S_{eff} this equation must have the following form:

$$\left\langle \frac{\partial}{\partial r} P_p \right\rangle = -\frac{\langle r^{1-d} \rangle}{D_{eff} S_{eff}} r^{d-1} q_r. \tag{3.47}$$

The product $D_{eff} S_{eff}$ must be equal to the ratio k_{eff}/η_{eff}. Assuming a homogeneous fluid saturating the rock we obtain $D_{eff} S_{eff} = k_{eff}/\eta$. In the original heterogeneous medium, $D(r)S(r) = k(r)/\eta$. Further, equations (3.46) and (3.47) are equivalent under the following condition:

$$(D_{eff} S_{eff})^{-1} = \langle r^{1-d} \rangle^{-1} \left\langle \frac{r^{1-d}}{D(r)S(r)} \right\rangle. \tag{3.48}$$

From here we obtain several simple permeability-upscaling rules for steady-state filtration in a heterogeneous medium. Assuming a homogeneous distribution of the viscosity in a 1D medium ($d = 1$), we obtain a well-known rule:

$$k_{eff}^{-1} = \left\langle \frac{1}{k(r)} \right\rangle. \tag{3.49}$$

Neglecting the elastic heterogeneity of the medium (especially in comparison with its hydraulic permeability) we can apply exactly the same rule for the diffusivity:

$$D_{eff}^{-1} = \left\langle \frac{1}{D(r)} \right\rangle. \tag{3.50}$$

In 1D statistically homogeneous structures the spatial averaging is independent of r_x. The averaging stabilizes with increasing L. Thus, the effective diffusivity (and, correspondingly, the permeability) indeed becomes a constant in the limit of large averaging lengths L.

The situation becomes less convenient in 2D and 3D media, where the effective parameters are explicit functions of the averaging length L and position r_x. Indeed, in 2D media we obtain

$$(D_{eff} S_{eff})^{-1} = 2L \left[\ln \frac{r_x + L}{r_x - L} \right]^{-1} \left\langle \frac{1}{r D(r)S(r)} \right\rangle. \tag{3.51}$$

In 3D media the analogous result is

$$(D_{eff} S_{eff})^{-1} = (r_x^2 - L^2) \left\langle \frac{1}{r^2 D(r)S(r)} \right\rangle. \tag{3.52}$$

Corresponding averaging rules in a piecewise-homogeneous medium composed of homogeneous layers of the thickness $h = 2L/N_h$ (here N_h is the total number of layers) and different diffusivities D_i ($i = 1, 2, \ldots, N_h$ is the number of a corresponding layer) are following. In 1D media we have:

$$D_{eff}^{-1} = N_h^{-1} \Sigma_i \frac{1}{D_i}. \tag{3.53}$$

In 2D media we obtain

$$D_{eff}^{-1} = \left[\ln \frac{r_x + L}{r_x - L} \right]^{-1} \Sigma_i \frac{1}{D_i} \ln \frac{r_x - L + ih}{r_x - L - h + ih}. \tag{3.54}$$

In 3D media the analogous result is

$$D_{eff}^{-1} = \frac{(r_x^2 - L^2)}{N_h} \Sigma_i \frac{1}{D_i} \frac{1}{(r_x - L + ih)(r_x - L - h + ih)}. \tag{3.55}$$

These results coincide with the rules derived in chapter 12 of Crank (1975).

However, in the case of a non-steady-state filtration these results are not necessarily applicable. To consider such situations we must return back to equation system (3.40) and (3.41). It is convenient to write it in a matrix form:

$$\frac{\partial}{\partial r} \begin{bmatrix} r^{d-1} q_r \\ P_p \end{bmatrix} = \begin{bmatrix} 0 & -r^{d-1} S(r) \frac{\partial}{\partial t} \\ -(D(r)S(r))^{-1} r^{1-d} & 0 \end{bmatrix} \begin{bmatrix} r^{d-1} q_r \\ P_p \end{bmatrix}. \tag{3.56}$$

The quantities $r^{d-1} q_r$ and P_p are continuous functions of r (even in the case of layer interfaces; note again that we assume a homogeneous fluid). Let us consider a sufficiently small d-dimensional spherical shell of thickness $2L \ll \delta_\alpha$, where $\delta_\alpha = \sqrt{Dt/\pi}$ is a distance reciprocal to the attenuation coefficient of the slow wave (2.105). Inside the $2L$-averaging length, both the pore pressure and the fluid flux will be nearly r-independent. Further, we are interested in filtration features on distances r significantly larger than $2L$. So r_x can be identified with r. The averaging on the scale $2L$ yields:

$$\frac{\partial}{\partial r} \begin{bmatrix} r^{d-1} q_r \\ P_p \end{bmatrix} = \begin{bmatrix} 0 & -r^{d-1} \langle S(r) \rangle \frac{\partial}{\partial t} \\ -\langle (D(r)S(r))^{-1} \rangle r^{1-d} & 0 \end{bmatrix} \begin{bmatrix} r^{d-1} q_r \\ P_p \end{bmatrix}. \tag{3.57}$$

The elements of the matrix in the right-hand side of this equation provide us with averaging rules. They imply that averaging of the diffusivity coinciding with (3.48). Therefore, features of a non-steady-state filtration on distances significantly larger than δ_α will be defined by effective parameters corresponding to averaging rules (3.49)–(3.50). These distances are of the order of several 10^{-1} parts of a slow-wave length or larger. These averaging rules are independent of the dimensionality of the medium. They correspond to a propagating slow plane wave in a local 1D heterogeneous structure. This is analogous to the eikonal approximation of the previous section. This approximation implies an assumption of locally plane wavefronts too. Note, however, that in contrast to the eikonal approximation, the radial scale of heterogeneity we consider here is assumed to be significantly smaller than δ_α.

However, averaging rules indicated in (3.57) and correspondingly in (3.49)–(3.50) do have a broader applicability. Let us assume that there exists a statistical ensemble of various realizations of our radially heterogeneous model of the medium. If the angular brackets denote the ensemble averaging (i.e. the statistical averaging) then its application to equation (3.56) will again provide equation (3.57) for the ensemble-averaged field quantities $r^{d-1}q_r$ and P_p. Then, for the statistical ensemble averaging rules (3.49)–(3.50) become exact. In reality we have one single realization of the medium only. However, in real media with random small-scale 3D spatial fluctuations of physical properties and with randomly distributed critical pore-pressure perturbations C, fitting smooth triggering-front envelopes of microseismic clouds is similar to constructing averaged surfaces of constant pressure (isobars). The ergodicity assumption (Rytov *et al.*, 1989) states then the equivalence of such a spatial averaging to the ensemble averaging. Hummel and Shapiro (2012) considered several numerical examples of non-steady-state filtration and demonstrated rather well the applicability of averaging rules (3.49)–(3.50).

3.7 Back front of seismicity

If the injection stops at time t_0 then induced earthquakes will gradually cease to occur. For any time moment after time t_0 a surface can be defined that describes spatial positions of a maximum pore-pressure perturbation at this time. This surface (it is a sphere in homogeneous isotropic rocks) separates the spatial domain, which is still seismically active, from the spatial domain (around the injection point), which is becoming seismically quiet. This surface was first described by Parotidis *et al.* (2004) and termed the back front of induced seismicity (see Figure 3.14).

In simple situations this surface can be analytically described. Let us approximate a fluid-injection borehole experiment by a point source of a constant strength q, and the duration t_0, i.e. a boxcar function of duration t_0. The source strength q (its physical unit is the watt) is related to the source term C_f from equations (2.219)–(2.221):

$$q = 4\pi C_f D. \tag{3.58}$$

Let $P_H(\mathbf{r}, t)$ be a solution for the pore pressure during the injection time. Note that this is a response of the medium to a Heaviside-source function. Because the system under consideration is linear, then its response $P_B(\mathbf{r}, t)$ to a boxcar-source function will be the following difference of the two Heaviside-source-caused responses:

$$P_B(\mathbf{r}, t) = P_H(\mathbf{r}, t) - P_H(\mathbf{r}, (t - t_0)). \tag{3.59}$$

Figure 3.14 shows a pore-pressure distribution in a 2D medium for two different points. One of them is close to the injection point. Another one is far from it. For

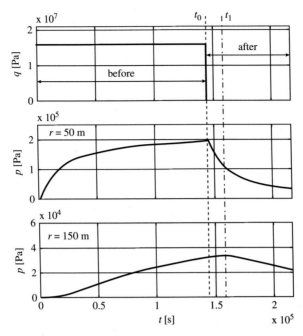

Figure 3.14 Pore-pressure evolution before and after injection termination. The back-front arrival at a given location corresponds to the maximum pore-pressure perturbation achieved at this location. (Modified from Parotidis *et al.*, 2004.)

small r, pore pressure rises immediately with the start of an injection and drops promptly after t_0. For larger r, pore pressure begins to increase after the beginning of the injection. This occurs until the time t_1, where the maximum pressure value is reached. The time difference $t_1 - t_0$ increases with the distance r. For a constant diffusivity D, t_1 depends solely on r. Assuming that events may be triggered for increasing pore-pressure values only, we expect no seismic activity after the maximum-pressure arriving time $t_1(r)$. Therefore, a termination moment of the seismic activity should correspond to the moment of the vanishing-time derivative of the pore pressure. This time moment is a solution of the following equation:

$$0 = \frac{\partial}{\partial t} [P_H(\mathbf{r}, t) - P_H(\mathbf{r}, (t - t_0))]. \tag{3.60}$$

Note, however, that a time derivative of a Heaviside-function response of a linear system is equal to its Green's function (we denote it here as $P_G(\mathbf{r}, t)$). Thus, the equation for the back-front surface becomes

$$0 = P_G(\mathbf{r}, t) - P_G(\mathbf{r}, (t - t_0)). \tag{3.61}$$

The 1D, 2D and 3D Green's functions for equation (3.1) in isotropic homogeneous media have the following form (see Carslaw and Jaeger, 1973, p. 262, and Landau and Lifshitz, 1991, section 51.7), respectively:

$$P_{G1}(\mathbf{r}, t) = \frac{q}{(4\pi Dt)^{d/2}} \exp\left(-\frac{r^2}{4Dt}\right). \tag{3.62}$$

Here $d = 1, 2, 3$ is the dimension of the space where the pressure diffusion occurs. For example, in the normal 3D space it is equal to 3. In a 2D fault it is equal to 2. In a 1D hydraulic fracture (later we will address such a situation) it is equal to 1. For given diffusivity D, and injection duration t_0, the solutions for $t(r)$ in equation (3.61) give time moments of pore-pressure maximums at the corresponding distance r. Writing this solution in a form of distance as function of time $r(t)$ yields the formulation of the radius $R_{bf} \equiv r(t)$ of the back front:

$$R_{bf}(t) = \sqrt{2d\,Dt \left(\frac{t}{t_0} - 1\right) \ln\left(\frac{t}{t - t_0}\right)}. \tag{3.63}$$

Along with the triggering front, the back front is also a kinematic signature of the pressure-diffusion-induced microseismicity. It is often observed on real data (see Figure 3.15). In situations where the injection has produced a very moderate or even zero impact on the permeability, the back front provides estimates of hydraulic diffusivity consistent with those obtained from the triggering front and approximately coinciding with the diffusivity of virgin rocks.

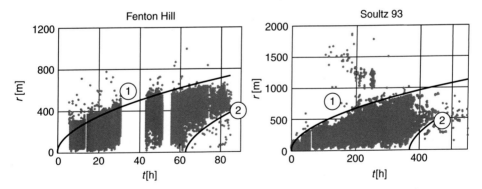

Figure 3.15 Examples of r–t plots with triggering fronts (1) and back fronts (2) of fluid-induced seismicity. The left-hand part and the right-hand part of the plot correspond to the case studies of geothermal borehole-fluid injections at Fenton Hill (an approximate estimate of the hydraulic diffusivity by the both curves is 0.14 m^2/s) and Soultz (an approximate estimate of the hydraulic diffusivity by the both curves is 0.5 m^2/s), respectively. (Modified from Parotidis *et al.*, 2004.)

3.8 Strength of pre-existing fractures

Besides hydraulic properties of rocks, the strength of pre-existing cracks is another important parameter that influences fluid-induced microseismicity. This strength corresponds to a value of the pore-pressure perturbation necessary to cause sliding along a pre-stressed fracture, i.e. to trigger an earthquake. This is the previously introduced criticality C of the medium. Thus, criticality is a pressure perturbation that, according to the Mohr–Coulomb criterion (1.66), results in rock failure. Variations of the criticality are caused by a combination of variations in the cohesion, friction coefficient and the pre-injection stress field. Langenbruch and Shapiro (2014) have recently shown that heterogeneities of elastic properties of rocks strongly contribute to fluctuations of C. If C is high in a given location, we consider this region to be stable. If C is low, we refer to this location as unstable, with a higher probability that events are triggered due to pore-pressure changes. In these terms, high criticality corresponds to high strength and low criticality indicates low strength, respectively.

Rothert and Shapiro (2007) found a way to estimate the probability density function $f(C)$ of the medium criticality from microseismic data. Here, we provide a short review of their approach. In general, the probability density function describes the frequency of a specific value of critical pore pressure in the corresponding statistical ensemble. This ensemble is replaced by a single realization of the medium under the ergodicity assumption.

3.8.1 Statistics of rock criticality

Let us assume the simplest possible distribution of the critical pore pressure, a uniform probability density function $f(C)$ for the criticality $C(\mathbf{r})$ at any point with position vector $\mathbf{r}$; C_{min} and C_{max} will denote the minimum and maximum possible criticality values, respectively. Then $f(C) = 1/\Delta C$, with $\Delta C = C_{max} - C_{min}$. The statistical properties of C are assumed to be independent of position. Therefore, $C(\mathbf{r})$ is a statistically homogeneous random field. A criticality field randomly distributed within a number of cells N_c in three dimensions is shown in Figure 3.16 (two examples of 2D distributions of C are shown in Figures 3.6 and 3.7). The gray scale corresponds to the criticality value at a given point. The probability density function (PDF) is given by the number of cells in the medium associated with a specific value of criticality normalized by the total number of cells.

We consider a point-injection source located at the origin of a Cartesian coordinate system. It causes pore-pressure perturbations $P_p(\mathbf{r}, t)$ in an infinite homogeneous isotropic continuum with hydraulic diffusivity D. The pore-pressure

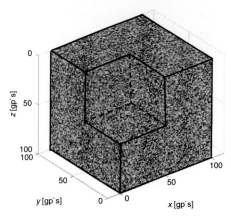

Figure 3.16 A synthetic distribution of the strength of pre-existing fractures (i.e. criticality C) in a 3D rock model. (Modified from Rothert and Shapiro, 2007.)

perturbation evolves in space outside the source according to the diffusion equation (3.1). A fluid injection starts at time $t = 0$ with a constant injection rate. We assume that for $t < 0$ the pore-pressure perturbation P_p is zero. Such a point source of injection with a time dependence given by a step function can be described by its duration t_0 and its strength q defined in (2.219)–(2.221) and (3.58). We will consider times less than t_0.

In the medium, the pore pressure $P_p(\mathbf{r}, t)$ rises monotonically as a function of $t \leq t_0$. The familiar solution of the diffusion equation (see equations (2.164) and (2.196)) satisfies the given initial and boundary conditions in three dimensions:

$$P_p(\mathbf{r}, t) = \frac{q}{4\pi D|\mathbf{r}|} \, \mathrm{erfc}\left(\frac{|\mathbf{r}|}{\sqrt{4Dt}}\right). \tag{3.64}$$

An induced microseismicity can be then modeled as described above in previous sections and shown in Figures 3.6–3.8.

To trigger a seismic event the following relationship between pore pressure and criticality must be fulfilled:

$$P_p(\mathbf{r_e}, t) = C(\mathbf{r_e}). \tag{3.65}$$

In the case of a homogeneous and isotropic medium it means

$$\frac{q}{4\pi D|\mathbf{r_e}|} \, \mathrm{erfc}\left(\frac{|\mathbf{r_e}|}{\sqrt{4Dt}}\right) = C(\mathbf{r_e}), \tag{3.66}$$

where $\mathbf{r_e}$ are all points where events have been triggered at time t.

To reconstruct the probability density function of C, we have to compute a number of events occurring in a spatio-temporal domain characterized by a criticality between C and $C + dC$. For analysis of a given microseismicity cloud using equation (3.66), the source strength q and hydraulic diffusivity D must be known. For the case of numerical data, q and D are known from the model set-up. For real injection experiments, the hydraulic diffusivity D can be obtained from independent measurements like borehole tests or core studies. Alternatively, it can be estimated from microseismicity according to the triggering front or back front approaches described above. The source strength q can be estimated by using the pressure at the borehole head (see equations (2.219)–(2.221) and (3.58)):

$$q = 4\pi p_0 D a_0. \tag{3.67}$$

A probability that a microseismic event occurs in a cell with criticality values between, for example, C_1 and $C_2 = C_1 + dC$ (dC is small) is given by the probability density function of criticality:

$$W_{C_{1,2}} = f\left(\frac{C_1 + C_2}{2}\right) \cdot dC. \tag{3.68}$$

On the other hand, this probability is equal to the fraction of cells with criticality between C_1 and C_2 in the spatial domain, where until the observation time t the pore pressure has already arrived in the range $C_1 \le P_p \le C_2$. This spatial domain is approximately a spherical volume with radius $(r_1 + r_2)/2$ such that $P_p(r_1, t) = C_1$ and $P_p(r_2, t) = C_2$.

To count the number of events triggered in the time interval from 0 to t and characterized by $C_1 \le C \le C_2$, it is convenient to introduce a new coordinate system (y, t). The variable y is defined as follows:

$$y = r_0 \equiv \frac{r}{\sqrt{4Dt}}, \tag{3.69}$$

where r is the distance from the injection point and t is the occurrence time of a microseismic event. We can compute the criticality at a point with coordinates (y, t) of a given event:

$$C = \frac{q}{4\pi D\sqrt{4Dt} \cdot y} \, \mathrm{erfc}(y). \tag{3.70}$$

Solving equation (3.70) for t we get

$$t = \frac{q^2}{64\pi^2 D^3 y^2 C^2} \cdot \mathrm{erfc}^2(y). \tag{3.71}$$

Given the source strength q and hydraulic diffusivity D, this relationship between t and y must be valid for each fluid-induced event characterized by the criticality C. The criticality C is just a parameter of a curve $t(y)$. Thus, the number of events with the criticality $C_1 \le C \le C_2$ is just a number of events between the corresponding curves $t(y, C_1)$ and $t(y, C_2)$.

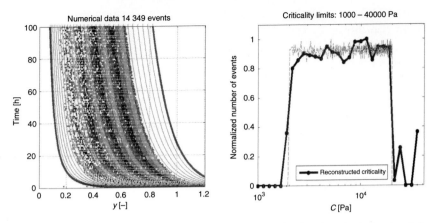

Figure 3.17 Left: $t(y)$ curves (3.71) for a numerically simulated microseismicity (dots) triggered using a criticality field with a uniform PDF. Right: a reconstructed PDF of criticality (logarithmic sampling). (Modified from Rothert and Shapiro, 2007.)

Therefore, the reconstruction of the criticality PDF can be carried out in the following way. The y–t domain is subdivided into a number of sub-domains with different criticalities (see the left-hand side of Figure 3.17). Note that equally spaced criticality values result in non-equally spaced domains (y, t). The criticality allocation can also be spaced logarithmically, resulting in a nearly equally spaced (y, t) domain. The numbers of events located between sequential curves $t_i(y)$ and $t_{i+1}(y)$ given by criticalities C_i and C_{i+1}, respectively, are calculated. The best choice of the maximum-observation time t_{max} is usually the complete injection time t_0. Usually, it provides the most complete statistics of events. At the maximum time t_{max} values y, being solutions of the equations $t_{max} = t_i(y)$ and $t_{max} = t_{i+1}(y)$, correspond to spherical surfaces of the radii r_i and r_{i+1} satisfying the equations $P_p(r_i, t_{max}) = C_i$ and $P_p(r_{i+1}, t_{max}) = C_{i+1}$. Thus, the counted numbers of events represent the number of locations with criticalities between C_i and C_{i+1} in the spatial ball of an approximate radius $(r_1 + r_2)/2$. Therefore, this number of events must be normalized by the volume of such an average ball and associated with the mean value of criticality $(C_i + C_{i+1})/2$. Of course, such a normalization is an approximation. The smaller the difference $C_{i+1} - C_i$, the better the approximation. Additionally, if criticalities are spaced non-uniformly, then the normalization procedure will, to some extent, also compensate for this. Figure 3.17 gives an example of how this algorithm works on a synthetic data set.

Often, hydraulic properties of rocks are anisotropic. In the case of a homogeneous anisotropic medium, the equation of pore-pressure diffusion (3.1) simplifies.

In the principal coordinate system of the diffusivity tensor we obtain equation (3.13). In coordinate system (3.14) it turns into equation (3.15) describing the relaxation of pore pressure in an isotropic medium with the scalar hydraulic diffusivity $D_{ii}/3$. The solution of the diffusion equation for the anisotropic case in analogy to equation (3.64) becomes

$$P_p(\mathbf{r}, t) = \left[\frac{4\pi}{q} (\frac{D_{ii}}{3})^{3/2} X_{el} \right]^{-1} \operatorname{erfc}\left(\frac{X_{el}}{\sqrt{4t}} \right), \qquad (3.72)$$

where we have introduced the following notation:

$$X_{el} = \sqrt{\frac{x_1^2}{D_{11}} + \frac{x_2^2}{D_{22}} + \frac{x_3^2}{D_{33}}}. \qquad (3.73)$$

We see that a hydraulic anisotropy influences the pore-pressure perturbation. Also pore-pressure-induced clouds of microseismic events will show anisotropic behavior. Microseismic clouds will tend to be ellipsoidal. Such ellipsoidal forms are clearly observed especially in the triggering-time normalized space (see Figure 3.11).

   Therefore, to apply the criticality-reconstruction approach, we have to transform a microseismic cloud obtained in an anisotropic medium into a cloud that occurs in an effective isotropic medium. The coordinate system must be rotated into a system parallel to the principal axes of the diffusivity tensor. After the rotation, all coordinates must be additionally scaled as shown in equation (3.14).

   Rothert and Shapiro (2007) reconstruct probability density functions of rock criticality for the two already-presented real data sets obtained during experiments at Hot Dry Rock (HDR) geothermal sites. Below we summarize and further interpret their results.

3.8.2 *Case study: Soultz-sous-Forêts*

The Soultz-sous-Forêts Hot Dry Rock site is located in the Rhine Graben in eastern France. During a geothermal stimulation experiment performed by SOCOMINE (France) and CSMA (UK) in 1993, about $25\,300$ m^3 of water were injected into crystalline rocks over a depth interval of 2850–3400 m of borehole GPK1 (Dyer *et al.*, 1994, Cornet, 2000). Flow logs indicate that major fluid loss (about 60%) occurred at a depth of approximately 2920 m (for details see Cornet *et al.*, 1997; Shapiro *et al.*, 1999). During 400 h after the start of the injection, about 9300 events were recorded and later located with a location error of approximately 20–80 m. The cloud of events is shown in Figure 3.4. Events recorded after the stop of the injection (approx. 120) are not considered.

Shapiro *et al.* (2003) estimated a tensor of hydraulic diffusivity with an approximate relation of the principal components of 1:2.5:7.5. The largest-component axis of the tensor of hydraulic diffusivity is close to vertical one. Its largest horizontal component has nearly N/S direction. Its orientation is close to the orientation of maximum horizontal stress (see Dyer *et al.*, 1994; Cornet *et al.*, 1997; Klee and Rummel, 1993).

Owing to the fact that the injection-pressure function for Soultz can be rather well approximated by a step function, equation (3.71) is approximately valid and the criticality estimation approach can be applied. To estimate the PDF of the criticality, Rothert and Shapiro (2007) proceeded as follows.

In equation (3.71), t_{max}= 400 h was used for the maximum time of pressure perturbation. They considered a pore-pressure perturbation along a cylindrical source of length l = 100 m and radius R_c = 0.2 m. Surface pressure, i.e. well-head pressure (which can be also taken as an approximation of the injection pressure), was approximately constant at the level of 12 MPa (see figure 1 in Cornet, 2000). To use the method of the previous section based on a simple solution for a spherical injection cavity, the following far-field approximation was applied. The real injection source was replaced by a surface-equivalent sphere. The radius of a surface-equivalent sphere is then a_0 = 3.2 m. According to equation (3.67) this results in a point source of strength $q \approx 25 \cdot 10^6$ W. After scaling of the cloud into one that would occur in an effective isotropic medium, the algorithm for criticality reconstruction was applied. The maximum principal component of the hydraulic diffusivity tensor of D =0.05 m^2/s was used. The normalized numbers of events within the criticality shells was computed. The result is shown in Figure 3.18. The reconstructed probability density function is shown on Figure 3.19 in logarithmic and linear scales. The probability density function for criticalities lower than 0.001 MPa is nearly vanishing. Then it increases, and for criticality values above 3 MPa it steeply drops off.

Therefore, criticality for Soultz seems to be distributed in a broad range (three orders of magnitude) between 0.001 MPa and approximately 3 MPa. The bounds of the criticality limits seem to be sharp. The lower bound (C_{min}) is possibly defined by the magnitude of tidal stresses, which are of the order of 0.001 MPa. These stresses are permanently occurring and relaxing in the Earth. Thus, rocks *in situ* seem to be accustomed to such perturbations and, thus, do not show significant seismic reactions to such or smaller stress changes. The probability density function of criticality has an asymmetric shape. It does not seem to be Gaussian. It is rather roughly uniform or a truncated Gaussian one.

The upper bound (C_{max}) is possibly defined by the injection conditions. Indeed, as a result of replacing an open-hole-interval source by a surface-equivalent spherical cavity this quite rough evaluation approach tends to underestimate the

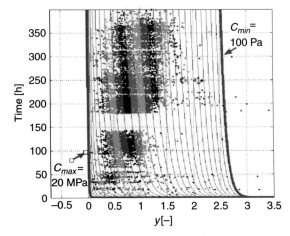

Figure 3.18   A y–t plot for seismicity induced in the Soultz 1993 geothermal experiment. (Modified from Rothert and Shapiro, 2007.)

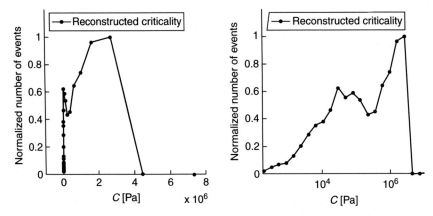

Figure 3.19   Reconstruction of the probability density function (PDF) of criticality for the Soultz 1993 data set. The PDF is given in logarithmic and linear scales. (Modified from Rothert and Shapiro, 2007.)

criticality. On the other hand it is clear that, even in the case of an exact pore-pressure modeling, the maximum C estimates cannot exceed the injection pressure of 12 MPa. To adjust these two values we must multiply the criticality range obtained above by approximately 4. Then we will obtain values of the criticality that are rather overestimated. This indicates that minimum pore pressures of 0.01 MPa are sufficient to seismically activate some pre-existing fractures at Soultz. This value is in agreement with the estimates following from the case study KTB considered at the beginning of this chapter. Necessary pressures for the majority of seismic events are approximately between 0.01 MPa and 10 MPa. However, the upper limit here is probably defined by the injection pressure. Thus, one cannot exclude that C_{max} can be even higher than 10 MPa.

3.8.3 Case study: Fenton Hill

Fenton Hill is located about 40 km west of Los Alamos on the west side of the Rio Grande Graben in New Mexico, USA. During a massive hydraulic stimulation experiment in 1983, about 21 600 m^3 of water were injected at a depth of approximately 3460 m for about 62 h (Fehler *et al.*, 1998). Fluid injection took place along an open-hole interval of approximately 20 m (House, 1987). Maximum bottom-hole pressure was approximately 48 MPa. During the experiment, 11 366 microseismic events were located with an accuracy better than 100 m (House, 1987). Approximately 9350 events occurred during the time interval of the injection and are shown in Figure 3.11.

Shapiro *et al.*, (2003) estimated the tensor of hydraulic diffusivity with an approximate relation of the principal components of 1:4:9. The largest axis of the tensor points approximately in a NNW direction and dips with an angle of about 72°. The medium component points SSE with a dip angle of 18°. The smallest component is about horizontal while pointing ENE. For the source strength q, Rothert and Shapiro (2007) considered pore-pressure perturbation along a cylindrical source of the length $l = 20$ m and the radius $R_c = 0.1$ m. Well-head injection pressure, i.e. overpressure, was approximately constant at 14 MPa (House, 1987). The radius of a surface-equivalent sphere is then $a_0 = 1$ m. The diffusivity was taken to be equal to $0.017 \, \text{m}^2/\text{s}$. According to (3.67) this approximately results in a source strength q of $35 \cdot 10^6$ W. After scaling in order to correct for anisotropy, the algorithm described above was applied to the data. The results are shown in Figures 3.20 and 3.21.

Comparable to the result for the Soultz data, the probability density function for criticalities lower than 0.001 MPa is vanishing, then it increases until 0.01 MPa, stays approximately constant until 0.1 MPa and then decreases towards 1 MPa. Criticality over a broad range characterizes the strength of pre-existing fractures. Again, the boundaries of the criticality limits seem to be sharp. The probability density function of criticality is again asymmetric (very roughly uniform or possibly truncated Gaussian).

As in the previous case study, the values of C are possibly somewhat underestimated. We can additionally constrain these estimates. To match the highest estimates of C to the injection pressure we must multiply all C values by approximately 14. Then we will obtain values of the criticality that are rather overestimated. Thus, as in the Soultz case study, this indicates that minimum pore pressures of 0.01 MPa are sufficient to seismically activate some pre-existing

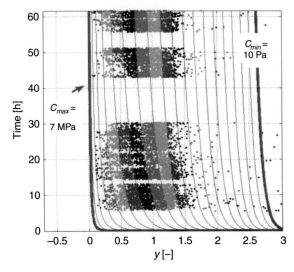

Figure 3.20 A y–t plot for seismicity induced in the Fenton Hill geothermal experiment. (Modified from Rothert and Shapiro, 2007.)

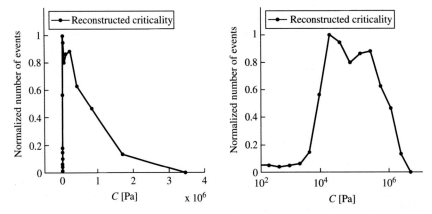

Figure 3.21 Reconstruction of the criticality probability density function (PDF) for the Fenton Hill data set. The PDF is given in logarithmic and linear scales. (Modified from Rothert and Shapiro, 2007.)

fractures at Fenton Hill. Necessary pressures for the majority of events are approximately between 0.01 MPa and 10 MPa. However, again one cannot exclude that C_{max} can be even higher than 10 MPa.

3.9 Spatial density of seismicity

Triggering front, back front and symmetry signatures of the fluid-induced seismicity considered above characterize temporal evolution and geometry of microseismic clouds. They describe kinematics of microseismic clouds. Already in the previous section, where we studied statistics of critical pore pressure, we worked implicitly with another type of seismicity characteristic that is related to its spatial density. Real data usually indicate that the density (i.e. probability) of seismic events is related to the magnitude of pore-pressure perturbations. Thus, we call the density of events, event rates and event magnitudes the dynamic signature of induced microseismicity.

We consider again a point source of a fluid injection in an infinite homogeneous and isotropic porous continuum. Furthermore, we assume the same triggering model of seismicity as in our previous discussions. The probability of an earthquake occurrence at a given point by a given time will be equal to $W(C(\mathbf{r}) \leq p_{max}(t, \mathbf{r}))$, which is the probability of the critical pressure being smaller than the maximum of the pore pressure $P_p(t, \mathbf{r})$ achieved at this point by the given time. If the pore-pressure perturbation caused by the fluid injection is a non-decreasing function of time (which is approximately the case for step-function-like borehole-injection pressures frequently used in reality) then this probability will be equal to

$$W = \int_0^{P_p(t,\mathbf{r})} f(C)dC, \qquad (3.74)$$

where $f(C)$ is the probability density function of the critical pressure. The pore pressure $P_p(t, r)$ is a solution of a diffusion equation describing pore-pressure relaxation.

The simplest possible PDF of the critical pore pressure is a uniform one, $f = 1/\Delta C$, where ΔC is a normalizing constant. In the previous section we saw that such a PDF can be accepted as a rough approximation of realistic criticality distributions. In this case $W = P_p(t, \mathbf{r})/\Delta C$. Therefore, in such a simple model the event probability is proportional to the pore-pressure perturbation.

Often realistic conditions of borehole fluid injections can be approximated by a point source of pore-pressure perturbation (3.67) of a constant strength q switched on at time 0. The corresponding solution of the diffusion equation then has the form given by (3.64). For infinite observation time this equation reduces to (see also Figure 3.22):

$$P_p(r, \infty) = q/(4\pi D r) = a_0 p_0/r. \qquad (3.75)$$

The event probability is proportional to the volumetric density of microseismic events. Therefore, a comparison of the volumetric event density with equation (3.64) can provide additional information on hydraulic properties of rocks. For a synthetic cloud of microseismicity we count the number of events in concentric spherical shells with the center at the injection point. Theoretically, the thickness

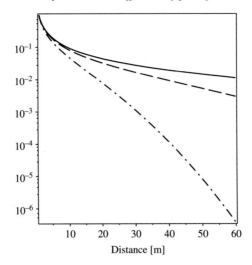

Figure 3.22 Normalized pore-pressure perturbation (3.64) in a medium with a given scalar diffusivity D due to a step-function point source of injection. For the solid line $D = 1 \text{ m}^2/\text{s}$ and $t = 10\,000$ s. For the dashed line $D = 10 \text{ m}^2/\text{s}$ and $t = 100$ s. For the dashed-dotted line $D = 1 \text{ m}^2/\text{s}$ and $t = 100$ s.

of these shells must be differentially small. However, in order to ensure sufficient statistics of events, the shells should not be too thin. In the case of real data, shell thickness can be of the order of a spatial event-location error. Owing to normalization, the central spherical volume defined by the internal spherical surface of the first shell is not considered. By normalizing event numbers to the shell volumes we obtain the event density. Then, we normalize all computed event densities by the event density in the first spherical shell taken into account. This allows us to work with non-dimensional quantities and to eliminate unknown proportionality factors.

To compare numerical results with predictions of equation (3.64), the analytical function $P_p(r, t)$ is also normalized by its value at the median radius of the first spherical shell taken into consideration. The time in the error function is the total period of the injection. Shapiro *et al.*, (2005a,b) provide examples of corresponding numerical simulations and show their agreement with (3.64).

Before such a comparison can be done for real data, the hydraulic anisotropy of rocks must be taken into account. In order to compare a spatial distribution of event density with analytical solution (3.64) we have to transform the microseismic cloud. Analogously to our considerations in the previous section of this chapter, this can be done by scaling an original event cloud along its principal axes and in relation of the inverse square roots of the principal components of the hydraulic diffusivity tensor (see equation (3.72)). Such a scaling procedure is a consequence of the fact that, after such a scaling, the diffusion equation in an anisotropic medium becomes equivalent to the diffusion equation in an isotropic medium. The hydraulic

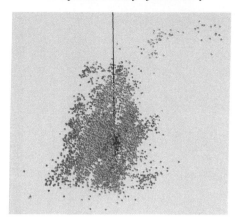

Figure 3.23   A 2D projection of the seismicity cloud of the Soultz case study. (Modified from Rothert and Shapiro, 2003.)

diffusivity in the resulting isotropic diffusion equation is equal to the arithmetic average of the principal components of the diffusivity tensor. After the scaling, a comparison between predictions of equation (3.64) and the event density is possible. It is completely analogous to the one in an isotropic medium (see Shapiro *et al.*, 2005a,b). Note that an inversion for a diffusivity tensor using a triggering front is able to provide orientations and relations between the tensor's principal components with a rather high precision. However, their absolute values (i.e. the magnitude of the arithmetic average of the principal components) are estimated less exactly. This is a consequence of the heuristic nature of proportionality factors in the triggering-front definition (3.6). This disadvantage can be compensated for by using the analysis of the spatial density of events we describe here.

As case histories we consider again the two data sets already addressed above, the microseismic clouds from Soultz and Fenton Hill. Figures 3.23 and 3.24 show 2D projections of the corresponding seismicity clouds.

As discussed above, the tensor of hydraulic diffusivity in Soultz is characterized by a significant anisotropy with an approximate relation of principal components in proportion 1:2.5:7.5. The average of the principal components of the hydraulic diffusivity D can then be fitted to match the data. A best-fit curve is shown in Figure 3.25. This fit provides an estimate of $D = 0.03$ m^2/s. We observe a very good agreement of the event density with the theoretical curve given by equation (3.64).

Independent methods of estimating hydraulic diffusivity at the same location yield rather similar values: the average principal component of the diffusivity tensor can be computed from results given by Shapiro *et al.* (1999) to be equal to $D = 0.023$ m^2/s. This estimate is also based on microseismicity but results from using the triggering front in separate angular sectors around the injection source.

Figure 3.24 A 2D projection of the seismicity cloud of the Fenton Hill case study. (Modified from Rothert and Shapiro, 2003.)

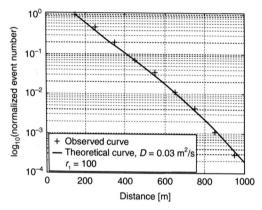

Figure 3.25 Event density versus theory for the Soultz 1993 case study. (Modified from Shapiro *et al.*, 2005a.)

The estimate of the apparent hydraulic permeability from the same location based on a borehole flow test and reported in Jung *et al.* (1996) also can be used to compute the hydraulic diffusivity. This approximately yields $D = 0.022$ m^2/s.

In the previous sections we have shown estimates of $D = 0.05$ m^2/s obtained from triggering fronts and back fronts fitted to the complete data set of the Soultz experiment (see Figure 3.15). These approaches assume that the medium is hydraulically isotropic. Therefore, they provide order-of-magnitude estimates only. Uncertainties of the event-density-based approach to estimating average diffusivity seems to be less, of the order of several tens percent (some numerical studies are presented by Shapiro *et al.*, 2005a,b).

Previously we have seen that, in the case of Fenton Hill, the tensor of diffusivity is characterized by an anisotropy with an approximate relation of the principal

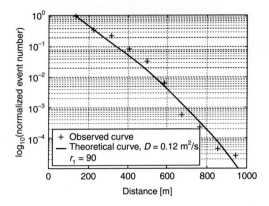

Figure 3.26 Event density versus theory for the case study Fenton Hill. (Modified from Shapiro *et al.*, 2005a.)

components as 1:4:9. A least-squares-fit of analytical function 3.64 gives an estimate of $D = 0.12$ m^2/s and is shown in Figure 3.26. Fitting of the triggering front and the back front of seismicity (see Figure 3.15) provide a very close result: $D = 0.14$ m^2/s.

Finally, Figure 3.26 shows some systematic deviations of the observed event density from the theoretical curve. This is possibly an indication of some non-linearity of the pore-pressure diffusion process in the Fenton Hill case study. The next chapter considers seismicity triggering by non-linear pressure diffusion in detail.

4

Seismicity induced by non-linear fluid–rock interaction

We have already seen that the method of passive seismic monitoring has significant potential for characterizing physical processes related to fluid stimulations of rocks. One of its important modern applications is spatial mapping of hydraulic fracturing. On the other hand, understanding spatio-temporal dynamics of microseismic clouds contributes to reservoir characterization. It helps to monitor and to describe hydraulic fractures.

We will start this chapter with a simple intuitive approach to the quantitative interpretation of hydraulic-fracturing-induced seismicity. The approach is based on a volume-balance model of the growth of long thin simple-geometry (nearly 1D) tensile hydraulic fractures. Then we will introduce a case study from a gas shale showing fracturing of a 3D rock volume. This motivates a more-general formulation of non-linear fluid–rock interaction by hydraulic stimulations of reservoirs. We show that linear pore-pressure relaxation and hydraulic fracturing are two asymptotic end members of a set of non-linear diffusional phenomena responsible for seismicity triggering. We formulate a general non-linear diffusion equation describing the pore-pressure evolution and taking into account a possibly strong enhancement of the medium permeability. Both linear pore-pressure relaxation and hydraulic fracturing can be obtained as special limiting cases of this equation.

From this formulation we derive an expression for the triggering front of fluid-induced seismicity, which is valid in the general case of non-linear pore-pressure diffusion. Our results are valid for an arbitrary spatial dimension of diffusion and a power-law time dependence of the injection rate. They show that, the larger the non-linearity of the fluid–rock interaction, the more strongly propagation of the triggering front depends on the mass of the injected fluid. Further, we investigate the nature of diffusivity estimates obtained from the triggering front of non-linear diffusion-induced seismicity. Finally, we introduce a model of anisotropic non-linear diffusion and show its application to the gas-shale data set mentioned earlier.

164

4.1 Seismicity induced by hydraulic fracturing

Here, we show that evaluation of spatio-temporal dynamics of a cloud of induced microseismic events can contribute to characterization of the hydraulic-fracturing process. Again, r–t plots (plots of distances from the injection source to hypocenters of seismic events versus their occurrence times) turn out to be a useful tool. For example, r–t plots show signatures of fracture volume growth, of fracturing fluid loss, as well as diffusion of the injection pressure into rocks and inside the fracture.

4.1.1 Triggering front of hydraulic fracturing

During hydraulic fracturing a treatment fluid is injected through a perforated domain of a borehole into a reservoir rock under a bottom-hole pressure larger than the minimum principal stress σ_3. Often this stress is called the fracture closure pressure, implying that uniaxial tensile strength from equation (1.79) is vanishingly small. Usually it is expected that a tensile fracture ($J = I$, see Section 1.2.4) will be opened (see Figure 4.1) so that its longest axis will be parallel to the maximum compressional tectonic stress σ_1 and its shortest axis will be parallel to the minimum compressional tectonic stress σ_3. Thus, depending on the tectonic stress regime, the hydraulic fracture can be vertical (e.g. a normal stress regime) or horizontal (e.g. a thrust stress regime). It can, of course, also be oblique in the case of tilted tectonic stresses.

In order to understand the main features of induced seismicity by hydraulic fracturing we apply a simple model of the fracture-growth process resulting from a consideration of the volume balance of an incompressible treatment fluid. The volume balance requires the volume $Q_c(t)$ of the injected fluid to be equal to a sum of the

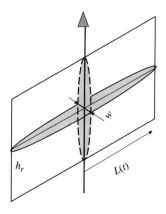

Figure 4.1 A geometrical sketch of a PKN hydraulic fracture.

stored-in-fracture fluid volume $Q_f(t)$ and of the lost-into-surrounding-formation fluid volume $Q_l(t)$:

$$Q_c(t) = Q_f(t) + Q_l(t). \tag{4.1}$$

A straight planar height-fixed fracture (usually a vertical one – this is the case for nearly all real-data examples given here) confined in a reservoir layer (see Figure 4.1) is considered. This is the so-called PKN (Perkins–Kern–Nordgren) model, which is very well known in the theory and practice of hydraulic fracturing (see Economides and Nolte, 2003, and references therein; see also Peirce and Detournay, 2008, for computational approaches to hydraulic fracture modeling).

Under these conditions, the half-length L of the fracture is approximately given as a function of the injection time t by the following expression:

$$L(t) = \frac{Q_I t}{4h_f C_L \sqrt{2t} + 2h_f w}, \tag{4.2}$$

where Q_I is an average steady-state injection rate of the treatment fluid, h_f is an average fracture height and w is an average fracture width. The first term in the denominator describes the fluid loss from the fracture into the rock. It is proportional to $\sqrt{t}$ (this is a result of the approximation proposed by Carter, 1957) and has a diffusion (filtration) character. The fluid-loss coefficient C_L depends on many factors, including the wall-building characteristics of the fracturing fluid, the relative permeability of the reservoir rock to the fracturing filtrate, the hydraulic diffusivity D of the medium, the injection pressure, etc. The fluid-loss coefficient is an important reservoir engineering parameter and is a subject of research. The second term, $2h_f w$, represents the contribution of the effective fracture volume and depends mainly on geometry of the vertical cross-section of the fracture.

In the case of hydraulic fracturing of a formation with a very low permeability (e.g. gas shale or tight gas sandstones) the fracture body represents the main permeable channel in the formation. A propagating fracture changes the effective stress state in its vicinity and activates mainly slip events (observations of tensile, implosive and CLVD events are also reported) in the critical fracture systems existing in surrounding rocks (Rutledge and Phillips, 2003). Thus, the fluid-induced microseismicity is concentrated in a spatial domain close to the hydraulic fracture (see also Warpinski *et al.*, 2001). Figure 4.2 schematically shows such a situation. Therefore, equation (4.2) can be considered as a 1D approximation for the triggering front of microseismicity in the case of a penetrating hydraulic fracture.

By the hydraulic fracturing of tight rocks (i.e. rocks with extremely low permeability) this equation replaces the triggering front (3.6). In the limit of a long-time

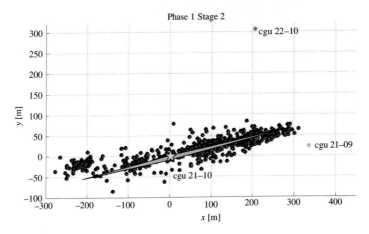

Figure 4.2 Schematic representation of a hydraulic fracture along with a cloud of microseismic events at one of treatment stages in tight gas sandstones of the Cotton Valley. (Modified from Shapiro, 2008, EAGE Publications bv.)

injection, the first (fluid-loss) term in the denominator of equation (4.2) becomes dominant and, therefore, equation (4.2) becomes identical to equation (3.6):

$$L \approx \sqrt{4\pi D_{ap} t},\qquad(4.3)$$

with apparent diffusivity D_{ap} given by:

$$D_{ap} = \frac{Q_I^2}{128\pi h_f^2 C_L^2}.\qquad(4.4)$$

Note, however, a completely different physical meaning of the apparent-diffusivity coefficient D_{ap} than the one of hydraulic diffusivity of rocks D from equation (3.6). The latter is a material property, being the hydraulic diffusion coefficient. On the other hand, D_{ap} is nearly inversely proportional to the hydraulic diffusivity of virgin rocks. It is a combination of geometrical factors and injection and rock parameters.

If the volume of fluid loss is insignificant (e.g. this can be the case during the initial phase of an injection), then the fracture length becomes close to a linear function of time:

$$L(t) = \frac{Q_I t}{2h_f w}.\qquad(4.5)$$

Equations (4.2)–(4.5) give us a simple model of the growth process of a hydraulic fracture during a fluid injection. They can be used as an approximation of the spatial growth of a related microseismic cloud in the direction of fracture penetration. In other words, they provide us with an approximate description of the 1D dynamics of the microseismic cloud during the injection phase.

4.1.2 Back front of hydraulic fracturing

What happens after the termination of the injection? To some extent the fracture can still continue to penetrate. However, the dominant part of the seismicity induced after an injection stop is located in the fractured domain (including the body of hydraulic fracture usually filled with a proppant). We assume that the seismicity is mainly triggered by the process of fluid-pressure relaxation in the fractured domain. This hypothesis is also very close to the established understanding of seismicity triggering by hydraulic fracturing (see Warpinski *et al.*, 2001). The process of pore-pressure relaxation can be approximately characterized by a system of equations describing the dynamics of fluid-saturated porous elastic solids (see Chapter 2). Such a description requires a more-precise treatment of fluid properties. For instance, in contrast to volume-balance considerations, the fluid compressibility must be taken into account.

We assume that the fractured domain is a thin planar infinitely long homogeneous strip of a porous elastic solid. The injection source is a thin straight line (a perforated borehole) crossing this strip normally to its bottom and top. The penetration depth of the treatment fluid and of the pore-pressure perturbation behind the fracture walls is assumed to be very small in comparison with the fracture length. Therefore, the geometry of the problem is effectively one dimensional. Let x be a coordinate along the fracture length.

As an approximation we can further assume a fluid injection with a constant fluid rate starting at time $t = 0$ at point $x = 0$ and terminating at time $t = t_0$. Assuming that events may be triggered only for increasing pore-pressure values, this would result in no later seismic activity at a point x, where $P_p(x, t)$ has reached its maximum at the time t. Therefore, (x, t) coordinates where the fluid pressure $P_p(x, t)$ reaches its maximum give the distance $x_{bf}(t)$ (we assume a symmetric fracture and consider the half-space with positive x) from the injection source to points at which the induced microseismicity is terminated after the end of fluid injection. This distance determines the so-called back front of the induced seismicity (Parotidis *et al.*, 2004).

In the following discussion we will accept for the back front a rough approximation of a linear diffusion of the pore pressure in the volume of the stimulated rock. In this case we obtain the following expression:

$$x_{bf} = \sqrt{2Dt\left(\frac{t}{t_0} - 1\right)\ln\left(\frac{t}{t - t_0}\right)}. \tag{4.6}$$

This equation is identical to equation (3.63) with $d = 1$.

Equations (4.2)–(4.5) and (4.6) provide us with approximate bounds of hydraulic-fracturing-induced microseismicity in the r–t domain.

4.1.3 Case study: Cotton Valley

We consider microseismic clouds (event locations courtesy of James Rutledge, LANL) induced by hydraulic fracturing experiments in two boreholes of the Carthage Cotton Valley gas field, East Texas, USA (see Figures 4.3 and 4.4). The gas reservoir is a low-permeability interbedded sequence of sands and shales. We will pay especial attention to the fracturing treatments called stages 2 and 3 of treatment well 21-10 (see Figures 4.5, 4.6 and the upper part of Figure 4.4) and described by Rutledge and Phillips (2003). Theses fracturing experiments are also described as the treatments B and A in (Rutledge *et al.*, 2004). There the hydraulic fracturing was conducted at depths of 2757–2838 m and 2615–2696 m, respectively. In both experiments the total perforation intervals were 24 m. About 990 and 650 microearthquakes, respectively, were located by using two borehole arrays of 3-C geophones.

For all located microearthquakes of stage 3, Figure 4.7 (bottom) shows event distances from corresponding nearest perforation points versus event occurrence

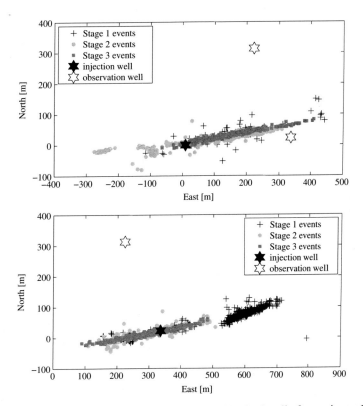

Figure 4.3 Map view of microseismicity induced by hydraulic fracturing at boreholes 21-09 (bottom) and 21-10 (top) in tight gas sandstones of the Cotton Valley. (Modified from Dinske *et al.*, 2010.)

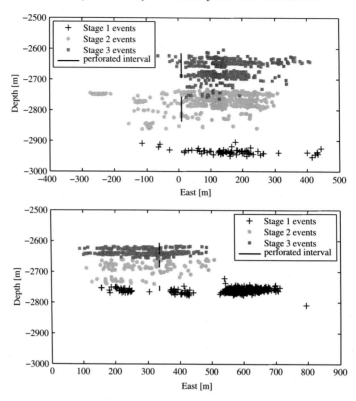

Figure 4.4   Vertical plane projection of microseismicity induced by hydraulic fracturing at boreholes 21-09 (bottom) and 21-10 (top) in tight gas sandstones of the Cotton Valley. (Modified from Dinske *et al.*, 2010.)

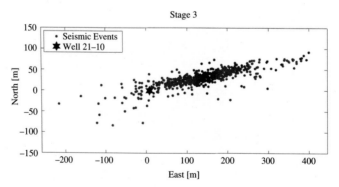

Figure 4.5   Map view of gel-fracturing stage 3 of Cotton Valley tight gas sandstone. (Modified from Dinske *et al.*, 2010.)

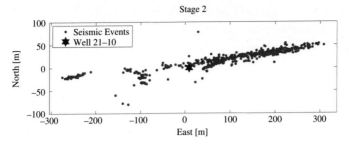

Figure 4.6 Map view of gel-fracturing stage 2 of Cotton Valley tight gas reservoir. Note an intersection of the hydraulic fracture of stage 2 with a pre-existing natural fracture between 30 and 100 m east. About 40% of induced earthquakes occurred along this natural fracture. Note also the difference in the event-location quality between Figure 4.2 and this figure (obtained after a precise event re-picking; data courtesy of James Rutledge, LANL). (Modified from Dinske *et al.*, 2010.)

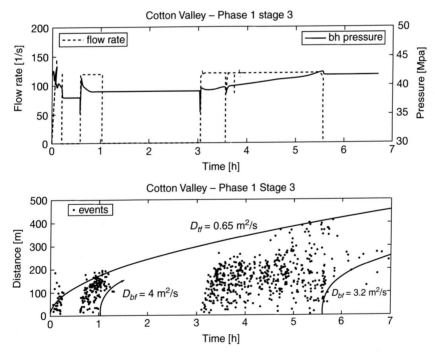

Figure 4.7 Hydraulic-fracturing treatment data (top) and *r–t* plot of induced microseismicity (bottom) for fracturing stage 3 of well 21-10 of the Carthage Cotton Valley gas field. (Modified from Shapiro *et al.*, 2006b.)

time. In the same time scale Figure 4.7 (top) shows the flow rate of the treatment fluid along with the measured bottom-hole pressure.

A parabolic envelope given by equation (4.3) seems to describe well the upper bound of the majority of microseismic events. One possible explanation is that during a large part of the fluid-injection period the fracture growth is mainly controlled by fluid-loss effects. Apparent hydraulic diffusivity $D_{ap} = 0.65$ m²/s characterizes these effects and can be further used in equation (4.4) in order to estimate the fluid-loss coefficient (C_L). Indeed, from this equation, we see that

$$C_L = \frac{Q_I}{8 h_f \sqrt{2\pi D_{ap}}}.$$ (4.7)

In the experiment under consideration the fluid rate during the injection intervals can be approximated as $Q_I = 0.12$ m³/s. The height of the fracture can be assumed to be approximately equal to the total length of the perforated borehole interval, $h_f = 24$ m. Thus, we obtain the following order-of-magnitude estimate. $C_L = 3 \times 10^{-4}$ m/s$^{1/2}$. It is known that, typically, the fluid-loss coefficient is of the order of 2×10^{-5} m/s$^{1/2}$ to 2×10^{-3} m/s$^{1/2}$ (see, for example, Economides and Nolte (2003)). Thus, we obtain a very realistic result.

The back front of the induced microseismicity is clearly seen after the stop of the fluid injection. The last (third) injection phase was completed with a cross-link gel and proppant. The back front (see equation (4.6)) provides a high diffusivity estimate $D_{bf} \approx 3.2$ m²/s. A possible interpretation is that this estimate mainly represents the hydraulic diffusivity along the propped hydraulic fracture saturated by the treatment fluid. Moreover, even after the end of the second injection phase (approximately 1 h after the injection start) a back-front signature can be seen in the r–t plot. An even higher diffusivity estimate, $D_{bf} \approx 4$ m²/s, is explained by the fact that during the first two phases of the injection no proppant but rather a low-viscosity fluid was used for the treatment. Thus, the fracture must have a somewhat larger permeability (before it has been completely closed) than in the case of a proppant-filled fracture. Also the fluid viscosity influences the diffusivity estimates.

Figure 4.8 shows a zoom of the first 1.5 h of the treatment. It indicates one more important signature of the process of hydraulic fracturing. During the first approximately 10 min of fracturing a quick, quasi-linear-with-time growth of the microseismic cloud up to 150 m in length can be observed. We can try to interpret these data by using equation (4.5), where the process of new fracture-volume creation is assumed to be dominant. For example, we can attempt to estimate the average width of the hydraulic fracture:

$$w = \frac{Q_I t}{2 h_f L(t)}.$$ (4.8)

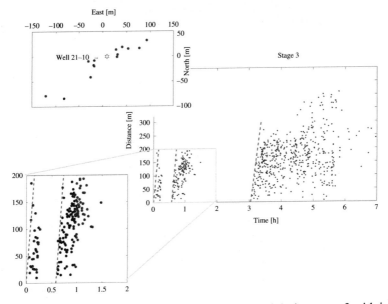

Figure 4.8 The r–t plot of the microseismicity induced during stage 3 with indicated features corresponding to the first opening and following reopening of the fracture. The inset in the upper left-hand corner of the figure shows a map view of locations of microseismic events from the vicinity of the straight line shown in the time range of first 10 min of the r–t plot. (Modified from Shapiro, 2008, EAGE Publications bv.)

The injection rate Q_I averaged over the first 10 min was about 0.08 m³/s. The average fracture height can again be approximated by the total perforation interval, i.e. 24 m. Figure 4.8 shows that during the first 10 min of the injection the fracture reached approximately 150 m half-length. Thus, the average fracture width w is approximately 7 mm. It is known that in low- to medium-permeability formations (and this is the case here – we work with a tight gas sand) the average hydraulic fracture width is in the range of 3–10 mm (Economides and Nolte, 2003). Again, we arrive at a very realistic result.

Interestingly, such quasi-linear growth of the microseismic cloud can be observed in the first 10–20 min of each of the three injection periods (see Figure 4.8 at 0 h, at approximately 40 min and 3 h after the start of the treatment). We interpret this as a signature of the fracture closing after each injection stop and its reopening at the start of following injection phases.

Very similar features (see Figures 4.9 and 4.10) can be also seen in the data of the hydraulic fracturing experiment corresponding to stage 2 of the gel treatment. For the fluid-loss coefficient we obtain $C_L = 3.7 \times 10^{-4}$ m/s$^{1/2}$ and the fracture average width is approximately of the order of 8 mm.

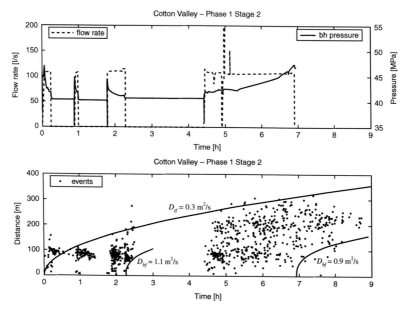

Figure 4.9 Hydraulic fracturing treatment data (top) and *r–t* plot of induced microseismicity (bottom) for the fracturing stage 2 in the well 21-10 of Carthage Cotton Valley gas field. (Modified from Shapiro *et al.*, 2006b.)

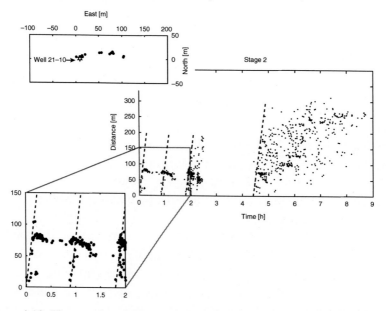

Figure 4.10 The *r–t* plot of the microseismicity induced during stage 2 with indicated features corresponding to the first opening and following reopening of the fracture. The inset in the upper left-hand corner of the figure shows a map view of locations of microseismic events from the vicinity of the straight line shown in the time range of first 10 min of the *r–t* plot. (Modified from Shapiro, 2008, EAGE Publication bv.)

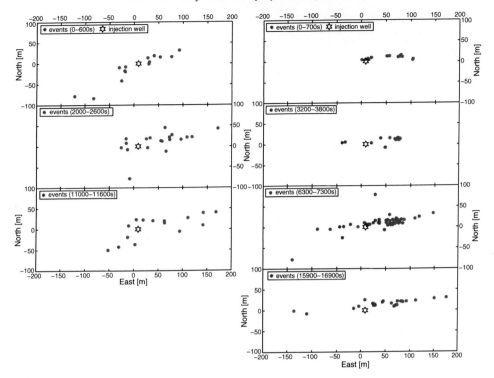

Figure 4.11 Temporal evolution of the hydraulic fractures of stages 3 (on the left) and 2 (on the right) of the Cotton Valley case study in terms of induced microseismic events (the map view).

Figures 4.8 and 4.10 help us to identify time intervals where the process of opening of hydraulic fractures was interrupted and even turned to their closing. Figure 4.11 shows microseismic images of the hydraulic fractures of both stages at the final moments of their multiple reopening. For the images, those events were selected that were used before for fitting straight lines on Figures 4.7–4.10.

In order to analyze a long-term opening process we consider composite microseismicity, which would correspond to a treatment of several hours' duration with an approximately constant injection rate. To do this, we eliminate time intervals corresponding to interruptions in the fluid injections at stages 2 and 3. Figure 4.12 shows composite r–t plots obtained in this way. We can clearly observe a predominantly diffusional character of the fracture growth. Such a corrected r–t plot can help to further improve estimates of the fluid-loss coefficient. Also, the back-front signatures can be nicely identified. How can we now use these signatures to characterize the reservoir?

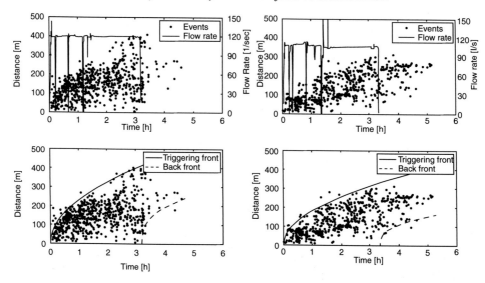

Figure 4.12 Composite r–t plots of stages 3 (to the left) and 2 (to the right) corresponding to several-hours-long fluid injection with an approximately constant injection rate. (Modified from Shapiro, 2008, EAGE Publication bv.)

4.1.4 Estimating permeability of virgin reservoir rocks

We can attempt to further interpret the fluid-loss coefficient neglecting near-fracture-surface effects. This means that we neglect the effect of a possible filter cake on the pressure difference Δp between fracture and far-field reservoir. In addition, we neglect the effect of the hydraulic-fracturing process on the pore space and pre-existing natural fractures in the vicinity behind the walls of the fracture. If all these effects are significant, then our estimate below can significantly deviate from the property of a virgin reservoir. It corresponds instead to an effective permeability.

As an approximation, we can use the following equation (see A9-13 of Economides and Nolte, 2003):

$$C_L \approx \sqrt{k\,\varphi\,c_t/(\pi\,\eta)} \cdot \Delta p, \tag{4.9}$$

where k and φ are the permeability and porosity of the reservoir, and c_t and η are the compressibility and viscosity of the reservoir fluid, respectively. Combining this equation with equation (4.7) we obtain

$$k \approx \frac{Q_I^2 \eta}{128 h_f^2 \Delta p^2 \varphi c_t D_{ap}}. \tag{4.10}$$

For example, we can use the following typical values for the Cotton Valley tight gas reservoir conditions (a hydrocarbon gas at temperature of 120 °C and pressure of 28 MPa): $\varphi = 0.1$; $c_t = 3.5 \cdot 10^{-8}$ Pa^{-1}; $\eta = 3 \cdot 10^{-5}$ Pa $\cdot$ s. Assuming a hydrostatic far-field reservoir pressure and taking for Δp an estimate of 15 MPa, we obtain for the virgin reservoir permeability an estimate of the order of 10^{-5} darcy. This is in a good agreement with values used for engineering simulations at Cotton Valley.

4.1.5 Estimating permeability of hydraulic fractures

The hydraulic diffusivity of a fractured domain can be approximately written as (see our equation (2.99) and its discussion for highly porous rocks; see also Economides and Nolte, 2003 chapter 12):

$$D \approx k_f/(c_f \, \eta_f \, \varphi_f), \tag{4.11}$$

where k_f is a permeability of the fractured domain, φ_f is its porosity, and c_f and η_f are the compressibility and viscosity of the treatment fluid, respectively. If we assume that the back-front diffusivity D_{bf} is equal to the diffusivity of the fracture, we will obtain:

$$k_f \approx D_{bf} c_f \eta_f \varphi_f. \tag{4.12}$$

For example, we use $D_{bf} \approx 3\,\text{m}^2/\text{s}$, $c_f \approx 1/(2.25 \cdot 10^9 \text{ Pa})$ (we assume the compressibility of the treatment fluid is of the same order as the compressibility of water), $\eta_f = 150 \cdot 10^{-3}$ Pa $\cdot$ s (viscosity of the cross-link gel, see Rutledge and Phillips, 2003) and $\varphi = 0.3$ (an estimate of the porosity of the fractured domain including the propped fracture). This provides us with an estimate of the effective permeability k_f of the hydraulic fracture of 70 darcy, which is a realistic value.

4.1.6 Case study: Barnett Shale

From the case studies of microseismicity we have considered so far we can conclude that linear pore-pressure relaxation and hydraulic fracturing are two asymptotic end members of a set of phenomena responsible for seismicity triggering. Later, we will show that both types of seismicity-inducing processes (on one hand a linear pore-pressure diffusion and, on the other, an opening of a new volume in rocks) are limiting cases of a non-linear pressure-diffusion process leading to an enhancing of the medium permeability. Such a non-linear process seems to be responsible for microseismicity triggering in shale. In particular, microseismicity of Barnett Shale shows signatures of non-linear-diffusion triggering with an extremely strong hydraulic permeability enhancement in a 3D stimulated volume of a reservoir. This indicates a process of a volumetric hydraulic fracturing via reopening of compliant pre-existing cracks and joints.

Features of hydraulic fracturing considered here are defined by specific properties of a particular Barnett Shale gas reservoir (Fisher *et al.*, 2002, 2004;

Maxwell *et al.*, 2006). Features of the tectonic stress there (which still remains to be better understood) are also significant. The Barnett Shale is a marine shelf deposit of the Mississippian age. It is an organic-rich black shale with extremely low permeability on the order of 0.1–0.5 microdarcy. Shapiro and Dinske (2009a) have analyzed a microseismic data set from Barnett Shale (courtesy of Pinnacle Technology). The data were obtained by hydraulic fracturing from a vertical bore-hole in the Lower Barnett Shale of the Fort Worth Basin. These data demonstrate a typical hydraulic fracture fairway network from a vertical well in the core area of Barnett (Fisher *et al.*, 2002).

One could propose the following scenario. We assume that the virgin reservoir rock is extremely impermeable. This leads to a vanishing fluid loss from opened fractures. We also assume that the treatment fluid is incompressible and that during its injection it deforms and opens weak compliant pre-existing fractures in a limited volume of the rock. We assume that the fractures can be opened if the pore pressure exceeds a given critical value. As soon as a fracture has been opened the permeability of the rock strongly increases, and the fluid can be further transported to open more fractures. Note that fracture opening (i.e. fracture width) is a function of the pore-pressure perturbation. However, in the case of extremely increasing permeability, one can assume that from the injection source up to the filtration front (where $P_p = 0$) radial variations of the bulk porosity filled by the treatment fluid are weak only. Then equation (4.5) should be replaced by

$$r(t) = A(Q_I t/\phi)^{1/3}. \tag{4.13}$$

Here $r(t)$ is a growing size of the fractured domain, ϕ is its porosity filled by the treatment fluid and A is a dimensionless geometric factor equal, for example, to 1 or to $(3/(4\pi))^{1/3}$ in the case of cubic or spherical fractured domain, respectively.

Therefore an r–t plot will show a $t^{1/3}$ parabolic envelope of corresponding microseismic clouds. Exactly this type of behavior is demonstrated by the Barnett Shale data set in Figures 4.13–4.15. Figure 4.13 shows the microseismic cloud induced by the hydraulic fracturing. The spatial evolution of microseismicity with injection time is also seen in the figure. Figures 4.14 and 4.15 show the injection pressure, the fluid rate and the r–t plot for microseismic events. One can see that the cubic-root type of the triggering front describes the data significantly better than a square-root parabola.

The cubic-root parabola of Figure 4.15 and equation (4.13) permits us to estimate the porosity opened and connected by the treatment fluid. An approximate estimate of this additional (volume-averaged) opened porosity is of the order of 0.01%.

Therefore, in this particular case study, microseismic features of hydraulic fracturing correspond to a non-linear pressure diffusion in a medium with permeability

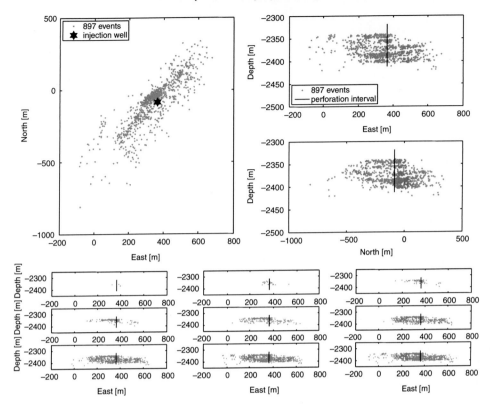

Figure 4.13 Hydraulic-fracturing-induced microseismicity in Barnett Shale (data courtesy of Shawn Maxwell, Pinnacle Technologies). Top: map and vertical-plane projections of the microseismic cloud. Bottom: spatial growth of the microseismic cloud with time on a vertical-plane projection (note that the vertical and horizontal scales are equal here; from Shapiro and Dinske, 2009a).

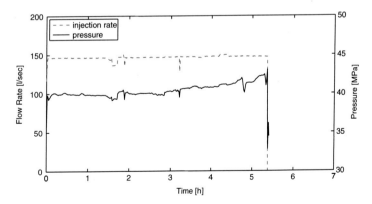

Figure 4.14 Borehole pressure (measured at the injection depth) and fluid flow rate for the Barnett Shale data set shown in Figure 4.13. (Modified from Shapiro and Dinske, 2009b.)

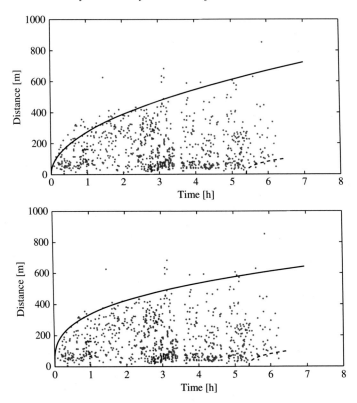

Figure 4.15 The r–t plot of induced microseismic events for the Barnett Shale data set shown in Figure 4.13. Different envelopes are shown. Top: a linear-diffusion-type approximation of the triggering ($t^{1/2}$); the dashed line gives a possible indication of the back front. Bottom: a cubic-root parabola ($t^{1/3}$) better matching the data. (Modified from Shapiro and Dinske, 2009b.)

that is strongly enhanced by the fluid injection. It seems that the volumetric tensile opening of pre-existing compliant fractures embedded into extremely imperme-able matrix is a dominant mechanism controlling the dynamics of the induced microseismicity here. This process can be denoted as a 3D volumetric hydraulic fracturing. The r–t plot shows a characteristic cubic-root parabolic behavior.

4.2 Seismicity induced by non-linear pressure diffusion

Above we have seen that spatio-temporal features of pressure-diffusion-induced seismicity can be found in a very natural way from the triggering front concept.

In a homogeneous and isotropic medium the triggering front can be defined corresponding to equation (3.6) in the limit of linear fluid–rock interaction. This approximation might be sometimes suitable for Enhanced Geothermal Systems

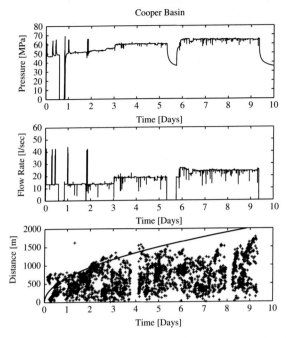

Figure 4.16 Injection pressure, flow rate (on the two upper plots) and an *r–t* plot of the corresponding fluid-injection-induced microseismicity at a geothermal borehole in the Cooper Basin, Australia. The data are courtesy of H. Kaieda. (After Shapiro and Dinske, 2009b.)

(EGS). In the previous chapter we considered the case studies Soultz and Fenton Hill in this approximation. Figures 4.16 and 4.17 show two more case studies of EGS with the triggering front constructed in the linear-diffusion approximation.

Another extreme corresponding to a strong non-linear fluid–rock interaction is hydraulic fracturing. Propagation of a hydraulic fracture is accompanied by opening new fracture volumes, fracturing fluid loss and its infiltration into reservoir rocks as well as diffusion of the injection pressure into the pore space of surrounding rocks and inside the hydraulic fracture. In order to understand the main features of the induced seismicity by such an operation, in the previous sections we applied a simple approximation of the process of the fracture growth. We used a volume-balance consideration for a single straight planar (usually vertical) fracture confined in a reservoir layer. Half-length L of the fracture is approximately given as a function of injection time t by expression (4.2). The fracture induces seismic events in its vicinity. Thus, equation (4.2) is an approximation of the triggering front in the case of a penetrating hydraulic fracture.

During the initial phase of hydraulic-fracture growth the contribution of the fracture volume in the fluid volume balance is often dominant. This can lead to a nearly

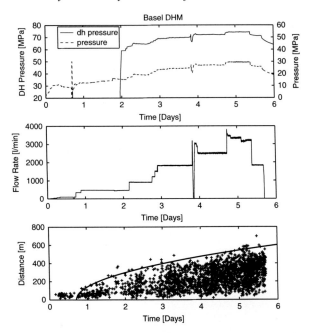

Figure 4.17 Injection pressure, flow rate (on the top two plots) and an *r–t* plot of the corresponding fluid-injection induced microseismicity at a geothermal bore-hole in Basel region of Switzerland. The data are courtesy of U. Schanz and M. O. Häring. (After Shapiro and Dinske, 2009b.)

linear expansion with time of the triggering front (see equation (4.5)). This is an example of 1D hydraulic fracturing. If the injection pressure drops, the fracture will close. A new injection of the treatment fluid leads to reopening of the fracture and, thus, to a repeated 1D hydraulic-fracture growth, i.e. repeated nearly linear (as function of time) propagation of the triggering front. Because of a pressure dependence of the fracture width and possible deviations of the fracture geometry from a simple nearly rectangular planar shape, deviations from a linear-with-time fracture growth can be observed. A long-term fluid injection leads to domination of diffusional fluid-loss processes. The growth of the fracture slows down and becomes approximately proportional to $\sqrt{t}$. Figure 4.18 shows an example of data obtained during hydraulic fracturing.

A process closely related to classical 1D hydraulic fracturing is the volumetric (i.e. 3D) hydraulic fracturing considered in the previous section. It involves a nearly incompressible treatment fluid, which during its injection deforms and opens weak compliant pre-existing fractures in a limited rock volume (see Figure 4.13). An *r–t* plot will show a $t^{1/3}$ parabolic envelope of corresponding microseismic clouds. This type of behavior is shown by the Barnett Shale data set we discussed in the previous section (see Figure 4.15).

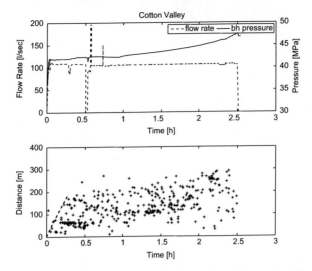

Figure 4.18 Hydraulic fracturing induced microseismicity at the Carthage Cotton Valley gas field. Top: borehole pressure (measured at the injection domain) and fluid flow rate. Bottom: *r–t* plot of induced microseismic events. The dashed line indicates approximately linear-with-time triggering front propagation corresponding to equation (4.5). This figure reproduces the data set of Figure 4.9 in the time period 4.5–7 h. (After Shapiro and Dinske, 2009b.)

Therefore, linear pore-pressure diffusion and non-linear fluid–rock interaction can show different signatures of seismicity triggering fronts. Shapiro and Dinske (2009a,b) proposed an approach unifying linear pressure diffusion and volumetric hydraulic fracturing, i.e. equations (3.6) and (4.13). Following them we derive a rather general non-linear diffusion equation describing the pore-pressure relaxation. This equation takes into account a possibility of strong enhancement of the medium permeability. Both asymptotic situations described above can then be obtained in an arbitrary spatial dimension. In the next chapter we will further use this formalism for a description of magnitude probabilities of induced events.

4.2.1 Non-linear diffusion and triggering fronts

Here we show that all types of scaling of microseismicity growth we reviewed in the previous section are particular cases of more-general non-linear diffusion dynamics. Under rather general conditions, including a possible strongly non-linear interaction of an injected fluid with a rock, the pore-pressure relaxation can be approximately described by a system of the two following differential equations.

The first one is the continuity equation expressing the fluid-mass conservation (see equation (2.65)):

$$\frac{\partial \phi \rho_f}{\partial t} = -\nabla \mathbf{q} \rho_f, \tag{4.14}$$

where ρ_f is the density of a pore fluid, $\mathbf{q}$ is its filtration velocity and ϕ is the rock porosity. Under realistic conditions (by neglecting irrotational skeleton deformations and stress dependences of elastic properties) the time-dependent part of quantity $\phi \rho_f$ is proportional to the pore-pressure perturbation P_p (see equation (2.113)), and can be substituted by $\rho_0 P_p S$, with ρ_0 being a reference fluid density. We recall that S is a storage coefficient depending on the bulk modulus of pore fluid K_f, the bulk moduli of grain material K_{gr} and of drained-rock skeleton K_{dr}, the P-wave modulus of drained skeleton $\lambda_{dr} + 2\mu_{dr}$ and the Biot–Willis coefficient $\alpha = 1 - K_{dr}/K_{gr}$ (see also equation (2.100)).

The second equation is the Darcy law, expressing a balance between the viscous friction force and the pore-pressure perturbation:

$$\mathbf{q} = -\frac{\mathbf{k}}{\eta} \nabla P_p, \tag{4.15}$$

where $\mathbf{k}$ is the tensor of hydraulic permeability of the rock and η is the dynamic viscosity of the pore fluid. In the case of non-linear fluid–rock interaction, the permeability can be a strongly pressure-dependent quantity (see Section 2.9.7). For simplicity we will neglect anisotropy and heterogeneity of rocks (a model of an anisotropic non-linear diffusion will be considered later in this chapter). We assume a spherical pore-pressure source switched on at time $t = 0$. Thus, we consider a spherically symmetric problem in a d-dimensional space and combine the two above equations in a corresponding d-dimensional spherical coordinate system with the origin at the injection point (see also equation (3.42)):

$$\frac{\partial r^{d-1} P_p}{\partial t} = \frac{\partial}{\partial r} D(P_p) r^{d-1} \frac{\partial}{\partial r} P_p, \tag{4.16}$$

where r is the radial distance from the injection point. We introduce a pressure-perturbation-dependent hydraulic diffusivity:

$$D(P_p) = \frac{k(P_p) \rho_f(P_p)}{S \eta \rho_0}. \tag{4.17}$$

The non-linear diffusion equation (4.16) must be completed by the initial condition of zero pore-pressure perturbation before the injection starts (i.e. $P_p = 0$ for $t < 0$) and by the two following boundary conditions. One of them gives the mass rate m_i of the fluid injection at the surface of an effective injection cavity of radius a_0:

$$m_i(t) = -A_d a_0^{d-1} \frac{k(P_p) \rho_f(P_p)}{\eta} \frac{\partial}{\partial r} P_p \mid_{r=a_0}, \tag{4.18}$$

where A_d will take one of the values $4\pi, 2\pi h, 2A_r$, if the dimension d of the space is $3, 2$ and 1, respectively. In particular, h denotes the height of a hypothetical homogeneous plane layer, where a cylindrically symmetric filtration takes place (i.e. $d = 2$). A_r denotes the cross-sectional area of an infinite straight rod in the case of a 1D filtration.

The second boundary condition states that a pore-pressure perturbation, along with its spatial derivatives, is vanishing at infinity faster than $1/r^{d-1}$.

Integrating equation (4.16) over distance r gives:

$$\frac{\partial}{\partial t} \int_{a_0}^{\infty} r^{d-1} P_p dr = D(P_p) r^{d-1} \frac{\partial}{\partial r} P_p \mid_{r=a_0} . \tag{4.19}$$

Integrating this over time and taking into account the boundary condition (4.18) at $r = a_0$ gives:

$$A_d \int_{a_0}^{\infty} r^{d-1} P_p dr = \int_0^t \frac{m_i(t)}{\rho_0 S} dt = \frac{m_c(t)}{\rho_0 S} \approx \frac{1}{S} \int_0^t Q_i(t) dt. \tag{4.20}$$

Here $m_c(t)$ denotes the cumulative mass of the fluid injected until time t. In the last approximate term we have neglected pressure dependency of the fluid density and we have introduced the volumetric rate of the fluid injection, $Q_i(t)$.

In the following we assume an injection source with a power-law injection rate $Q_i(t) = S(i + 1) A_d Q_0 t^i$. For example, in the case of a constant-rate injection, $i = 0$ and $Q_i = S A_d Q_0$. The dimensional constant Q_0 determines the strength of the injection source (of course, along with the exponent i).

Further, we assume a power-law dependence of the diffusivity on pore-pressure perturbations (see also equations (2.283)–(2.285)):

$$D = (n + 1) D_0 P_p^n. \tag{4.21}$$

If n is large, the hydraulic diffusivity (and permeability, respectively) will depend strongly on the pressure. If $n = 0$, the pressure relaxation will be described by a linear diffusion equation. The exponent n can be considered as a measure (an index) of non-linearity of the fluid–rock interaction.

Finally, we assume that the radius a_0 of the injection cavity is vanishingly small (compared to distances under consideration).

Under these assumptions, equations (4.16) and (4.20) take the following forms, respectively:

$$\frac{\partial r^{d-1} P_p}{\partial t} = D_0 \frac{\partial}{\partial r} r^{d-1} \frac{\partial}{\partial r} P_p^{n+1}, \tag{4.22}$$

$$\int_0^{\infty} r^{d-1} P_p(t, r) dr = Q_0 t^{i+1}, \tag{4.23}$$

where the proportionality constants D_0 and Q_0 are defined by properties of the medium and of the injection source.

To understand features of the solution of the problem stated by equations (4.22)–(4.23) we will apply the Π-theorem of the dimensional analysis (Barenblatt, 1996). The two equations above show that the pressure perturbation depends on the following quantities: r, t, D_0 and Q_0. They have the following physical dimensions:

$$[r] = L, \quad [t] = T, \quad [D_0] = \frac{L^2}{T P^n}, \quad [Q_0] = \frac{P L^d}{T^{i+1}}, \tag{4.24}$$

where, L, T and P denote physical dimensions of length, time and pressure, respectively. It is clear that three of the quantities r, t, D_0 and Q_0, arbitrarily chosen, have independent physical dimensions. The physical dimension of one of them (for example r) can be expressed in terms of three others. In other words, only one dimensionless combination, θ, can be constructed:

$$\theta = \frac{r}{(D_0 Q_0^n t^{n(i+1)+1})^{1/(dn+2)}}. \tag{4.25}$$

The following combination of the quantities t, D_0 and Q_0 has the dimension of pressure:

$$\left(\frac{Q_0^2}{D_0^d t^{(d-2i-2)}}\right)^{1/(dn+2)}. \tag{4.26}$$

The Π-theorem of the dimensional analysis states then that the pressure must have the following form:

$$P_p = \left(\frac{Q_0^2}{D_0^d t^{(d-2i-2)}}\right)^{1/(dn+2)} \Phi(\theta), \tag{4.27}$$

where Φ is a dimensionless function, which must be found by solving the problem formulated in (4.22)–(4.23) and the initial condition $P_p = 0$ for $t < 0$.

We see that the spatial distribution of P_p is completely defined by the dimensionless parameter θ. If θ is large enough (i.e. small times and large distances from the source) we expect an insignificant pressure increase. If θ is small (large times and small distances) a strong change of pressure must occur. Thus, a constant value of θ denotes a front of changing pressure. From an analysis of similar non-linear equations (for example, analogous equations for very intensive thermal waves were considered in Barenblatt, 1996, pp. 76–79, for the case $d = 3$) it is known that a sharp front separating a zero-perturbation domain from non-vanishing perturbation values can occur (see also our discussion at the end of this section and Figure 4.19). Thus, a constant value of θ defines the triggering front R_t of induced seismicity. From equation (4.25) we obtain a general result:

$$R_t \propto (D_0 Q_0^n t^{n(i+1)+1})^{1/(dn+2)}. \tag{4.28}$$

Let us first assume that the diffusivity is pressure independent. Then $n = 0$ and

$$R_t \propto \sqrt{Dt}.$$ (4.29)

This corresponds to equation (3.6). The triggering of seismic events is controlled by a linear pore-pressure relaxation (i.e. a linear diffusion process). Let us further assume that n is large (this corresponds to an extremely non-linear diffusion process). Then

$$R_t \propto (Q_0 t^{(i+1)})^{1/d}.$$ (4.30)

In this case the triggering front is completely defined by the volume of the injected fluid. In other words, equation (4.30) is an expression of the volume balance for an incompressible fluid. The opening of a new pore (or fracture) volume occurs in the spatial domain between injection source and triggering front. This volume is equal to the volume of the injected fluid. For example, in the case of a hydraulic fracture growth with $d = 1$ and a constant injection rate, $i = 0$,

$$R_t \propto Q_0 t.$$ (4.31)

This is in perfect agreement with equation (4.5), which describes the linear-with-time opening of a single hydraulic fracture. The dashed line in Figure 4.18 indicates such a process.

In the case of $d = 3$ and a constant injection rate, $i = 0$, we obtain:

$$R_t \propto (Q_0 t)^{1/3}.$$ (4.32)

This situation corresponds to equation (4.13) and Figure 4.15.

A special case of equation (4.22) with $n = 1$ corresponds to a classical equation describing the water head in an unsaturated medium. This equation was first proposed and solved by the French scientist J. Boussinesq (Barenblatt et al., 1990; Barenblatt, 1996). The Boussinesq equation with $d = 2$ and various values of i corresponds to unsaturated flow to or from a borehole in a reservoir layer. The parameterization $n = 1, d = 1$, and $i = (3\alpha_i - 1)/2$ corresponds to the initial condition problem of the Boussinesq equation formulated by Barenblatt et al. (1990). Particularly, the case of $i = -1$ (i.e. $\alpha_i = -1/3$) corresponds to a constant (i.e. time-independent) amount of fluid instantaneously injected into the medium. This yields the classical law $R_t \propto (D_0 Q_0 t)^{1/3}$ of propagation of the saturation head in a 1D medium. In the case of $i = -1$ and a very strong non-linear diffusion ($n \gg 1$), equation (4.28) yields:

$$R_t \propto Q_0^{1/d} t^{1/(dn+2)}.$$ (4.33)

Another form of the triggering front can be obtained by substituting the cumulative mass of the injected fluid (equation (4.20)) into equation (4.28):

$$R_t \propto \sqrt[dn+2]{D_0 t \left(\frac{m_c(t)}{S\rho_0 A_d}\right)^n}. \tag{4.34}$$

Again we see that in the case of linear diffusion (i.e. $n = 0$) the injected mass does not influence propagation of the triggering front. However, the stronger non-linearity, the stronger is the effect of the injected fluid mass. In the limit of large n we obtain:

$$R_t \propto \sqrt[d]{\frac{m_c(t)}{S\rho_0 A_d}}. \tag{4.35}$$

Therefore, we conclude that, depending on parameter n, equation (4.22) describes a broad range of phenomena: from a linear pore-pressure relaxation to hydraulic fracturing.

Finally, we will attempt to analyze a full solution of problems (4.22)–(4.23). Substituting P_p given by equation (4.27) into equations (4.22) and (4.23) we obtain:

$$\frac{\partial^2 \Phi^{n+1}}{\partial \theta^2} + \frac{(d-1)}{\theta}\frac{\partial \Phi^{n+1}}{\partial \theta} + \frac{n(i+1)+1}{dn+2}\theta\frac{\partial \Phi}{\partial \theta} + \frac{d-2(i+1)}{dn+2}\Phi = 0, \tag{4.36}$$

$$\int_0^\infty \theta^{d-1}\Phi(\theta)d\theta = 1. \tag{4.37}$$

This equation system must be completed by the condition $\Phi(\infty) = 0$.

For the set of constants $d = 3, i = -1$ (corresponding in our case to an instantaneous injection of a finite volume of fluid into a 3D poroelastic medium), equations (4.36)–(4.37) reduce exactly to the equation system (2.44)–(2.45) of Barenblatt (1996). Thus, for our somewhat more-general problem, we try to find a solution in a binomial form similar to those of Barenblatt (his equations (2.46)–(2.50)). Using a direct trial we obtain that, under the condition $i = -1$, system (4.36)–(4.37) has the following solution:

$$\Phi = J(\theta_t^2 - \theta^2)^{1/n}, \text{ for } \theta < \theta_t, \text{ and } \Phi = 0, \text{ for } \theta > \theta_t, \tag{4.38}$$

where $J = [n/(2(n+1)(dn+2))]^{1/n}$, and θ_t is another constant that can be found from substituting solution (4.38) and consequent integrating in (4.37). For example, the case of $d = 1, n = 1, i = -1$ corresponds to the Boussinesq equation analyzed by Barenblatt *et al.* (1990). For this case they obtained $\theta_t^2 = 8$.

In this example the condition $\theta = \theta_t$ defines a pressure front and, respectively, a seismicity triggering front (exactly in the sense of equation (4.28)). From (4.38)

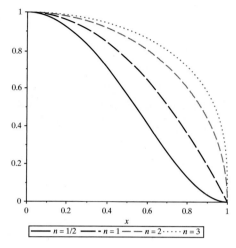

Figure 4.19 Pressure (represented by the vertical axis and normalized to its maximal value) as function of distance x (normalized to the triggering front) corresponding to equation (4.39) for different values of n.

the following scaling can be obtained for the pressure on distances smaller than the triggering front size (see Figure 4.19):

$$p(t, r) = p(t, 0) \left(1 - \frac{r^2}{R_t^2} \right)^{1/n} . \tag{4.39}$$

The pressure becomes vanishing small for larger distances. Moreover, we see that, in the case of finite $n \geq 1$, there exist a pressure front that is characterized by a discontinuous spacial derivative of the pressure. The triggering front of seismicity can be directly associated with this pressure front. The pressure profile has an abrupt character at the front in the case of a significant non-linearity. Simultaneously, the pressure distribution behind the front becomes closer to a rectangular one. We conclude that the stronger the non-linearity of the fluid–rock interaction, the more uniform the pressure distribution behind the triggering front.

4.2.2 Triggering fronts and diffusivity estimates

In this section we follow Hummel and Shapiro (2012) to investigate what kind of diffusivity estimates will be obtained from fitting triggering fronts on r–t plots in the case of non-linear pressure diffusion.

Earlier, Hummel and Müller (2009) analyzed synthetic clouds of 1D and 2D microseismicity triggered by non-linear pore-pressure diffusion using an exponential dependence of the diffusivity on the pore pressure (see also equations (2.286)–(2.287)):

$$D(P_p) = D_0 \exp(\kappa P_p). \qquad (4.40)$$

Here D_0 denotes diffusivity at vanishing pore pressure, and κ is the permeability compliance. At the limit of vanishing κ the pressure relaxation will be described by a linear diffusion equation. For the permeability compliance, Hummel and Müller (2009) reported values from different literature sources ranging from 0 to 40 GPa^{-1} for sandstones and values up to 500 GPa^{-1} for fractured rocks. Millich *et al.* (1998) state that for the KTB site the evolution of the *in situ* permeability with effective pressure can be described by an exponential relation with a permeability compliance value of 140 GPa^{-1}.

Hummel and Shapiro (2012) numerically considered power-law and exponential pressure dependencies of the diffusivity (4.21) and (4.40), respectively. They restricted their computations to radially symmetric structures. They found that, for a weak fluid–rock interaction (permeability enhancement at the injection source is less than one order of magnitude), the fracturing domain diffusivity does not differ significantly from the effective diffusivity empirically estimated by using the seismicity behavior in r–t plots. If non-linearity is large, all three values, the initial medium diffusivity, the fracturing domain diffusivity and the diffusivity obtained from the triggering front using the r–t-plot approach are significantly different. However, in an order-of-magnitude approximation their results show that the r–t-plot approach provides diffusivity estimates after the hydraulic stimulation of the rock.

Hummel and Shapiro (2012) used boundary and initial conditions corresponding to a point-like (or a small-radius injection cavity) pressure source of magnitude p_0 applied instantaneously at $t = 0$ and remaining constant for all times $t > 0$. This corresponds to a Dirichlet-type boundary condition (a constant injection pressure). They used $p_0 = 10$ Pa and correspondingly renormalized the realistic permeability compliances κ to the range of values between 0 and 0.5 Pa^{-1}, so that the injection impact on the permeability would be equal to the one for an injection with $p_0 = 10$ MPa (a typical order of magnitude in the case of a reservoir stimulation). This corresponds to values of the product κp_0 less than 5. For the index of non-linearity n they used $n = 0, 1, 2$ and 3. In both exponential and power-law-diffusion models they used the numerical value 1 for the quantity D_0. This corresponds to 1 m^2/s for the exponential model and 1 m^2/(sPa^n) for the power-law diffusion.

Note that the initial and boundary condition problem we consider in this section differs from previous problems. Here we define a constant-pressure boundary problem. In the previous section we considered boundary conditions corresponding to pre-defined fluid-injection rates. To explore the behavior of the triggering front we again start with the dimensional analysis.

For exponential diffusion model (4.40), the pore-pressure relaxation radius R_t (we will again identify this radius with the radius of the triggering front) depends on the quantities p_0, t, D_0, A_0 and κ. Here A_0 denotes the surface area of an effective injection source cavity. The dimensions of significant physical variables and parameters of the problem are:

$$[p_0] = P, \quad [t] = T, \quad [D_0] = \frac{L^2}{T}, \quad [A_0] = L^2, \quad [\kappa] = \frac{1}{P}, \tag{4.41}$$

where, L, T and P denote physical dimensions of length, time and pressure, respectively. We can construct two dimensionless quantities, Θ_{e1} and Θ_{e2}:

$$\Theta_{e1} = \frac{D_0 t}{A_0}, \quad \Theta_{e2} = \kappa p_0. \tag{4.42}$$

Then, the relaxation radius can be written in the following form:

$$R_t = \sqrt{D_0 t}\, f_e(\Theta_{e1}, \Theta_{e2}), \tag{4.43}$$

where f_e is a function of variables Θ_{e1} and Θ_{e2}.

In the case of power-law-diffusion model (4.21), the relaxation radius of the pore-pressure perturbation depends on quantities p_0, t, D_0 and A_0. The physical dimensions of these quantities are

$$[p_0] = P, \quad [t] = T, \quad [D_0] = \frac{L^2}{T P^n}, \quad [A_0] = L^2. \tag{4.44}$$

We can construct one dimensionless quantity only:

$$\Theta_{p1} = \frac{D_0 t p_0^n}{A_0}. \tag{4.45}$$

Hence, the relaxation radius has the form

$$R_t = \sqrt{D_0 t p_0^n}\, f_p(\Theta_{p1}), \tag{4.46}$$

where f_p is a function of Θ_{p1}.

For both the exponential and power-law diffusivity models, the strength of the injection source is given by a surface-force density. This is the injection pressure applied to the surface of the injection cavity. In one dimension, the injection source represents a surface of a half-space. The pressure is distributed along an infinite vertical plane intersecting the coordinate axes at the origin. Therefore, surface area A_0 of the effective injection source becomes infinitely large. As a result, the quantity Θ_{e1} goes to zero. Thus, for the exponential diffusion model, pressure fronts (and, therefore, triggering fronts) are given by

$$R_t = \sqrt{D_0 t}\, f_e(0, \Theta_{e2}). \tag{4.47}$$

For the power-law-diffusion model, triggering fronts in one dimension are given by

$$R_t = \sqrt{D_0 t p_0^n}\, f_p(0). \tag{4.48}$$

In equations (4.47) and (4.48) we assume that, for $\Theta_{e1} \to 0$ and $\Theta_{p1} \to 0$, the function limits $f_e(0, \Theta_{e2})$ and $f_p(0)$ are finite. For both models describing non-linear fluid–rock interaction relations (4.47) and (4.48) show a $\sqrt{t}$-dependence of the triggering front.

In two dimensions, the situation is similar. The pressure is distributed across the surface of an infinitely extended cylinder so that A_0 becomes again infinite. This cylinder is normal to the plane in which pore-pressure diffusion takes place. As a result, $\Theta_{e1} \to 0$ and $\Theta_{p1} \to 0$. Thus, triggering fronts for both the exponential-diffusion model and the power-law-diffusion model are also described by relations (4.47) and (4.48), respectively.

In contrast, in the case of pore-pressure diffusion in three dimensions, the injection pressure is distributed across the surface area of an effective injection-source cavity of a finite radius a_0. In this case, the surface area is limited and does not approach infinity. Moreover, we are interested in a solution for a point-like injection source, that is, $A_0 \to 0$. Therefore, now we are looking for asymptotic behavior of both functions f_e and f_p in the limits $\Theta_{e1} \to \infty$ and $\Theta_{p1} \to \infty$. For such situations, it is more difficult to derive a relation for the triggering front. According to Barenblatt (1996), it is possible that the limits of f_e and f_p, if they exist, will become either constant or power-law dependent. Hummel and Shapiro (2012) numerically showed that, in the case of the exponential diffusivity model, the triggering front still have a $\sqrt{t}$-dependence. Therefore, the limit of f_e is a constant. However, for the power-law diffusivity model they observed a more general power law of the temporal dependence of the triggering front. This indicates a power-law dependence of f_p (see also equation (4.46)):

$$f_p(\Theta_{p1}) \sim \left(\frac{D_0 t p_0^n}{A_0} \right)^{b_n - 0.5} , \tag{4.49}$$

where the exponent b_n depends on the injection source and properties of the medium.

Numerical results of Hummel and Shapiro (2012) (see Figures 4.20 and 4.21) support these results of the dimensional analysis. They first computed the pressure distribution for several numerical models. In the case of non-linear diffusion ($\kappa > 0$ and $n > 0$) the hydraulic diffusivity strongly increases with pressure. As a result, the pore-pressure profiles (Figure 4.20) penetrate deeper into the medium compared to the linear diffusion. Owing to the geometrical spreading in models with higher spatial dimension, the pore-pressure profiles are characterized by a large pressure drop in the vicinity of the source. With increasing influence of the non-linear fluid–rock interaction, the shape of the pore-pressure profiles tends to change from a concave one to a more convex one. For the exponential diffusion

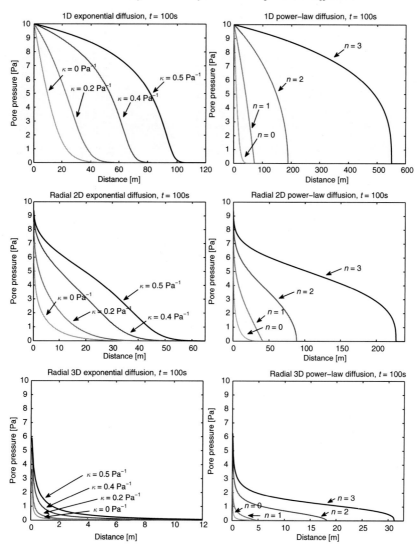

Figure 4.20 Snapshot profiles of the pore-pressure diffusion (at the moment 100 s after the injection start with $p_0 = 10$ Pa) in numerical models with exponential (left-hand column) and power-law (right-hand column) pressure dependence of the hydraulic diffusivity (equations (4.40) and (4.21), respectively). The profiles are shown for various κ and n. From the top to bottom, respectively, the profiles correspond to 1D, 2D and 3D media. (Modified from Hummel and Shapiro, 2012.)

model, the pressure profiles are characterized by a smooth transition at the pressure heads (as in the linear-diffusion case). For the power-law model, the pressure drops to zero abruptly.

Further, Hummel and Shapiro (2012) have numerically simulated microseismic clouds in their numerical models of pore-pressure diffusion. Figure 4.21 shows

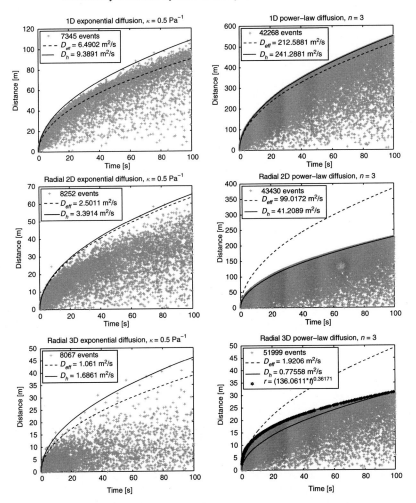

Figure 4.21 The r–t plots of synthetic microseismicity together with the results of the diffusivity analysis for the exponential (left-hand column) and power-law (right-hand column) diffusion models in 1D (top panels), 2D (middle panels) and 3D (bottom panels) media. One can see that for 1D and 2D media, the $\sqrt{t}$-dependent triggering front (solid parabola) is the envelope of the microseismic event clouds. This is also the case for the 3D diffusion in an exponential model. In the 3D power-law model the hexagram-marked curve represents a function $r \propto t^{b_n}$ describing the envelope of seismicity. (Modified from Hummel and Shapiro, 2012.)

selected r–t plots from their publication. It is seen that for the exponential diffusion model the envelope of the microseismic clouds is a $\sqrt{t}$ parabola. This behavior of the triggering front is influenced neither by the non-linearity nor by the spatial dimension. The $\sqrt{t}$-dependence agrees with the results obtained from our dimensional analysis. Figure 4.21 also shows several results of Hummel and

Shapiro (2012) for the power-law model. The results are in a good agreement with our dimensional analysis too. Indeed, envelopes of 1D and 2D synthetic micro-seismic clouds also have $\sqrt{t}$-dependence. For the 3D case the situation changes. A $\sqrt{t}$-parabola (solid curve) does not represent an envelope of the microseismic cloud. We observe a more general power-law time dependence t^{b_n} (see also equations (4.46) and (4.49)). Hummel and Shapiro (2012) obtained the following exponents: for $n = 1$ $b_n \approx 0.40$, for $n = 2$ $b_n \approx 0.37$ and for $n = 3$ $b_n \approx 0.36$. These results agree very well with equation (4.28) from the previous section. Indeed, for a constant fluid-injection rate ($i = 0$) in three dimensions ($d = 3$) this equation provides a power-law temporal behavior of the triggering front t^{b_n} with very close exponents:

$$b_n = \frac{n+1}{3n+2}.$$

(4.50)

Hummel and Shapiro (2012) attempted further to use synthetic microseismic clouds for estimating hydraulic diffusivity in corresponding models. They computed heuristic diffusivity by using equation (3.6) and compared it with approximate estimates of the effective diffusivity of the stimulated medium. To compute the effective diffusivity they used the following approach.

The fluid stimulation leads to a hydraulic diffusivity, which increases with time and is heterogeneously distributed in space. At the last moment of the injection a monotonic (non-decreasing) stimulation creates a maximum enhanced hydraulic diffusivity. An effective-diffusivity value represents such an enhanced diffusivity. The replacement of a heterogeneously distributed diffusivity by a single effective-diffusivity value is designed in such a way that the position of a triggering front in the replacing homogeneous medium would approximately coincide with its actual position (see also Section 3.6.3).

For a particular synthetic microseismic event cloud, one subdivides a distance r_{fe} from the injection point to the farthest event at a given time t_{fe} into equidistant intervals dr. For time t_{fe} and all distance samples r, one computes the corresponding pore-fluid pressure $p(r; t_{fe})$. This pore pressure is used to compute the stimulated diffusivity at the point r and time t_{fe}. Then, corresponding to equation (3.50), an estimate of the effective diffusivity can be obtained by the following integration:

$$D_{eff}(r_{fe}; t_{fe}) = r_{fe} \left[\int_0^{r_{fe}} \frac{dr}{D(r; t_{fe})} \right]^{-1}.$$

(4.51)

This estimate taken at the last moment of the injection is assumed to be the effective diffusivity of the stimulated structure. Hummel and Shapiro (2012) compared the heuristic diffusivity estimates obtained from the triggering front to the effective diffusivity values computed by using equation (4.51). They observed that the heuristic diffusivity estimates and effective diffusivity values show good correlation for indices of non-linearity $n < 2$ in one dimension and $n < 3$ in two dimensions. For larger values of n, the heuristic estimates are approximately half as much as the effective values. However, the order of magnitude of the estimates is still the same compared to the effective values. Even in the case of very strong non-linear fluid–rock interaction triggering front (3.6) still provides reasonable diffusivity estimates. This can be also seen from the parabolic curves corresponding to the effective and heuristic diffusivity estimates shown in Figure 4.21.

4.3 The model of factorized anisotropy and non-linearity

Now we can try to use the power-law non-linear diffusion model for an interpretation of the Barnett Shale microseismic data set we considered in Section 4.1.6. Shale and most other rocks associated with hydrocarbon reservoirs can be strongly elastically and hydraulically anisotropic. The anisotropy controls the distribution of microseismicity. Below we follow Hummel and Shapiro (2013) and discuss their approach to account for non-linear anisotropic pore-pressure diffusion.

We make several strong assumptions. We first assume that the principal directions of the pressure-dependent tensor of hydraulic diffusivity remain unchanged during the stimulation. We consider the pressure-dependent diffusivity tensor D_{ik} in its principal coordinate system. Then, this tensor is represented by a diagonal matrix with the principal components D_{11}, D_{22} and D_{33}. Furthermore, we assume that the pressure dependence of any of these components can be expressed by the same function $f_D(P_p)$. For example, for both models considered in the previous section this function is given by an exponential law or a power law. Thus, we assume that the tensor of diffusivity can be described by the following decomposition of its pressure-dependent and anisotropic parts:

$$\mathbf{D}(P_p) = \begin{bmatrix} D_{011} & 0 & 0 \\ 0 & D_{022} & 0 \\ 0 & 0 & D_{033} \end{bmatrix} f_D(P_p). \tag{4.52}$$

The pressure-dependent part describes an impact of the hydraulic stimulation on the permeability. In general, the function $f_D(P_p)$ can express an arbitrary functional dependence on pressure. Hummel and Shapiro (2013) assumed a power-law diffusivity: $f_D(P_p) = (n+1)P_p^n$. The matrix in (4.52) controls the directivity of the stimulated hydraulic transport. Then we substitute (4.52) into (3.1) and obtain the following form of the non-linear diffusion equation:

$$\frac{\partial P_p}{\partial t} = D_{011} \frac{\partial}{\partial x_1} f_D(P_p) \frac{\partial}{\partial x_1} P_p$$
$$+ D_{022} \frac{\partial}{\partial x_2} f_D(P_p) \frac{\partial}{\partial x_2} P_p + D_{033} \frac{\partial}{\partial x_3} f_D(P_p) \frac{\partial}{\partial x_3} P_p. \tag{4.53}$$

Now we can apply the transformation approach described in Section 3.5. We use here a slightly different normalization:

$$X_1 = x_1\sqrt{\frac{D_n}{D_{011}}}, \quad X_2 = x_2\sqrt{\frac{D_n}{D_{022}}}, \quad X_3 = x_3\sqrt{\frac{D_n}{D_{033}}}, \quad (4.54)$$

where

$$D_n = \sqrt[3]{D_{011}D_{022}D_{033}} \quad (4.55)$$

is a reference diffusivity. Note that the choice of D_n is a matter of convenience. For example, in Section 3.5 we used $D_n = D_{ii}/3$.

In this coordinate system, equation (4.53) becomes

$$\frac{\partial P_p}{\partial t} = D_n \left(\frac{\partial}{\partial X_1} f_D(P_p)\frac{\partial}{\partial X_1} + \frac{\partial}{\partial X_2} f_D(P_p)\frac{\partial}{\partial X_2} + \frac{\partial}{\partial X_3} f_D(P_p)\frac{\partial}{\partial X_3} \right) P_p.$$
$$(4.56)$$

Thus, in this coordinate system for a spherically symmetric source (e.g. a point source or a spherical cavity) we can assume radial symmetry of the solution and write this equation in spherical coordinates (see also (4.16)):

$$\frac{\partial r^2 P_p}{\partial t} = D_n \frac{\partial}{\partial r} f_D(P_p) r^2 \frac{\partial}{\partial r} P_p. \quad (4.57)$$

This model accounts for permeability variations due to the reservoir stimulation. However, it assumes that the functional type of the pressure dependence of the permeability is identical for any direction at any location. Under these conditions a cloud of induced seismicity in the normalized coordinate system X_1, X_2, X_3 will stay spherical at all times that elapse after the start of the stimulation.

The model of a factorized anisotropy and non-linearity can be applied to real data in the following way. We first should compensate for the directivity of the hydraulic transport. This can be done by re-normalization of the cloud from its originally nearly ellipsoidal shape to a spherical shape. Characteristic scales of the microseismic cloud can be estimated at the injection final moment t_0. It is convenient to rotate the event cloud into its principal coordinate system and then to estimate the lengths of its principal axes. For the particular Barnett Shale case study they are $L_x = 1121$ m, $L_y = 393$ m and $L_z = 80$ m. In accordance with (4.54) and (4.55), the scaling of the coordinates of seismic events is:

$$X_1 = x\frac{L_n}{L_x}, \quad X_2 = x_2\frac{L_n}{L_y}, \quad X_3 = x_3\frac{L_n}{L_z}, \quad (4.58)$$

where

$$L_n = \sqrt[3]{L_x L_y L_z}. \quad (4.59)$$

Hummel and Shapiro (2013) applied scaling (4.58) to the cloud shape corresponding to several times elapsed after the injection start. Figure 4.22 shows that

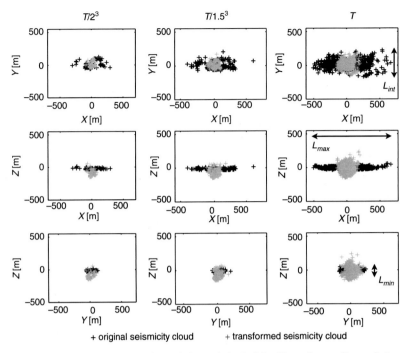

Figure 4.22 Temporal evolution of the original (black) and transformed (gray) event cloud. Vertical and horizontal scales are equal. Note that, for all three snapshots (here T denotes the complete injection duration and x, y and z are the two horizontal and one vertical dimensions of the microseismic cloud, respectively), scaling (4.58) has not changed. (Modified from Hummel and Shapiro, 2013.)

such scaling results in the spherical shape of the transformed cloud at all elapsed times. This implies equivalent functional pressure dependence of the stimulated permeability in different directions and thus supports the model of the factorized anisotropy and non-linearity, i.e. equation (4.52).

Now one can use the triggering front to analyze spatio-temporal features of the transformed event cloud. For such an analysis, Hummel and Shapiro (2013) applied the approach that we discussed in the previous section. They considered a power-law fitting function $r(t) = (a_n t)^{b_n} * 1$ m for the triggering front. They obtained the following estimates of the parameters: $a_n \approx 180$ s^{-1} and $b_n \approx 0.36$. The corresponding parabola $r(t)$ approximates satisfactory the triggering front in the scaled coordinate system. The values of the parameter b_n approximately correspond to the non-linearity index n in the range of 5–7. Then they simulated $P_p(r, t)$ in the case of a pressure source. For this they assumed a constant injection pressure equal to the time average of the one measured in the experiment (approx. 8.4 MPa; see Figure 4.23). In the numerical modeling this pressure was applied to the surface of an injection cavity with radius 0.5 m. Then they adjusted the parameter D_0 so that

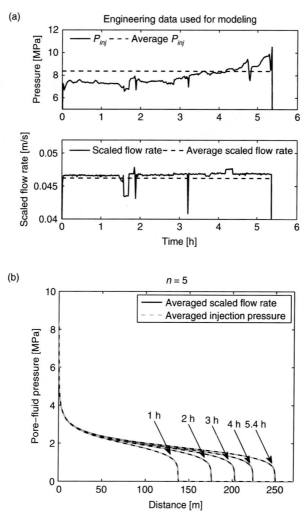

Figure 4.23 (a) Injection pressures and filtration velocities (calculated from the injection rate for the spherical source of the radius of 0.5 m) from the hydraulic fracturing of the Barnett Shale case study. (b) Numerically simulated time-dependent pressure profiles based on the averaged injection pressures and on the flow rates. Estimated values $D_0 = 3.4 \cdot 10^{-32}$ m^2/(s Pa^5) (note that this is just a normalizing constant depending on n; it should not be interpreted independently; here $n = 5$) and $S^{-1} = 8.4$ GPa correspond to the pore-pressure perturbation penetrating up to 250 m into the medium. (After Hummel and Shapiro, 2013.)

the diffusion front reaches the size of the seismicity cloud at the time $t_0 = 5.4$ h elapsed after the injection start (see Figure 4.23). The radius of the transformed cloud is approximately 250 m.

Afterwards, the boundary condition was changed to a fluid-flow source. This is the Neumann boundary condition. In other words, on the surface of the injection

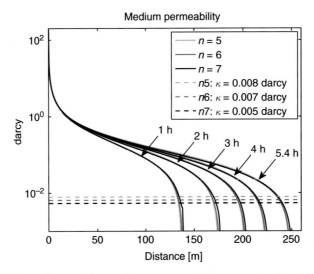

Figure 4.24 Snapshots of the medium permeability computed for the pressure profiles like those shown in Figure 4.23. The permeability of the stimulated rock (excluding the singular source vicinity) is of the order of 10^{-3}–10^{-1} darcy. In comparison, the permeability (dashed lines) based on the diffusivity estimates using the triggering front are in the order of $5 - 8 \cdot 10^{-3}$ darcy. The permeability of the virgin rock is in the range of nano-microdarcy. (After Hummel and Shapiro, 2013.)

cavity the quantity $-DS\nabla P_p$ was given. Corresponding to the Darcy law, in an isotropic medium, this quantity is equal to the filtration velocity **q**. Giving the averaged filtration velocity at the injection cavity surface approximately equal to 0.046 m/s (Figure 4.23) and adjusting now the value of the unknown storage coefficient S so that the pressure profile coincides with the one in the previous simulation, one obtains $S^{-1} = 8.4$ GPa (this value corresponds to $n = 5$; see Figure 4.23).

Now we have all necessary parameters to reconstruct the spatio-temporal evolution of the stimulated permeability. Examples of such reconstructions are shown in Figure 4.24. Here we neglected possible pressure dependences of the storage coefficient and of the density of the treatment fluid. Hummel and Shapiro (2013) show some indications of these dependencies in the case study under consideration. Nonetheless, this modeling example shows the possibility of a reasonable characterization (which is in agreement with engineering and microseismic data) of the stimulated reservoir permeability and of the storage coefficient using the non-linear diffusion approach.

5

Seismicity rate and magnitudes

Magnitudes M of fluid-induced seismicity are usually in the range $-3 < M < 2$. Nevertheless, especially for long-term injections with durations of months or years, earthquakes with larger magnitudes ($M = 4$ or even larger) have been observed (Ake et al., 2005; Majer et al., 2007). Fluid-induced earthquakes with M from 3 to 4 occurred at several Enhanced Geothermal Systems (EGS) like those of Basel, Cooper Basin, The Geysers Field and Soultz (Giardini, 2009; Majer et al., 2007; Häring et al., 2008; Dyer et al., 2008; Baisch et al., 2009). Smaller but still perceptible events can be also observed by hydraulic fracturing of hydrocarbon reservoirs.

Induced seismic hazard is a topic of significance in the shale-gas industry (National Research Council, 2013). Its understanding is of a considerable importance for mining of deep geothermic energy. It is of significance for CO_2 underground storage (see Zoback and Gorelick, 2012) and possibly also for other types of geo-technological activities (see Avouac, 2012).

Identifying parameters that control magnitudes and their statistics is a key point for evaluating the seismic hazard of fluid injections. Similarly to the tectonic seismicity, statistics of the induced seismicity can be rather well described by the Gutenberg–Richter frequency–magnitude distribution (Shapiro et al., 2007, 2010, 2011; Shapiro and Dinske, 2009a,b; Dinske and Shapiro, 2013). However, large-magnitude events deviate from it (Shapiro et al., 2011). In this chapter we analyze the influence of fluid injections on the frequency–magnitude statistics of induced events. We start with a model of point-like independent seismic events. This model describes the statistics of numerous small-magnitude earthquakes well. The model allows us to formulate a simple description of the seismicity rate and to introduce parameters quantifying the seismo-tectonic state of a fluid-injection site. One such useful parameter is the seismogenic index. This helps to predict the probability of given-magnitude events. However, the model of point-like events tends to overestimate the probability of significant magnitudes. In the second half of the chapter we consider statistics of large-magnitude events. We take the finiteness of rupture surfaces of such earthquakes into account and consider the influence of the finiteness of stimulated-rock volumes on this statistics. We also address the issue of the maximum-magnitude induced earthquake.

5.1 The model of point-like induced events

Shapiro *et al.* (2007) proposed a model that allows us to calculate the expected number of events with a magnitude larger than a given magnitude value M. This model will enable us to identify the main factors that affect the magnitude probabilities. Later we will see that this model is rather successful for describing the statistics of numerous small-magnitude events. However, it overestimates the number of large-magnitude earthquakes.

The model assumes a non-linear pore-pressure relaxation as a triggering mechanism of induced seismicity. We continue to consider a point-like pressure source in an infinite, homogeneous, permeable, porous continuum and repeat our argumentation for equation (3.74). As a result of a fluid injection and the consequent process of pressure relaxation, the pore pressure P_p will change throughout the pore space. We assume that a random set of pre-existing cracks (defects) is statistically homogeneously distributed in the medium and is characterized by the volume concentration N. For simplicity we assume that the cracks do not mutually interact. Each of these cracks is characterized by an individual critical value C of the pore pressure necessary in accordance with the Mohr–Coulomb failure criterion for the occurrence of an earthquake along such a defect. This critical pressure C is randomly distributed on a set of pre-existing cracks. If C is high we speak about a stable pre-existing fracture. If C is low we mean a fracture close to its failure. Statistical properties of C are assumed to be independent of spatial locations (i.e. $C(\mathbf{r})$ is a statistically homogeneous random field). If at a given point $\mathbf{r}$ of the medium (with a pre-existing crack there) pore pressure $P_p(t, \mathbf{r})$ increases with time, and at time t_0 it becomes equal to $C(\mathbf{r})$, then this point will be considered as a hypocenter of an earthquake occurring at this location at time t_0. For simplicity, we assume that no earthquake will be possible at this point again. This is equivalent to an assumption that phenomena leading to recharging critical cracks, such as thermal relaxation, stress corrosion, viscoelastic tectonic deformation and other effects that may be also related to the rate- and state-dependent friction, described by Dieterich (1994), Segall and Rice (1995), Segall *et al.* (2006) and mentioned in our Section 1.2.2, are much slower than the diffusional process of the pore-pressure relaxation.

Under these assumptions the probability of an earthquake occurring in the time interval from the injection start till a given time t and at a defect located at a given point $\mathbf{r}$ will be equal to $W_{ev}(C(\mathbf{r}) \leq Maximum\{p(t, \mathbf{r})\})$. This is the probability of the critical pressure being smaller than the maximum pore pressure at **r** in the time interval from the injection start till the time t. If the pore-pressure perturbation caused by the fluid injection is a non-decreasing function, which is the case for non-decreasing borehole injection pressures, then this probability will be given by equation (3.74). Rothert and Shapiro (2007) have shown that C is usually of the order of 10^3–10^6 Pa (see also Section 3.8), and $f(C)$ can be roughly approximated by a boxcar function $f(C) = 1/(C_{max} - C_{min})$ if $C_{min} \leq C \leq C_{max}$ and $f(C) = 0$ elsewhere (see Section 3.8). Usually C_{max} is of the order of several MPa. It is orders of magnitude larger than C_{min}. Thus, $f(C) \approx 1/C_{max}$. We also assume that C_{max} is larger than the pressure perturbation (excluding maybe a small vicinity of the source). Then,

$$W_{ev} = p(t, \mathbf{r})/C_{max}. \tag{5.1}$$

Therefore, the event probability is proportional to the pore-pressure perturbation. In turn, a spatial density of events is proportional to the event probability. Thus, the spatial event density must be also proportional to the pore-pressure perturbation. Such a distribution of microearthquake spatial density is indeed observed in reality (see Section 3.9).

5.1.1 Event number and event rate during a monotonic injection

It is apparent that the probability of an event with a magnitude larger than a given value is an increasing function of the total event number. Thus, first we compute the complete number of events induced from the injection start till a current time moment t. The product of the crack bulk concentration N and a spatial integral of probability (5.1) will give us the necessary quantity:

$$N_{ev}(t) = \frac{A_d N}{C_{max}} \int_{a_0}^{\infty} p(t, r) r^{d-1} dr. \tag{5.2}$$

Here the spatial integration is similar to the one of equations (4.18)–(4.19). From equation (4.20) we further obtain:

$$N_{ev}(t) = \frac{N m_c(t)}{C_{max} \rho_0 S}, \tag{5.3}$$

where $m_c(t)$ is a fluid mass injected until time t and S is the storage coefficient. This result is a consequence of the conservation of the fluid mass. Thus, it is valid in the very general case of non-linear fluid–rock interaction. We see that the number of events induced is proportional to the injected fluid mass. Note that an important validity condition for this statement is the non-decreasing pore-pressure perturbation in the whole space. If we neglect possible changes of the fluid density then this result will simplify correspondingly:

$$N_{ev}(t) = \frac{N}{C_{max}S} \int_0^t Q_i(t)dt = \frac{A_d N}{C_{max}} Q_0 t^{i+1}, \tag{5.4}$$

where in the last expression a power-law injection rate (parameterized after equation (4.20)) was assumed.

Finally, in the case of a constant injection rate $Q_i(t) = Q_I = SA_d Q_0$ we obtain

$$N_{ev}(t) = \frac{Q_I N t}{C_{max} S}. \tag{5.5}$$

This result shows that the cumulative event number is growing proportionally to the injection time. In other words, if the injection rate is constant, the events rate $N'_{ev}(t)$ will also be a constant equal to

$$N'_{ev}(t) \equiv \frac{\partial N_{ev}}{\partial t} = \frac{Q_I N}{C_{max} S}. \tag{5.6}$$

The quantity

$$F_t = \frac{C_{max}}{N} \tag{5.7}$$

depends on seismotectonic properties of the injection region only. Following Shapiro *et al.* (2007) we will address this quantity as a "tectonic potential." In our simple model, the tectonic potential is defined by a bulk concentration of pre-existing cracks (i.e. defects or, in other words, potential rupture surfaces for induced events) and the upper limit of their critical pressure C_{max}. Note that $1/C_{max}$ defines an average probability density of critical pressures. Note also that the tectonic potential has physical units of energy. Such a relatively simple expression of the tectonic potential is due to simplicity of our model. In reality, F_t can be a more complex function of stress state, tectonic history, rheology, lithology, hit flow, seismicity and possibly other parameters of the injection site.

According to (2.200), (2.220) and (2.221), by a constant injection rate and a linear pressure diffusion, we have $Q_I/S = 4\pi D p_0 a_0$, where p_0 is the injection pressure. This relation provides us with an alternative interpretation of equation (5.5):

$$N_{ev}(t) = 4\pi D p_0 a_0 t / F_t. \tag{5.8}$$

Equations (5.5) and (5.8) show that the probability of seismicity increases with the volume of the injected fluid, with the reciprocal tectonic potential, with the stiffness $1/S$ of fluid saturated rocks (for example, with decreasing porosity, increasing fluid stiffness, increasing rock consolidation, etc.) and with the diffusivity of the medium.

To compute the number of events with magnitudes larger than a given one, we must introduce some assumptions about the magnitude statistics. We assume that the probability $W_{\geq M}$ of an event having a magnitudes larger than M is independent of the number of events. This means that the ratio of a number of events with magnitudes larger than a given one to the complete number of events N_{ev} is, in average, a constant independent of the value of N_{ev}. Then, we obtain from (5.3):

$$N_{\geq M}(t) = W_{\geq M} \frac{m_c(t)}{F_t \rho_0 S}. \tag{5.9}$$

This rather general result has the following interesting consequence. In the logarithmic scale we obtain (note that here we are using the International Standard notation $\lg \equiv \log_{10}$ and $\ln \equiv \log_e$)

$$\lg N_{\geq M} = \lg W_{\geq M} + \lg m_c - \lg F_t \rho_0 S = \lg W_{\geq M} + \lg \frac{A_d Q_0}{F_t} + (i+1)\lg t, \tag{5.10}$$

where the last equality is obtained for a power-law injection rate considered in the previous chapter. This equation means that, in a logarithmic plot, curves of functions $N_{\geq M}(t)$ for different M must be mutually parallel. This is a consequence of the assumption of independence of magnitude probability of event numbers. Figures in the following section will confirm this statement. Moreover, in the case of power-law injections, on bi-logarithmic plots $N_{\geq M}$ versus t, theoretical curves $N_{\geq M}(t)$ are straight lines with proportionality coefficients $i + 1$.

To be more specific, we assume further that the induced seismicity obeys Gutenberg–Richter statistics (Gutenberg and Richter, 1954). Broad seismological experience shows that these statistics describe global as well as local regional tectonic seismicity (see, for example, Shearer, 2009; and Turcotte *et al.*, 2007). The Gutenberg–Richter statistics of an induced seismicity might be a consequence of a power-law-type size distribution of pre-existing defects (Langenbruch and Shapiro, 2014; see also our Section 5.2.4). The Gutenberg–Richter magnitude scaling means that the probability $W_{\geq M}$ of events with magnitudes larger than M is given as follows:

$$\lg W_{\geq M} = a - bM, \tag{5.11}$$

where a and b are the so-called Gutenberg–Richter a- and b-parameters (or values). They are regional seismicity constants. Note also that $a = \lg W_{\geq 0}$. Thus, from (5.9), we obtain

$$N_{\geq M}(t) = 10^{a-bM} \frac{m_c(t)}{F_t \rho_0 S} = 10^{a-bM} \frac{A_d}{F_t} Q_0 t^{i+1}. \tag{5.12}$$

This result clarifies which characteristics of rocks and of injections define magnitudes of induced earthquakes. The probability of significant events increases with an injection duration, with the strength of an injection source and with the reciprocal tectonic potential of an injection site (i.e. with the concentration of critical cracks multiplied by an average of critical-pressure probability density).

The number of events with magnitude larger than a given one is proportional to the cumulative mass of the injected fluid. This equation also shows a temporal scaling of this number, t^{i+1}. It is controlled by the scaling i of the injection rate. Note also a simple rule for the event rate $N'_{\geq M}(t) \equiv \partial N_{\geq M}(t)/\partial t$:

$$N'_{\geq M}(t) = 10^{a-bM} \frac{N m_i(t)}{C_{max} \rho_0 S} = 10^{a-bM} \frac{Q_i(t)}{S F_t} = 10^{a-bM} \frac{A_d(i+1)}{F_t} Q_0 t^i. \quad (5.13)$$

Therefore, the event rate is proportional to the mass rate of the injection or, approximately, to the volumetric injection rate. A temporal scaling of $N_{\geq M}$ is convenient to consider in a bi-logarithmic scale. For the logarithm of the event number, equation (5.12) results in the following:

$$\lg N_{\geq M}(t) = \lg \frac{m_c(t)}{S \rho_0 F_t} - bM + a = \lg \frac{A_d Q_0}{F_t} + (i+1)\lg t - bM + a. \quad (5.14)$$

Again, in a bi-logarithmic coordinate system in the case of an injection rate scaled as t^i, $N_{\geq M}$ as a function of time tends to be a straight line with a proportionality coefficient $i+1$.

5.1.2 Seismicity rate after termination of an injection

In the following sections we will compare our rather general conclusions on the seismicity rate of non-decreasing injections with real data. In this section we first briefly comment on the seismicity decay after a terminated injection.

Predicting the seismicity rate after the termination of an injection requires significantly more complete knowledge (or more assumptions) about the seismicity triggering process. Assuming that linear pore-pressure diffusion is the governing mechanism of seismicity triggering and assuming a constant injection rate into a homogeneous porous medium, Langenbruch and Shapiro (2010) formulated an analytical solution and calculated numerically the seismicity rate after an injection stop. They found that the decay rate of the induced seismicity is similar to the Omori law, which describes the decay rate of aftershock activity after tectonically driven earthquakes. For natural earthquakes it is usually claimed that the Omori exponent is close to 1. However, values in the range 0.3–2 have also been reported. Two models frequently used to explain such behavior of aftershocks are the rate- and state-dependent behavior of the friction coefficient and the stress corrosion of sub-critical cracks (Kanamori and Brodsky, 2004).

Langenbruch and Shapiro (2010) found this exponent to be larger than 1 for the case of the fluid-induced seismicity. Let t_0 denote the duration of a fluid injection and let $\tau_0 = t/t_0$ denote a normalized time after the termination of the injection ($\tau_0 \geq 1$). Langenbruch and Shapiro (2010) propose the following approximation of the seismicity rate after the termination of the injection (at least for $\tau_0 = O(1)$):

$$N'_{ev}(t) \approx N'_{ev}(t_0)\tau_0^{-p_d}, \tag{5.15}$$

where the Omori exponent p_d characterizes the decay rate, and (see also (2.220) and (2.221))

$$p_d - 2 \propto \frac{C_{min}\sqrt{Dt_0}}{C_f}. \tag{5.16}$$

In other words, if the minimum critical pressure is very low and $\tau_0 = O(1)$ then the Omori exponent for the seismicity induced under conditions described above will approach 2. Further, their analytic results show that, for $\tau_0 \gg 1$, the Omori exponent approaches 1. However, their estimations for the case studies Fenton Hill and Soultz (in the range of τ_0 of the order of 1) provided as high values of the Omori exponent as 7.5 and 9.5, respectively. This corresponds to C_{min} of the order of 5000 Pa. Thus, the decay rate of induced seismicity depends on the fracture strength. Moreover, the values show that the presence of unstable fractures (with the vanishing criticality C) results in an increase of the seismicity rate shortly after an injection stop (compare Figures 5.1 and 5.2). This also leads to a probability increase for large-magnitude events shortly after the injection termination. Later in this chapter we will observe that this effect can be additionally enhanced by the growing scale of the finite stimulated volume with the time elapsed after the start of the injection. In the case of $C_{min} = 0$, one obtains

$$N'_{ev}(t_0) \approx \nu_I \equiv \frac{4\pi DC_f}{F_t}. \tag{5.17}$$

We call this quantity the reference seismicity rate.

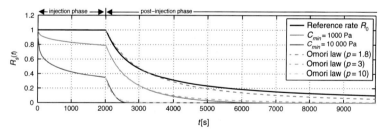

Figure 5.1 Seismicity rate in the case of stable pre-existing fractures (C_{min} is finite and significant). The rate is normalized to the reference seismicity rate ν_I defined in (5.17). Parameters of the model are $D = 1$ m^2/s, $t_0 = 2000$ s, $p_0 = 1$ MPa, $a_0 = 4$ m, $C_{max} = 1$ MPa. The solid lines correspond to $C_{min} = 0$ Pa, $C_{min} = 1000$ Pa and $C_{min} = 10\,000$ Pa (from the upper to the lower curve). The dashed lines show the modified Omori law with $p_d = 1.8$, 3.0 and 10.0. (After Langenbruch and Shapiro, 2010.) A black and white version of this figure will appear in some formats. For the color version, please refer to the plate section.

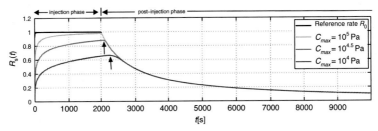

Figure 5.2 Seismicity rate in the case of unstable pre-existing fractures ($C_{min} = 0$). The rate is normalized to its maximum value, given by the reference seismicity rate ν_I. Parameters of the model are $D = 1$ m^2/s, $t_0 = 2000$ s, $p_0 = 1$ MPa, $a_0 = 4$ m. The solid lines correspond to $C_{max} = 10^6$ Pa, $C_{max} = 10^5$ Pa, $C_{max} = 10^{4.5}$ Pa and $C_{max} = 10^4$ Pa (from the upper to the lower curve). The arrows denote the time of maximum probability to induce an event with significant magnitude. (After Langenbruch and Shapiro, 2010.) A black and white version of this figure will appear in some formats. For the color version, please refer to the plate section.

Note that all these observations can change significantly in the case of a non-linear fluid–rock interaction. In contrast, features of the seismicity rate during non-decreasing injection time intervals are governed mainly by the mass conservation law and thus are more universal. In the next section we return back to seismicity rate of active injection phases.

5.1.3 Case studies of magnitude distributions

Here we discuss the behavior of $N_{\geq M}$ as a function of injection time for some borehole injections: at Ogachi (Japan), in Paradox Valley (USA), in Cooper Basin (Australia), at Basel (Switzerland), in Cotton Valley (USA) and in Barnett Shale (USA). Note that for our discussion we use only those time intervals of these experiments which approximately correspond to power-law temporal dependencies of their injection rates, t^i.

We restrict our analysis to magnitude ranges that are possibly weakly influenced by features of observation systems, of registration and of processing. The so-called completeness magnitude M_c is an important characteristic of seismicity catalogs. It is defined in the following way. It is assumed that all events with magnitudes $M \geq M_c$ are recorded by the observation system and registered in data sets.

First, we compare the number of events as a function of time as predicted by equation (5.14) with observations from data sets at injection sites Ogachi (a geothermal system) and in Paradox Valley (a salt-water disposal). Both injections correspond approximately to the assumptions made above, and induced a sufficiently large number of earthquakes (larger than 100). During an experiment at

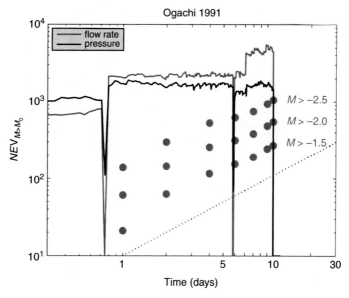

Figure 5.3 $N_{\geq M}$ as functions of injection time for the Ogachi 1991 experiment. The points are observed cumulative numbers of earthquakes with magnitudes larger than the indicated ones. The straight line has the proportionality coefficient 1, predicted by equation (5.14). The curves show the injection pressure (the lower line in the time range 1–10 days) and the injection rate (the upper line in the time range 1–10 days). (Modified from Shapiro *et al.*, 2007.) A black and white version of this figure will appear in some formats. For the color version, please refer to the plate section.

the Ogachi geothermal site in 1991, a volume of more than $10\,000$ m^3 of water was injected at a depth of 1000 m into hard rocks (granodiorite). The pressure remained relatively stable throughout the experiment (see Figure 5.3). A microseismic event cloud of about 500 m thickness and 1000 m length with nearly 1000 detected events was stimulated (Kaieda *et al.*, 1993). The magnitudes were determined by measuring velocity amplitudes and, alternatively, seismogram oscillation durations (Kaieda and Sasaki, 1998).

Magnitude statistics has been biased by the performance of the observation system and by processing in the magnitude ranges $M < -2.5$ and $M > -1.5$. Here $M_c = -2.5$ is an estimate of the completeness magnitude. However, there were systematic errors in magnitude measurements above $M = -1.5$. When the injection pressure is nearly constant, the functions $N_{\geq M}(t)$ are nearly linear in the bi-logarithmic plot. The distances between lines corresponding to different magnitudes M are regularly distributed and time independent. These two features are as predicted by equation (5.14). The steps between the lines can be used to estimate the b value of the Gutenberg–Richter law.

At Paradox Valley, the injection was carried out in several phases between 1991 and 2004 in order to reduce salinity in the Colorado River (Ake *et al.*, 2005). The brine was injected into a fractured limestone formation at a depth of more than 4 km. The injection became regular in 1996, but still showed some fluctuations and a 20-day shut down every six months since the year 2000. The microseismic event cloud extended to more than 15 km from the injection borehole. About 4000 events with a magnitude larger than -0.5 were induced. The largest event had a magnitude 4.3. The completeness magnitude M_c was close to 0.5. Though being one order of magnitude larger on the spatial scale and two orders of magnitude larger on the temporal scale, the microseismic activity in Paradox Valley shows the same features as described above for the Ogachi case study (see Figure 5.4).

The magnitude distributions observed in Figures 5.3 and 5.4 agree rather well with (5.14), which predicts a linear relation between $\lg N(M, t)$ and $\lg(t)$. Significant deviations seen in Figure 5.4 are explained by a strong irregularity of the injection pressure at the Paradox Valley site.

Our model and its main consequence, equation (5.14), also provide a convenient frame for comparison seismicity induced in different experiments. Figure 5.5 shows a common plot of magnitude frequencies as functions of time for the both regions and three injection experiments. We can clearly see that the most important

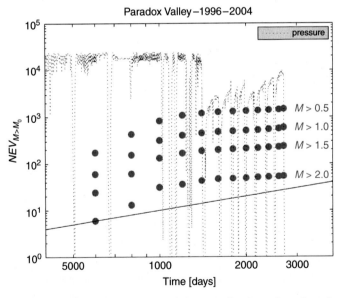

Figure 5.4 Pressure (irregular black line) and distribution of earthquake magnitudes for the Paradox Valley brine injection experiment. The straight line shows the theoretical proportionality coefficient 1 given by (5.14) for a constant-rate continuous-in-time fluid injection. (Modified from Shapiro *et al.*, 2007.)

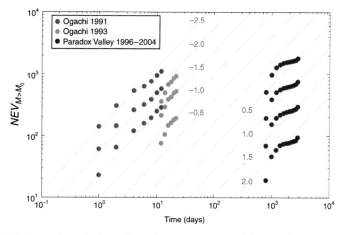

Figure 5.5 A combined plot of numbers of events with magnitudes larger then given ones as functions of injection durations at Ogachi and at Paradox Valley. Thin dashed lines correspond to equation (5.14) with $i = 0$. (Modified from Shapiro *et al.*, 2007.) A black and white version of this figure will appear in some formats. For the color version, please refer to the plate section.

factor influencing the appearance of the seismicity on this plot is the sensitivity of observational systems. For example, the surface-based registration system of Paradox Valley is not able to detect small-magnitude events. However, the influence of this factor can be taken into account by extrapolation and construction of additional lines corresponding to magnitudes of interest. The predictive potential of such a plot can be seen. For example, it is clear that injection durations at the Ogachi site are at least two orders of magnitude shorter than injection durations necessary to induce events of $M = 2$ and larger. It is also seen that the experiment of Ogachi 1991 has approximately the same tectonic potential as the Paradox Valley injection. Figure 5.5 also shows the data from another injection experiment at Ogachi performed in 1993. The magnitude distributions exhibit a similar behavior as for the experiment in 1991. However, one can see that the later experiment is characterized by a slightly larger tectonic potential. This is possibly due to criticality releasing by the first injection experiment, which led to an effective decrease of the concentration of critical cracks (one can also consider this as an increase in C_{max}). Thus, the model proposed here can help to optimize a design of injection experiments. For example, it can help to decide which parameters should be changed (e.g. the injection rate or duration) in order to reduce the seismogenic risk of the stimulation (see also the next section).

Further, we compare the data related to EGS projects in Australia at the Cooper Basin and in Switzerland at Basel. Cooper Basin is located in South Australia. We consider here the microseismicity induced in the borehole Habanero-1 (Soma *et al.*, 2004), which was drilled down to a depth of 4421 m. In its last 754 m it penetrated granite. The open-hole section was below the depth of 4115 m. However, the main outflow zone was at a natural fracture at the depth of 4254 m.

Basel is located in the Upper Rhine Graben (Dyer *et al.*, 2008). The borehole Basel 1 was drilled till the depth of approximately 4800 m and penetrated more than 2 km of granite in its lower part. The open-hole section corresponds approximately to the last 400 m of the borehole.

Figures 4.16 and 4.17 show *r–t* plots of microseismic events induced in the Cooper Basin and Basel experiments. Solid parabolic lines provide linear diffusion approximations of microseismic cloud expansions (4.3) with hydraulic diffusivities of the order of 0.4 m^2/s and 0.06 m^2/s, respectively. These approximations are quite rough. However, in both case studies the estimates of hydraulic diffusivities provided by linear diffusion approximations are still very realistic and correspond well to hydraulic properties of naturally fractured granite.

We see (Figure 5.6) that the Cooper Basin injection period can be characterized by a roughly constant flow rate. In contrast, the injection at Basel is characterized by an increasing fluid flow. In the first case we can approximate the injection rate exponent *i* by 0. In the second case, we can roughly approximate *i* by 1. This corresponds to a smoothing of the real stepwise-increasing flow rate. Therefore, corresponding to equation (5.14), we expect an approximately linear dependence between $\lg N_{\geq M}$ and $\lg t$ in both cases. However, the proportionality coefficient of this dependence is equal to $i + 1$. Thus, in the case of the Cooper Basin injection this linear dependence should be characterized by the proportionality coefficient 1. In contrast, the Basel injection should be characterized by the proportionality

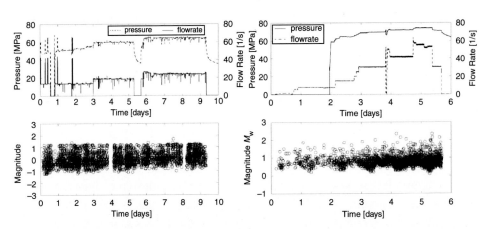

Figure 5.6 Water injection rate and pressure as functions of injection time as well as observed microearthquake magnitudes in crystalline rocks of Cooper Basin (left; the data are courtesy of H. Kaieda, CRIEPI) and of Basel (right; the data are courtesy of U. Schanz and M. O. Häring, Geothermal Explorers). (After Shapiro and Dinske, 2009b.)

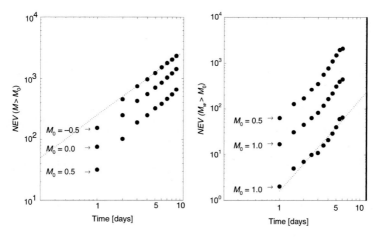

Figure 5.7 Left: numbers of events with magnitudes larger than given ones as functions of injection durations in the Cooper Basin. The thin dotted line has proportionality coefficient 1. Right: numbers of events with magnitudes larger than given ones as functions of injection durations in Basel. The thin dotted line has proportionality coefficient 2. Note that only slopes of the dotted lines (rather then their exact locations) are essential here. (After Shapiro and Dinske, 2009b.)

coefficient 2 of the $\lg N_{\geq M}$ versus $\lg t$ dependency. Figure 5.7 shows that this is approximately the case.

Two other case studies are hydraulic fracturing injections into tight gas reservoirs of Cotton Valley sandstones and of Barnett Shale. Their r–t diagrams (see equations (4.5) and (4.13), respectively) correspond to the growth of a single, millimeters-thick dominant hydraulic fracture in the case of the Cotton Valley reservoir stimulation (Figure 4.18), and to the growth of a 3D fracturing domain (with additional compliant porosity of the order of 0.01%) in the case of the Barnett Shale example (Figure 4.15). Figure 5.8 shows magnitude distributions in time for both case studies. In both cases the injection rate is approximately constant. Therefore, a linear dependence between $\lg N_{\geq M}$ and the $\lg t$ in both cases should be characterized by the proportionality coefficient 1. This is in agreement with Figure 5.9.

The fluid–rock interaction at the geothermal boreholes we considered above was, to first approximation, linear (i.e. only small injection-induced alterations of hydraulic diffusivity on the reservoir scale). The fluid–rock interaction in the case of hydraulic fracturing of tight gas reservoirs was definitely strongly non-linear (i.e. due to hydraulic fracturing the hydraulic diffusivity was strongly – by several orders of magnitude – increased; this corresponds to a large index of non-linearity n and, consequently, is strongly non-linear equation (4.22)).

In spite of this principal difference between examples describing stimulations of the geothermal and tight gas reservoirs we observe a very similar character

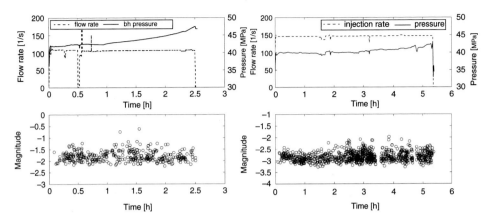

Figure 5.8 Water injection rate and pressure as functions of injection time and observed microearthquake magnitudes in Cotton Valley tight sands (left; the data are courtesy of J. Rutledge, LANL) and Barnett Shale (right; the data are courtesy of S. Maxwell, Pinnacle Technology). (Modified from Shapiro and Dinske, 2009b.)

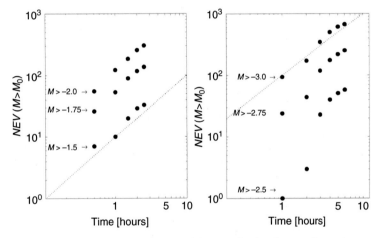

Figure 5.9 Left: numbers of events with magnitudes larger than given ones as functions of injection durations in the Cotton Valley tight sands. The thin dotted line has proportionality coefficient 1. Right: numbers of events with magnitudes larger than given ones as functions of injection durations in Barnett Shale. The thin dotted lines have the proportionality coefficient 1, correspondingly to equation (5.14) with $i = 0$. Note that only slopes of the dotted lines (rather then their exact locations) are essential here. (Modified from Shapiro and Dinske, 2009b.)

of magnitude distributions in time. This is in good agreement with the theoretical model developed above. Indeed, our equations (5.3) and (5.9) show that the cumulative event number $N_{ev}(t)$ induced till time t, as well as the number of events with the magnitude larger than a given one, $N_{\geq M}$, are independent of the index of non-linearity, n. Therefore, the magnitude distribution in time is independent of the non-linearity grade of the fluid–rock interaction. These event numbers as functions of time are only controlled by the cumulative mass of the injected fluid. The larger this cumulative mass, the larger is the number of events induced at a given location. The larger the number of such events, the higher is the probability of large-magnitude events. Equally larger will be the total seismic moment of events induced. This is also in good agreement with the observation of McGarr (1976) that the total seismic moment of induced earthquakes increases with the volume perturbations in rocks. In addition to this observation, our model shows that the magnitude distribution in time is a function of the spatio-temporal redistribution of the pore pressure in rocks caused by a fluid-mass injection. Such a pore-pressure redistribution also continues after the termination of the injection. Thus, the model above is able to predict magnitude distributions after the injection stops. However, such a prediction requires more-specific knowledge about the pressure-relaxation laws. The mass-conservation principle alone is not sufficient for such a task. In addition, all four considered case studies support our hypothesis on the independence of the magnitude probability density of event numbers. It is seen from the fact that, for different magnitudes M, the lines of $\lg N_{\geq M}$ as functions of time are nearly mutually parallel.

A significant difference in the geothermal and tight-gas case studies is observed in the b-values characterizing the microseismicity. The b-values are equal to the differences between the lines of $\lg N_{\geq M}$ divided by the differences of corresponding magnitudes. The approximate values for the Cooper Basin, Basel, Cotton Valley and Barnett Shale examples are 0.75, 1.5, 2.5 and 2.5, respectively. We observe approximately the following tendency: the larger the non-linearity, the larger the b-value. Therefore, hydraulic fracturing is characterized by a much stronger dominance of small earthquakes in the common number of induced seismicity than geothermal stimulations. Later in this chapter we will see that this is likely to be the geometric effect of finiteness of the stimulated rock volume.

5.1.4 Seismogenic index

In the previous section we found that the rate of induced seismicity is controlled by the rate of the injected fluid mass. Our observations supported the theoretical prediction that by monotonic-rate (non-decreasing-rate) injections the number of events with magnitudes larger than a given one increases with time as the injected fluid mass does. This rule seems to work independently of the non-linearity grade of the triggering process. Here we show that, by using the seismicity rate of such events and the fluid-injection rate, a parameter (seismogenic index) can be derived that quantifies the seismotectonic state of the injection site. This index is independent of injection parameters and depends only on tectonic features of the injection location. It can be used to quantitatively compare tectonic situations at injection sites in terms of a potential risk to induce an event of a significant magnitude. Along with injection parameters the seismogenic index permits us to estimate the occurrence probability of a given number of such events during a given time period. We will also estimate this index for several injection experiments.

We rewrite equation (5.14) in the following form (we neglect possible variations of the fluid density and take the first part of the right-hand side of the equation):

$$\lg N_{\geq M} - \lg Q_c(t) + bM = a - \lg(F_t S), \tag{5.18}$$

where Q_c is the cumulative injected volume and M is an arbitrary event magnitude. The quantity on the left-hand side of this equation is experimentally measurable. It depends on the injection parameters and on the induced seismicity. We are concentrating now on the quantity Σ defined by the right-hand side of this equation:

$$\Sigma \equiv a - \lg(F_t S) = a + \lg \frac{N}{C_{max} S}. \tag{5.19}$$

This quantity is independent of the injection time. Neither does it depend on any other injection characteristics. It is completely defined by the seismotectonic features of a given location. We will call it the seismogenic index. The larger this index, the larger a probability of significant magnitudes. If we know the Σ- and b-values, we will be able to predict the temporal distribution of magnitudes of injection-induced events. In the case of a monotonic injection, for an arbitrary injected cumulative volume $Q_c(t)$ the expected number of events with magnitudes larger than M is

$$\lg N_{\geq M} = \Sigma + \lg Q_c(t) - bM. \tag{5.20}$$

It is difficult to theoretically calculate Σ because of some unknown parameters (e.g. N). However, Σ can be estimated by using equation (5.18) and parameters of the seismicity induced by an injection experiment at a given location (see also Shapiro *et al.*, 2007, 2010, 2011; Shapiro and Dinske, 2009a,b; Dinske and Shapiro, 2013):

$$\Sigma = \lg N_{\geq M} - \lg Q_c(t) + bM. \tag{5.21}$$

Equation (5.20) can be further rewritten in a more conventional form of the Gutenberg–Richter law but with the a-value being time-dependent:

$$\lg N_{\geq M}(t) \approx a_t(t) - bM, \tag{5.22}$$

with

$$a_t(t) = \lg Q_c(t) + \Sigma. \tag{5.23}$$

Thus, for fluid injections, the *a*-value becomes a function of the time that has elapsed since the start of the injection. It is given by a sum of the seismogenic index and the time-dependent cumulative volume of the injected fluid.

Finally, one more useful reformulation of equation (5.20) can be proposed. Let us introduce a specific magnitude M_Σ defined as follows:

$$M_\Sigma = \frac{\Sigma}{b}. \tag{5.24}$$

Note that, as with the seismogenic index, the specific magnitude is completely defined by seismotectonic features of the injection site. The larger the specific magnitude, the larger the probability of significant induced events. Using the specific magnitude, the Gutenberg–Richter law modified for fluid injections (5.20) can be written in the following form:

$$\lg N_{M_\Sigma} \approx \lg Q_c(t). \tag{5.25}$$

Therefore, the number of events with $M \geq M_\Sigma$ is given just by the cumulative volume of the injected fluid. Recently, Dinske and Shapiro (2013) observed this type of behavior in real data. Note that both Σ and M_Σ depend on the choice of the metric system for the injected volume. We work in the System International and measure Q_c in m^3. Furthermore, we address moment magnitudes (see equation (1.135)) in this book. One usually observes Σ in the range from -8 to 1, and M_Σ in the range from -4 to 1.

The statistical model of induced seismicity we summarized above can be equally applied for non-monotonic injections. In such a case the probability (3.74) is given by a minimum monotonic majorant of the pore pressure (see Parotidis *et al.*, 2004) and numerical computations are then required to predict $\lg N_{\geq M}(t)$ for each particular situation.

The seismogenic index is a convenient quantity for a quantitative comparison of seismotectonic activity at different locations. Here we compute the seismogenic index at several borehole injection locations. Some of them are geothermal locations. Another part represents hydraulic fracturing of hydrocarbon reservoirs. For our discussion we use time periods of the injections approximately corresponding to non-decreasing fluid rates. We try to restrict our analysis to magnitude ranges that are possibly weakly influenced by features of observation systems, of registration and of processing. Further, we attempt to use moment magnitudes. Where moment magnitude are not given, our estimates of the seismogenic index will be biased by the order of difference between the local magnitudes and the moment magnitudes in corresponding magnitude ranges. At all locations shown here this bias will be insignificant for our conclusions (see also comments below).

We consider familiar case studies. These are four geothermic locations: Cooper Basin, Basel, Ogachi and Soultz. We also include non-geothermal sites: KTB (Jost *et al.*, 1998; Shapiro *et al.*, 2006b), Paradox Valley (Ake *et al.*, 2005), Cotton Valley (Rutledge and Phillips, 2003) and Barnett Shale case study (Maxwell *et al.*, 2006). The last two locations correspond to hydraulic fracturing of gas reservoirs. The German Continental Drilling Program Site (KTB) was a place of several borehole fluid-injection experiments, the results of two of them corresponding to different depths and conditions we show here.

Dinske and Shapiro (2013) estimated the seismogenic index Σ for all locations mentioned above by using (5.21). The values of the seismogenic index are shown in Figure 5.10. They demonstrate a reasonable stability (i.e. time independence). They are stable regardless of the injection duration or the cumulative injected fluid volume. Statistical errors and temporal fluctuations of the obtained estimates are restricted in the interval of ± 0.5. This is also the order of the bias

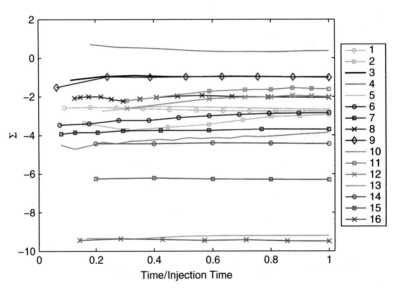

Figure 5.10 Seismogenic index computed for different locations of Enhanced Geothermal Systems, hydraulic fracturing in hydrocarbon reservoirs, and other injection locations (injection times are given in parentheses). 1: Ogachi 1991 (11 days), 2: Ogachi 1993 (16 days), 3: Cooper Basin 2003 (9 days), 4: Basel 2006 (5.5 days), 5: Paradox Valley (2500 days), 6-9: Soultz 1996 (48 hr), 1995 (11 days), 1993 (16 days) and 2000 (6 days), 10: KTB 2004/05 (194 days), 11–12: KTB 1994 (9 hr) [upper and lower bound, calculated for two *b*-values], 13: Barnett Shale (6 hr), 14–16: Cotton Valley Stages A (2.5 h), B (2.5 h) and C (3.5 h) (Modified from Dinske and Shapiro, 2013.) A black and white version of this figure will appear in some formats. For the color version, please refer to the plate section.

caused by differences between local and moment magnitudes. This fact is indirectly confirmed by comparison of different magnitude studies for the same locations; see, for example, publications of Mukuhira *et al.* (2008) and Deichmann and Giardini (2009) on the Basel experiment.

We observe the following general tendency. The seismogenic index of hydraulic fracturing sites is lower than the index of geothermal sites. In our examples of hydrocarbon reservoirs it is lower than −4. It is larger than −3 for the geothermal sites. Significantly higher Σ has been estimated for the Cooper Basin data set. It is of the order of −1.2 to −0.9. The largest index is obtained for the Basel data set: 0.1 to 0.7. The seismogenic index and *b*-value derived from seismicity in the sandstone and shale gas reservoirs in Cotton Valley and Barnett differ significantly from corresponding quantities in geothermal reservoirs. The gas reservoirs are characterized by a low seismogenic index, $-10 < \Sigma < -4$, and a high *b*-value. The rather large diversity of Σ at Cotton Valley has the following reasons. In Stage B (also called stage 2 in the previous chapter), the propagating hydraulic fracture intersected and consequently opened a natural fracture system as shown in Figure 4.6 (Dinske *et al.*, 2010) that may have resulted in the highest value of seismogenic index for the Cotton Valley reservoir. The fracturing stages were not only performed in different boreholes but also with different treatment fluids: a gel–proppant mixture was injected in Stage A (also called stage 3 in the previous chapter) and B, whereas water was used in Stage C. Note that, for the fracturing in Barnett shale, water was also injected. In both situations the value of Σ is of the same order.

For reservoir locations where multiple fluid injections had been carried out, such as in Soultz or at the KTB, we observe several different indices. This is due to the fact that the fluid was injected either in different wells and/or at different depths. For example, at the KTB site (see Section 3.1) the 1994 injection was performed at the depth of about 9 km using the main borehole to stimulate the SE1 fault zone, whereas in 2004/2005 the fluid was injected at the depth of ∼4 km using the pilot borehole. In this latter experiment seismicity occurred along the less-prominent fault zone SE2. Additionally, there was a long-term fluid-extraction phase (one year) preceding the last injection. Also in Soultz the various injections have stimulated different parts of the geothermal reservoir. A possible reason for the differences in the derived seismogenic indices are different properties of the fracture systems. Also multiple injections from the same borehole source can influence estimates of the index. As we have mentioned above, this can happen due to reducing the bulk concentration of critically stressed defects (or equally, due to elevating C_{max}) by preceding injections. In general, geothermal reservoirs and the KTB as well as the Paradox Valley reservoirs are not only characterized by a higher seismogenic index but also by a lower *b*-value (if compared to hydrocarbon case

studies). These observations lead to the conclusion that fluid injections in geothermal reservoirs have a higher potential to induce an earthquake having a significant
magnitude. This is in agreement with the observed earthquake magnitude ranges
(Majer *et al.*, 2007).

Dinske and Shapiro (2013) analyzed possible correlations between the seismogenic indices and other parameters of injections, reservoirs and seismic hazard
estimates. In spite of the low data statistics due to the limited number of catalogs, they observed some interesting general tendencies. The seismogenic index
tends to be higher the greater the depth of a reservoir. They also observed a significant positive correlation between the seismogenic index and the peak ground
acceleration, which is the seismic hazard indicator for naturally occurring seismicity. Dinske and Shapiro (2013) found a correlative linear relationship between the
seismogenic index and the seismic hazard indicator specified as the peak ground
acceleration with 10% probability of exceedance in 50 years, corresponding to a
return period of 475 years (Giardini *et al.*, 2003).

Summarizing, the seismogenic index helps us to characterize the level of seismic activity one should expect by fluid injections into rocks. The index Σ could
be estimated from a seismicity induced by a preliminary small-scale and short-
term injection test at sites where long-term and large-scale rock stimulations are
planned. Correlations between the seismogenic index and other parameters, such
as seismic hazard indicators, for example, can be also used to roughly estimate Σ.

5.1.5 Occurrence probability of events with given magnitudes

A higher seismogenic index leads to a higher probability of significant events. Let
us first assume a constant injection rate Q_I. Then, according to equations (5.5) and
(5.17), the cumulative event number at elapsed time t will be

$$N_{ev}(t) = \frac{N}{C_{max}S}Q_I t = v_I t, \qquad (5.26)$$

i.e. the cumulative event number is just proportional to the event occurrence time,
where v_I is a constant temporal event rate. Our model and assumptions described
above imply that the events occur independently from each other. In other words,
in our model the induced seismicity is assumed to be a Poisson process (like, for
example, a radioactive decay).

The Poissonian nature of seismicity processes can be tested by analyzing statistic
properties of waiting times between two successive events (the inter-event time).
Langenbruch *et al.* (2011) performed such an analysis. Here we briefly summarize their main results. In a Poisson process with a constant expected event rate
v_I (the corresponding process is called a homogeneous Poisson process, HPP)

the probability $W_n(\nu_I, t)$ of having n events in the time interval $[0, t]$ is given by (Hudson, 1964):

$$W_n(\nu_I, t) = \frac{(\nu_I t)^n}{n!} \exp(-\nu_I t). \qquad (5.27)$$

The probability that, in a time interval Δt after a given event, no events will occur is:

$$W_0(\nu_I, \Delta t) = \exp(-\nu_I \Delta t). \qquad (5.28)$$

This probability can be interpreted as the probability $W_{>\Delta t}$ that the waiting time is longer than Δt. The probability $W_{<\Delta t}$ that the waiting time is shorter than Δt is then:

$$W_{<\Delta t} = 1 - \exp(-\nu_I \Delta t). \qquad (5.29)$$

Differentiating this expression in respect to Δt provides us with the probability density function of this quantity, $PDF_{\Delta t}$:

$$PDF_{\Delta t} = \nu_I \exp(-\nu_I \Delta t). \qquad (5.30)$$

Probability (5.29) can be further rewritten in the following form:

$$W_{<\Delta \tau} = 1 - \exp(-\Delta \tau), \qquad (5.31)$$

where $\Delta \tau \equiv \nu_I \Delta t$ is the normalized waiting time. Therefore, we obtain the following relation for the PDF of $\Delta \tau$:

$$PDF_{\Delta \tau} = \exp(-\Delta \tau). \qquad (5.32)$$

Langenbruch *et al.* (2011) analyzed the inter-event time distributions of injection-induced earthquakes for different catalogs collected at geothermal injection sites at Soultz and Basel. These catalogs are rather complete and contain large numbers of events. They found that the distributions of waiting times during phases of constant seismicity rate coincides quite well with the exponential distribution of the homogeneous Poisson process (Figure 5.11).

Furthermore, Langenbruch *et al.* (2011) analyzed the waiting times for the complete event catalogs and found that induced earthquakes are distributed according to a non-homogeneous Poisson process in time. Moreover, they replaced the time scale by the injected-volume one. They observed that in the fluid-volume scale the seismicity process again becomes a homogeneous Poisson process. These results strongly indicate that corresponding seismicity is directly induced by the fluid injection.

Sequence of events with magnitude larger than M is also a Poisson process. The corresponding event rate, ν_M is equal to

$$\nu_M = \nu_I W_{\geq M}. \qquad (5.33)$$

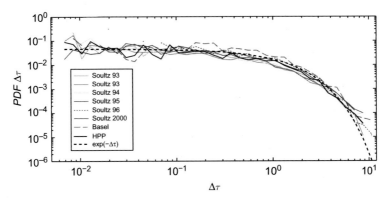

Figure 5.11 Estimated probability density functions of the normalized inter-event time for stationary periods of injections at Soultz and Basel. (After Langenbruch *et al.*, 2011.) A black and white version of this figure will appear in some formats. For the color version, please refer to the plate section.

Using the Poissonian distribution we obtain the occurrence probability $W_{n,M}(t)$ of n events of magnitude larger than M in the time interval $(0, t)$:

$$W_{n,M}(t) = \frac{(\nu_M t)^n}{n!} \exp(-\nu_M t). \qquad (5.34)$$

Let us now return to an arbitrary non-decreasing injection rate. Equation (5.10) suggests that the induced seismicity is still a Poisson process but in respect to the cumulative injected fluid mass rather than to the time. Moreover, it is a Poissonian process in respect to the expected cumulative event number $N_{\geq M}$:

$$W_{n,M}(t) = \frac{\left(N_{\geq M}(t)\right)^n}{n!} \exp\left(-N_{\geq M}(t)\right). \qquad (5.35)$$

In the case of a constant injection rate the process reduces to a standard homogeneous Poissonian process in time. Its distribution is given by equation (5.34).

Of practical importance is the probability of event absence. It is given by:

$$W_{0,M}(t) = \exp\left(-N_{\geq M}(t)\right) = \exp\left(-Q_c(t) 10^{\Sigma - bM}\right), \qquad (5.36)$$

where in the last part of the equation we have substituted the Gutenberg–Richter magnitude distribution.

For an illustration we consider an injection experiment with parameters close to the Basel case study. We assume that the seismogenic index Σ is equal to 0.25. We also assume that, during 5.5 days of the stimulation, 11 570 m³ of fluid have been injected into the rock. We assume further that the b-value is approximately 1.5. Equation (5.36) helps to answer the following question. What is the probability that during this injection events with magnitude larger than $M = 2.5$ will occur? This probability is given by substituting the injection parameters into (5.36), computing the value of $W_{0,M}(t)$, computing the value of $1 - W_{0,M}(t)$ and then multiplying the result by 100%. We obtain approximately 97%. The probability that events with magnitude larger than 3.5 will occur is approximately 11% (see also Figure 5.12).

Let us assume that we want to exclude occurrence of events with magnitude larger than M with a probability $W_{0,M}(t)$ (e.g. 90%, i.e. $W_{0,M}(t) = 0.9$). Then (5.36) provides us with an estimate of a tolerable maximal cumulative injected volume:

$$Q_c(t) = -10^{bM-\Sigma} \ln W_{0,M}(t). \qquad (5.37)$$

Let us assume that we want to exclude the occurrence of events with magnitude larger than M with a very high probability. This means that $W_{0,M}(t)$ is very close to 1. In other words, $\delta W_{0,M}(t) = 1 - W_{0,M}(t)$ is much smaller than 1. In this case the previous equation can be expanded in a Taylor series and we approximately obtain:

$$\delta W_{0,M}(t) = Q_c(t)10^{(\Sigma-bM)}. \qquad (5.38)$$

This equation provides the possibility of estimating the injection time, which is allowed to exclude the occurrence of events with magnitude larger than M with a

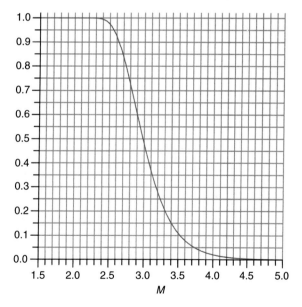

Figure 5.12 The probability that events with magnitude larger than a given one would occur during a fluid injection. The parameters of the injection and of the seismotectonic state are given in the text. They are close to those of the Basel geothermal-stimulation experiment. (After Shapiro *et al.*, 2010.)

given probability $1 - \delta W_{0,M}$ (for instance, $\delta W_{0,M} = 0.001$ means the absence of non-desired events with a probability of 99.9%). Note that the condition of validity of equation (5.38) is

$$Q_c(t)10^{(\Sigma-bM)} \ll 1. \qquad (5.39)$$

We recall that volumes are measured in m^3 in all equations of this book.

If we assume a constant injection rate Q_I, then $Q_c(t) = Q_I t$, and a requirement to exclude events with magnitude larger than a given one will reduce to the restriction for the injection duration:

$$t < \frac{\delta W_{0,M}}{Q_I}10^{(bM-\Sigma)}. \qquad (5.40)$$

Therefore, the seismogenic index and the fluid injection rate are two key parameters controlling the Poisson statistics of induced seismicity.

5.2 Statistics of large magnitudes

Fluid-induced seismicity results from stimulations of finite rock volumes. The finiteness of stimulated volumes influences frequency–magnitude statistics. We observe that fluid-induced large-magnitude events are frequently under-represented in comparison with the Gutenberg–Richter law. This is an indication that the events are more probable on rupture surfaces contained within the stimulated volume. Following Shapiro *et al.* (2013) we will introduce different possible scenarios of event triggering: rupture surfaces located completely within or intersecting only the stimulated volume. These scenarios correspond to the lower and upper bounds of the probability to induce a given-magnitude event. We will show that the bounds depend strongly on the minimum principal axis of the stimulated volume. We will further compare the bounds with the seismicity induced in several case studies. The observed frequency–magnitude curves mainly follow the lower bound. Fitting the lower bound to the frequency–magnitude distribution provides estimates of the largest expected induced magnitude and the characteristic stress drop, in addition to improved estimates of the Gutenberg–Richter a- and b-parameters.

5.2.1 Observations

From the previous sections we have seen that, during an active fluid injection with non-decreasing injection pressure, the Gutenberg–Richter law modified for injections (5.20) describes well the number $N_{\geq M}$ of induced earthquakes with magnitudes larger than M as a function of the injection time t.

Usually, large events have poor statistics (their number is usually less than 10), which is not representative for a single-injection experiment. However, one can identify systematic behavior by considering several injection experiments. Shapiro *et al.* (2011) observed systematic deviations of large-magnitude statistics from equation (5.20). The number of such events is significantly smaller than predicted, especially for short injection-time intervals.

Let us first consider the geothermal fluid-injection experiment at Basel (Häring *et al.*, 2008; Dyer *et al.*, 2008; Deichmann and Giardini, 2009). Figure 5.13 shows observed and theoretical curves of $N_{\geq M}(t)$. The simultaneous curve of the injected-water volume $Q_c(t)$ is shown. Immediately after the injection termination, the curve $Q_c(t)$ starts to decrease because of the outflow of the injected water. On a bilogarithmic scale, theoretical curves $N_{\geq M}(t)$ corresponding to (5.20) are given by a time-independent shifting of the curve of the injected-fluid volume downwards. Note that (5.20) is applicable for injection periods only.

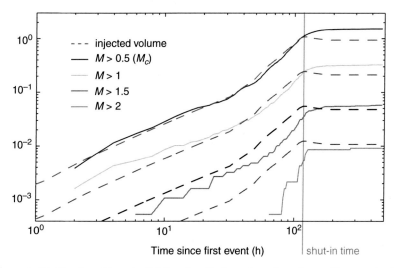

Figure 5.13 Number $N_{\geq M}$ (solid lines) of induced earthquakes with magnitudes M larger than indicated values as functions of the time t elapsed from the time of the first event in the catalog (nearly the injection start) at the Basel borehole (Häring *et al.*, 2008). The plot also shows the injected water volume $Q_c(t)$ (the upper dashed line). The quantities $Q_c(t)$ and $N_{\geq 0.5}(t)$ (the upper solid line; 0.5 is approximately a completeness magnitude) are normalized to their values at the moment of the maximum injection pressure (several hours before the injection termination). Immediately after the injection termination the curve $Q_c(t)$ starts to decrease because of an outflow of the injected water. Theoretical curves $N_{\geq M}(t)$ corresponding to (5.20) are given by lower dashed lines. They are constructed by a time-independent shifting of the curve of the injected-fluid volume. The lower solid lines shows the observed quantities $N_{\geq M}(t)$ normalized by the same value as the quantity $N_{\geq 0.5}(t)$. (Modified from Shapiro *et al.*, 2011.) A black and white version of this figure will appear in some formats. For the color version, please refer to the plate section.

In a range of small magnitudes, the theoretical and observed curves are in good agreement. However, the numbers of large-magnitude events are lower than theoretically predicted ones. This is seen from the fact that curves of $N_{\geq M}(t)$ for large magnitudes are no longer parallel to such curves for small magnitude ranges. The large-magnitude curves deviate significantly downwards (especially for short times elapsed after the injection start; note that corresponding theoretical predictions give curves that are just parallel to the $Q_c(t)$ curve in the time period of the injection).

We observe the same tendency at other locations. Figure 5.14 shows examples of distributions $N_{\geq M}$ as functions of the injected-fluid volume for several case studies. In all examples, an approximate agreement of equation (5.20) with numbers for small-magnitude events has been demonstrated in the previous section. This

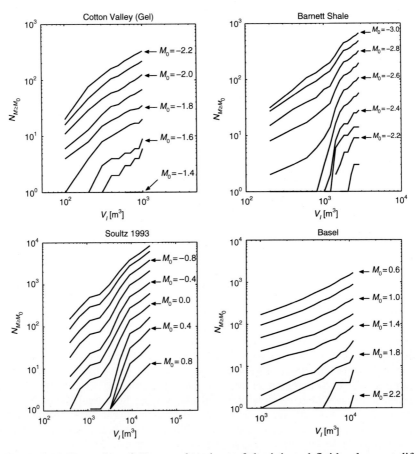

Figure 5.14 Examples of $N_{\geq M}$ as functions of the injected fluid volume at different sites. All curves correspond to the injection periods only. (After Shapiro *et al.*, 2013.)

is also seen in Figures 5.14. Indeed, according to equation (5.20) in a range of not-too-large magnitudes, lines of $\lg N_{\geq M}(t)$ are nearly mutually parallel. Theoretically, they all should be parallel to the function $\lg Q_c(t)$. These lines are nearly regularly spaced. Theoretically they should be equally spaced with an increment given by the product of the b-value with the increment of the magnitude. However, the number of large events is systematically smaller than a regular spacing of the lines $\lg N_{\geq M}(t)$ would imply.

The fact that induced small-magnitude events obey the Gutenberg–Richter statistics but large-magnitude events deviate from it indicates that the probability of an earthquake on the corresponding rupture surface depends on the geometric relation between this surface and the stimulated volume. Indeed, large-magnitude events correspond to large ruptures. Such events are less common than small-magnitude events. The statistics of rupture surfaces in rocks must correspond to the classical Gutenberg–Richter statistics of earthquakes, which is a common observation. However, potential rupture surfaces of induced events (we consider a potential rupture surface as the equivalent of a pre-existing crack) must intersect injection-stimulated volumes. For a given rupture surface the probability to intersect a stimulated volume will be higher the larger this rupture surface is. Thus, if we consider the proportion of large potential rupture surfaces to small ones, we will observe that it is significantly higher for the surfaces intersecting the stimulated volume than for rupture surfaces in rocks in general. This suggests that large-magnitude events should be over-represented in comparison to the Gutenberg–Richter statistics for small-magnitude events. However, on the contrary we observe an under-representation of large-magnitude events.

Thus, we will consider the two following extreme scenarios. In the first scenario (for reasons that will become evident below, we call it the upper-bound scenario, or just u-scenario) we assume that, to induce an event with a given rupture surface, it is enough to stimulate an arbitrary small spot of this surface. In induced seismicity, the portion of large-magnitude events in respect to the portion of small-magnitude events should be higher than the proportion of large-scale rupture surfaces in relation to small ones in rocks generally. Therefore, large-magnitude events should be over-represented in comparison to the expectations based on the Gutenberg–Richter statistics. This scenario seems to contradict the observations.

In another scenario (we call it the lower-bound scenario, or just l-scenario) we assume that, to induce an earthquake on a rupture surface, this surface must be stimulated completely. This is in agreement with the following formulation of the Mohr–Coulomb failure criterion: to enable an earthquake along a given interface, an interface-integrated tangential stress must overcome the total friction force. As soon as the largest part of a potential rupture surface remains unperturbed, the

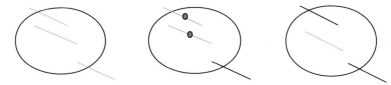

Figure 5.15 A sketch of three different scenarios of inducing earthquakes on given rupture surfaces. The ovals symbolize the stimulated volume. The light-gray straight segments denote rupture surfaces of induced seismic events. The black straight segments denote potential rupture surfaces, on which no earthquakes were induced. The small circles on two light-gray segments in the middle sketch denote rupture centers located within the stimulated volume.

probability of an earthquake remains low. Therefore, to enable an earthquake, a significant part of the corresponding rupture surface (we assume: the complete surface) should belong to the stimulated volume. This scenario seems to be more adequate than the previous one.

Figure 5.15 schematically shows both the scenarios we discussed above. In addition, one more triggering situation is shown. The left-hand sketch corresponds to the first considered scenario. Any intersection of a rupture surface with the stimulated volume suffices for inducing the earthquake along this surface. The right-hand sketch corresponds to the last considered scenario. An earthquake will occur only if its rupture surface is located completely within the stimulated volume (it is the only light-gray segment shown on the sketch). The figure shows one more situation. Earthquakes are induced on those rupture surfaces, whose centers (denoted in the figure by the small circles on the corresponding segments) are located within the stimulated volume. Note that the geometrical center of a rupture surface can be replaced here by any characteristic point inside of the surface. Such a point can be considered as a nucleation point of the earthquake. We can clearly see that the number of the induced events decreases from the left to the right (from three to one light-gray segment, respectively).

To quantify the scenarios shown in Figure 5.15 and to compare them with the frequency-magnitude statistics of induced events we must modify the probability $W_{\geq M}$ from Section 5.1. The modification of this probability must take into account the effect of the finiteness of the stimulated volume and of rupture surfaces. Note that our scenarios are not directly related to the stimulation physics of potential rupture surfaces. They just describe different possible statistical patterns of the phenomenon of induced seismicity. Therefore, they are applicable to the seismicity induced by elastic stress- or pore-pressure perturbations as well as to the seismicity induced by other processes like rate- and state-dependent friction alterations. On the other hand, a clear preference of the seismogenic process to follow one of these scenarios can provide us with useful seismotectonic information.

5.2.2 Statistics of earthquakes with finite rupture surface

Let us assume that a finite volume V of rocks (the stimulated volume) has been somehow altered so that seismogenic conditions in it have been changed sufficiently to produce seismicity. In practice we assume that the stimulated volume is approximately defined by an outer envelope of a cloud of hypocenters of induced seismicity (see Figure 5.16). We consider the following simplified abstract model. A stimulated volume is an ellipsoid or a cuboid that can grow with time (for example, due to a fluid injection). Potential rupture surfaces (we identify them with pre-existing cracks or faults) are randomly (or preferentially) oriented planar circular discs (penny-shaped inclusions with vanishing thickness). The spatial distribution of centers of the discs is random and statistically homogeneous with the bulk concentration N.

We introduce first the probability

$$W_f(X) = f_X(X)dX, \tag{5.41}$$

which is the probability of a given potential rupture surface in the unlimited medium having diameter X. Note that $f_X(X)$ is a PDF of a rupture surface of the size X. For example, later, we will assume a power-law $f_X(X)$ and show that it

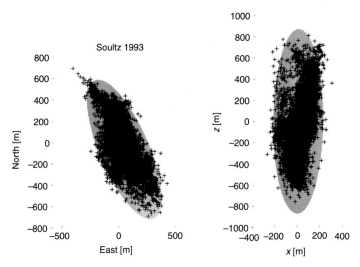

Figure 5.16 Two projections of the microseismic cloud of the case study Soultz (1993) and of an approximating ellipsoid of the aspect ratio 43:33:10 (we use such an ellipsoid for numerical simulations of geothermal stimulated volumes). The left-hand-side projection is a map view. It corresponds approximately to the plane of the intermediate and minimum principal axes (the L_{int}, L_{min}-plane). The right-hand-side projection corresponds approximately to the plane of the maximum and minimum principal axes (the L_{max}, L_{min}-plane). The minimum principal axis of the ellipsoid is nearly parallel to the x-axis. (After Shapiro et al., 2011.)

corresponds to the Gutenberg–Richter statistics of earthquakes. Further, let $W_c(X)$ denote a probability that the center of the rupture surface (of diameter X) belongs to the stimulated volume under the condition that this surface intersects the volume. Let us also for an instant assume that all ruptures have the size X. It is clear then that the product of $W_c(X)$ with the number of all rupture surfaces intersecting the stimulated volume gives the number of rupture surfaces having their centers in the volume. On the other hand, this number is equal to the product NV. Thus, the ratio $NV/W_c(X)$ is the number of the all rupture surfaces intersecting the volume. Recalling now that the rupture surfaces are statistically distributed over their size we conclude that the ratio $W_f(X)/W_c(X)$ gives the probability of a rupture surface intersecting the stimulated volume having diameter X.

Now we can formulate the probability $W_E(X)$ of an induced seismic event to have a rupture surface of diameter X:

$$W_E(X) = W_s(X)W_f(X)/W_c(X) = G_w(X)W_f(X), \qquad (5.42)$$

where in the second part on the right-hand side we have introduced one more notation: $G_w(X) \equiv W_s(X)/W_c(X)$. W_s denotes the probability of the corresponding rupture surface being sufficiently stimulated to produce the event. This probability is a conditional one. It implies that the rupture surface has something to do with the stimulated volume. At least, the rupture surface intersects the volume (intersecting includes also touching). The presence of $W_c(X)$ in equation (5.42) shows that there are many more rupture surfaces intersecting the stimulated volume than just the product NV. Note that the left-hand sketch of Figure 5.15 shows in light-gray all the rupture surfaces intersecting the stimulated volume. The middle sketch shows in light gray the rupture surfaces corresponding exactly to the product NV. In other words their number is equal to the product of the bulk concentration of rupture surfaces (which is the bulk concentration of the centers of the potential ruptures in rocks) with the stimulated volume V.

The quantity $G_w(X) = W_s(X)/W_c(X)$ describes the influence of the geometry of the stimulated volume. G_w takes also into account rupture surfaces intersecting the stimulated volume. Depending on W_s, it can be larger or smaller than 1. Corresponding to the two scenarios considered in the previous section, this will lead to $W_E > W_f$ or $W_E < W_f$, respectively. Let us now concentrate on its part, the probability $W_s(X)$.

Let us consider all potential rupture surfaces intersecting or located within the stimulated volume. We recall that $W_c(X)$ is the probability of such a surface of diameter X having its center within the stimulated volume. Let us also define a probability of a seismic event along this whole rupture surface, $W_{e1}(X)$, under the condition that the rupture center is located within the stimulated volume but the rupture surface is not entirely contained in the volume. Respectively, $W_{e2}(X)$ will denote the event probability for such a rupture surface under the condition that its center is outside of the stimulated volume. We further compute the probability $W_s(X)$ of a seismic event on a rupture of diameter X. Such an event occurs on a rupture located completely within the stimulated volume. Its probability is given by the product $W_{vol}(X)W_c(X)$, where $W_{vol}(X)$ is the probability for a rupture of size X completely belonging to the stimulated volume under the condition that its center is located within the volume. Such an event can also occur on a rupture with the center inside the volume but intersecting the volume only. The corresponding probability is given by $(1 - W_{vol}(X))W_c(X)W_{e1}(X)$. Finally, such an event can occur on a rupture intersecting the stimulated volume and having its center outside the volume. The corresponding probability is given by $(1 - W_c(X))W_{e2}(X)$. Therefore, the probability of a seismic event along a rupture of diameter X is given by the sum:

$$W_s(X) = W_{vol}(X)W_c(X) + (1 - W_{vol}(X))W_c(X)W_{e1}(X)$$
$$+ (1 - W_c(X))W_{e2}(X). \tag{5.43}$$

In a general case considering any potential rupture surface intersecting the stimulated volume (i.e. no pre-conditions for locations of rupture centers), the lower bound W_{sl} of the probability $W_s(X)$ will be given by $W_{e1}(X) = W_{e2}(X) = 0$:

$$W_{sl}(X) = W_c(X)W_{vol}(X). \tag{5.44}$$

This equation corresponds to the l-scenario (see the right-hand part of Figure 5.15). For the upper bound of $W_s(X)$, $W_{su}(X)$, several alternatives can be considered. The largest and simplest one is $W_{e1}(X) = W_{e2}(X) = 1$, and thus,

$$W_{su}(X) = 1. \tag{5.45}$$

This corresponds to the u-scenario discussed in the previous section (see the left-hand part of Figure 5.15), where stimulation of an arbitrary small spot of a potential rupture surface is enough for a corresponding seismic event. Equation (5.45) means an over-representation of the large-magnitude events with respect to the standard Gutenberg–Richter distribution. It can be seen from equation (5.42), where $W_c(X)$ becomes especially small for large X.

The next simple assumption would be $W_{e1}(X) = 1$ and $W_{e2}(X) = 0$. This assumption means that, for triggering an event, the center of its potential rupture surface must be within the stimulated volume (see the middle part of Figure 5.15). Such a restriction is a reasonable formalization of the intuitive requirement that a "significant part" or a "nucleation spot" of the rupture surface must be within the stimulated volume (note also a topological equivalence between a disc center and any other "nucleation center" placed inside the rupture). It leads to the following estimate:

$$W_{su0}(X) = W_c(X). \tag{5.46}$$

According to equation (5.42) this would mean that the statistics of induced events should be given by $W_f(X)$, i.e. given by the standard Gutenberg–Richter distribution.

Three bounds (5.44)–(5.46) represent three different scenarios of the development of induced seismicity. Their comparison with real data will clearly show which of the scenarios is preferable for the induced seismogenesis.

The relation between the different estimates of W_s is

$$W_{sl} < W_{su0} < W_{su}. \tag{5.47}$$

Corresponding to the bounds of the probability W_s, we obtain the bounds for the quantity $G_w = W_s/W_c$:

$$G_{wl} < G_{wu0} < G_{wu}, \tag{5.48}$$

where

$$G_{wl}(X) = W_{vol}(X), \tag{5.49}$$

$$G_{wu0}(X) = 1, \tag{5.50}$$

and, finally,

$$G_{wu}(X) = 1/W_c(X). \tag{5.51}$$

In the following section we discuss the probabilities $W_c(X)$ and $W_{vol}(X)$ associated with geometry of the stimulated volume.

5.2.3 Rupture-surface probability and geometry of stimulated volumes

Shapiro *et al.* (2011) investigated the probability $W_{vol}(X)$. This is the probability that a disc of diameter X is completely contained within a given stimulated volume under the condition that its center belongs to the volume. They found an exact expression of the probability $W_{vol}(X)$ for a spherical stimulated volume of the diameter L (see Appendix 1 in Section 5.3 of this chapter). In this case $W_{vol}(X) = W_{sp}(X/L)$, where the function $W_{sp}(X/L)$ is defined as follows:

$$W_{sp}\left(\frac{X}{L}\right) \equiv \left(1 + \frac{1}{2}\left(\frac{X}{L}\right)^2\right)\left(1 - \left(\frac{X}{L}\right)^2\right)^{1/2} - \frac{3\pi}{4}\frac{X}{L} + \frac{3}{2}\frac{X}{L}\arcsin\left(\frac{X}{L}\right). \tag{5.52}$$

The subscript "*sp*" indicates a spherical stimulated volume. This function is quickly decreasing with increasing X/L. Therefore, $W_{vol}(X)$ can strongly influence the statistics of induced earthquakes as soon as the size X approaches or exceeds a

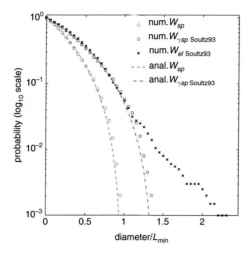

Figure 5.17 Probability W_{vol} for a spherical and an ellipsoidal (geometry similar to the case study Soultz 1993, see Figure 5.16) stimulated volumes shown as functions of disc diameters normalized to minimum principal axes of the volumes. The lines are given for spherical volumes (i.e. the exact functions W_{sp}). The lower line was computed for a spherical volume with a diameter equal to L_{min}. The upper line was computed for a spherical volume with diameter γ (see equation (5.53)). The corresponding numerical results are given as circles and squares, respectively. The crosses denote results of numerical modeling for the ellipsoid. (After Shapiro *et al.*, 2011.)

characteristic scale of the stimulated volume. We denote such a characteristic scale as Y. It is clear that, in the case of a spherical stimulated volume, $Y = L$.

Shapiro *et al.* (2011) have numerically computed $W_{vol}(X)$ for ellipsoidal volumes with principal axes $L_{min} < L_{int} < L_{max}$. For numerical computations they approximate a disc by a regular polygon with 16 sides (a hexadecagon). The centers of the polygons are distributed in a given stimulated volume with a given bulk concentration. These polygons have random orientations. For a given X they compute the number of polygons with all vertices located within the stimulated volume. This serves as an approximate criterion that the discs are located completely within the volume. Then, they normalize the result by the number of all centers. This gives an approximation of $W_{vol}(X)$ (see Figure 5.17).

Shapiro *et al.* (2011) found that

$$\gamma = \left[\frac{1}{3} \left(1/L_{min}^3 + 1/L_{int}^3 + 1/L_{max}^3 \right) \right]^{-1/3} \qquad (5.53)$$

usually provides a reasonably good estimate of a characteristic scale Y such that $W_{vol}(X)$ becomes very small ($W_{vol}(X) \ll 0.1$) for $X > Y$. If L_{min} is sufficiently small, then it will provide a dominant contribution to γ. Thus, $Y = O(\gamma)$.

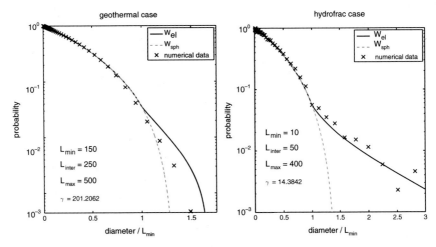

Figure 5.18 A comparison of numerically computed (crosses) and theoretically estimated (lines) probabilities W_{vol} as functions of disc diameters normalized to the minimum principal axes of the volume. To the left: for an ellipsoidal geothermal-type stimulated volume. The dashed line represents the approximation $W_{sp}(X/\gamma)$ (see the text below equation (5.53)). The solid line shows the result of equation (5.87). To the right: the same, but for a hydraulic-fracture like ellipsoid. The parameters of the ellipsoids and resulting size γ are given on the plots. (After Shapiro *et al.*, 2013.)

The function $W_{sp}(X/\gamma)$ will give a good approximation if the axes of the stimulated ellipsoid are close to each other. Frequently (especially in the case of hydraulic fracturing) one of the axes is extremely small: $L_{min} \ll L_{int} < L_{max}$. In this case, further corrections are required to approximate $W_{vol}(X)$ (Shapiro *et al.*, 2013). Equation (5.87) from Appendix 2 (Section 5.4) defines a function $W_{el}(X)$ providing such an approximation (see also Figure 5.18).

Shapiro *et al.* (2013) have also considered the situation with rupture inclined by an angle $\pm\phi$ to the plane of the maximum and intermediate tectonic stresses. From Chapter 1 we know that this angle is defined by the friction coefficient and is usually close to $\pm 30°$ (see Section 1.2). Thus, the rupture surfaces have a larger angle to the minimum stress axis. To simplify the consideration, Shapiro *et al.* (2013) assumed that the all potential rupture surfaces have these tilts. They approximated the stimulated volume by a cuboid with sides L_{min}, L_{int} and L_{max} and obtained:

$$W_{vol}(X) = W_{cub}(X) \equiv \left(1 - \frac{X}{L_{min}}|\sin\phi|\right)\left(1 - \frac{X}{L_{max}}|\cos\phi|\right)\left(1 - \frac{X}{L_{int}}\right).$$
$$(5.54)$$

For example, if $\phi = 0$, all potential rupture surfaces will belong to the same plane. Such geometry seems to be less relevant for the seismicity induced by fluid stimulations of rocks. However, it is more adequate for aftershocks of earthquakes in subduction zones.

The probability $W_{vol}(X)$ is important for the lower bound of the probability $W_E(X)$ of a seismic event with a rupture surface of the size X (see equations (5.42), (5.44) and (5.49)). Another important probability, which can influence the statistics of induced earthquakes is $W_c(X)$ (the upper bound of the probability W_E especially can be affected; see equations (5.46) and (5.51)) This is the probability of a rupture surface of the size X having its center within the stimulated volume, given that this rupture surface intersects the volume. Shapiro *et al.* (2013) investigated this probability $W_c(X)$. Here we summarize their results.

Let us first consider chaotically oriented rupture surfaces and a spherical stimulated volume of diameter L. The sought-after probability $W_c(X)$ (Figure 5.19) is given by the ratio of the total number of rupture surfaces with the centers within the stimulated volume to the total number of all rupture surfaces having any intersections with (or completely located within) this volume (see Appendix 3 in Section 5.5):

$$W_c(X) = W_{csp}\left(\frac{X}{L}\right) \equiv \left(1 + \frac{3}{2}\frac{X^2}{L^2} + \frac{3\pi}{4}\frac{X}{L}\right)^{-1}. \qquad (5.55)$$

Figure 5.19 shows this result along with results of a numerical modeling.

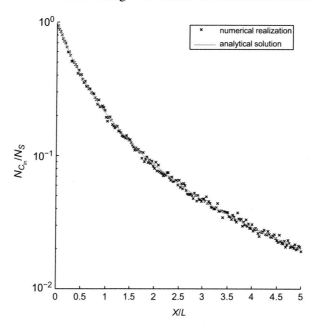

Figure 5.19 Probability $W_c(X)$ of a rupture surface of the diameter X having its center within a spherical stimulated volume of the diameter L: analytical and numerical results. (After Shapiro *et al.*, 2013.)

Shapiro *et al.* (2013) have numerically investigated this probability for ellipsoidal volumes with principal axes $L_{min} < L_{int} < L_{max}$. Numerical results show that substituting into equation (5.55) instead of L the following quantity

$$\gamma_c = \left[\frac{1}{3} \left(1/L_{min}^{3/2} + 1/L_{int}^{3/2} + 1/L_{max}^{3/2} \right) \right]^{-2/3} \qquad (5.56)$$

usually provides a good estimate of W_c. Note, again, that if L_{min} is sufficiently small, then it will provide a dominant contribution to γ_c.

For rupture surfaces tilted by an angle $\pm\phi$ to a plane of the maximum and intermediate tectonic stresses, Shapiro *et al.* (2013) approximate the stimulated volume by a cuboid with sides L_{min}, L_{int} and L_{max}, rather than by an ellipsoid. They proposed the following approximation for W_c:

$$W_{cc}(X) \equiv \left[\left(1 + \frac{X|\sin\phi|}{L_{min}} \right) \left(1 + \frac{X|\cos\phi|}{L_{max}} + \frac{X}{L_{int}} \right) + \frac{\pi X^2 |\cos\phi|}{4L_{max}L_{min}} \right]^{-1} .$$
$$(5.57)$$

Also here, if $\phi = 0$ then all potential rupture surfaces belong to the same plane. Equation (5.57) then provides probability W_c for a stimulated area of a rectangular form. Such a geometry may be relevant for aftershocks of tectonic earthquakes.

5.2.4 Distributions of magnitudes and the Gutenberg–Richter law

We consider again the probability of an induced seismic event to have a rupture surface of diameter X, $W_E(X)$. This probability is formulated by equation (5.42). Substituting equation (5.41) into (5.42) gives

$$W_E(X) = W_s(X)W_f(X)/W_c(X) = G_w(X)f_X(X)dX. \qquad (5.58)$$

Therefore, the quantity $G_w(X)f_X(X)$ is the probability density function of an induced event with rupture surface of the size X. Now we must relate the probability density of magnitudes to the probability density $G_w(X)f_X(X)$. Below, we will closely follow the work of Shapiro *et al.* (2013).

The spatial scale of the rupture surface controls the magnitude of a corresponding earthquake. A relationship between the rupture size X and the earthquake magnitude M can be found by combining equations (1.133) and (1.135). This yields the moment magnitude (Lay and Wallace, 1995; Shearer, 2009; Kanamori and Brodsky, 2004)

$$M = 2\lg X + \left[\lg \Delta\sigma - \lg C_0 \right]/1.5 - 6.07 \qquad (5.59)$$

for seismic moments measured in Nm. In the last part of the equation we conventionally assume that the slip displacement scales as a characteristic length X of the slipping surface (this is a result of the linear elastic theory of the fracture mechanics; see also equation 9.26 and table 9.1 from Lay and Wallace, 1995, and our equation (1.98)). The quantity $\Delta\sigma$ is usually defined as a static stress drop, and C_0 is a geometric constant of the order of 1. We will use a shorter form of equation (5.59):

$$M = 2\lg(X/C_\sigma), \tag{5.60}$$

where we have introduced a convenient notation C_σ for the cubic root of the reciprocal stress drop: $C_\sigma = 1084 C_0^{1/3}/\Delta\sigma^{1/3} \approx 10^3 \Delta\sigma^{-1/3}$.

In what follows we are interested in the statistic of magnitudes. Thus, we have to consider M, X and C_σ as random variables. Equation (5.60) defines the magnitude M as a function of two random variables, X and C_σ. It can be also written in the following form:

$$X = C_\sigma 10^{M/2}. \tag{5.61}$$

This equation defines the rupture length X as a function of two random variables, M and C_σ.

There are two transformation equations relating the pair of random variables $(M; C_\sigma)$ to another pair $(X; C_\sigma)$. The first relation, $X(M, C_\sigma)$, is given by equation (5.61). The second relation is given by the trivial statement: $C_\sigma = C_\sigma$. These two relations define a coordinate transformation from the coordinate system $(M; C_\sigma)$ to the system $(X; C_\sigma)$. The Jacobian of this transformation is equal to $\partial X(M, C_\sigma)/\partial M$. This Jacobian and the transformation equation (5.61) yield the PDF of magnitudes, f_M:

$$f_M(M) = 1.151 f_X(C_\sigma 10^{M/2}) f_C(C_\sigma) C_\sigma 10^{M/2}, \tag{5.62}$$

where we accepted $\ln 10^{1/2} \approx 1.151$ and assumed that the random variables X and C_σ are statistically independent. Furthermore, we introduced the following notations: $f_X(X)$ is a PDF of the rupture length and $f_C(C_\sigma)$ is a PDF of C_σ. Thus, a probability $W_{\geq M}$ of events with the magnitude larger than an arbitrary M is equal to:

$$W_{\geq M} = 1.151 \int_0^\infty \int_M^\infty f_X(C_\sigma 10^{M/2}) f_C(C_\sigma) C_\sigma 10^{M/2} dM dC_\sigma. \tag{5.63}$$

Let us first assume that the following factorization is possible:

$$f_X(C_\sigma 10^{M/2}) = f_1(C_\sigma) f_2(10^{M/2}), \tag{5.64}$$

where f_1 and f_2 are two independent functions. This will be the case if $f_X(X)$ is a power-law function. Then

$$W_{\geq M} = A_C \int_M^\infty f_2(10^{M/2}) 10^{M/2} dM \tag{5.65}$$

with the proportionality coefficient

$$A_C = 1.151 \int_0^\infty f_1(C_\sigma) f_C(C_\sigma) C_\sigma dC_\sigma. \tag{5.66}$$

Thus, under the factorizing assumption for f_X, the randomness of the stress drop influences the distribution of magnitudes by modifying its proportionality factor (5.66) only.

Let us assume further a power-law PDF of a size of potential rupture surfaces in an unlimited medium: $f_X(X) \approx A_X X^{-\beta}$ (here $\beta > 0$ and A_X is a proportionality constant). Note that such a PDF cannot be exactly valid because of a possible integration singularity at $X = 0$. We assume that a power-law function is a good approximation of a real PDF of potential rupture surfaces above a certain very small size (which corresponds to a magnitude significantly smaller than M under considerations). Thus, a PDF $f_X(X)$ strongly decreases with X. Power-law size distributions are typical for natural fractal-like sets (Scholz, 2002; Shapiro and Fayzullin, 1992; Shapiro, 1992). This type of self-similarity has been already related to the Gutenberg–Richter frequency–magnitude distribution of earthquakes (Shearer, 2009; Turcotte *et al.*, 2007; Kanamori and Brodsky, 2004).

Indeed, a power-law PDF $f_X(X)$ with $\beta = 2b + 1$ along with equations (5.60) and (5.64) gives the following functions f_1 and f_2:

$$f_1(C_\sigma) = A_X C_\sigma^{-2b-1}, \tag{5.67}$$

and

$$f_2(10^{M/2}) = 10^{-bM} 10^{-M/2}. \tag{5.68}$$

Then equation (5.65) provides the Gutenberg–Richter law (we have previously introduced it in the form of equation (5.11)):

$$W_{\geq M} = 10^{a-bM}, \tag{5.69}$$

where

$$a = \lg \left[\frac{A_X}{2b} \int_0^\infty f_C(C_\sigma) C_\sigma^{-2b} dC_\sigma \right] \tag{5.70}$$

and

$$b = (\beta - 1)/2. \tag{5.71}$$

Therefore, a power-law size distribution of rupture surfaces leads to the Gutenberg–Richter magnitude distribution (5.68) in a rather general case of an arbitrary statistically distributed stress drop. In the following we will include in our model a possibility of events of different size by assuming that the Gutenberg–Richter magnitude distribution is the result of a power-law size distribution of all rupture surfaces spanned by earthquake events. This assumption is supported by the recent study of Langenbruch and Shapiro (2014). They showed that Gutenberg–Richter relation results from the power-law size distribution of spatial fluctuations of the Mohr–Coulomb failure stress caused by the fractal nature of elastic heterogeneity of rocks.

5.2.5 *Lower and upper bounds for magnitude distributions*

In order to account for the finiteness of the stimulated volume we must include the influence of the geometry into our considerations. Consequently, we must include the quantity G_w (see equation (5.42)) as a factor into the PDF of magnitudes. Taking into account (5.58) we see that we must replace $f_X(X)$ by $G_w(X)f_X(X)$ in equation (5.63). Taking into account equation (5.64) and results (5.67)–(5.71) we obtain:

$$W_{\geq M} = 1.151 A_X \int_0^\infty \int_M^\infty 10^{-bM} G_w(C_\sigma 10^{M/2}) C_\sigma^{-2b} f_C(C_\sigma) dM dC_\sigma. \quad (5.72)$$

Note that the quantity G_w is usually a function of a ratio of the rupture scale X and the characteristic scale Y of the stimulated volume. For example, in equations (5.52)–(5.57) $Y = L, \gamma, L_{min}$, respectively. Thus, the explicit dependence of G_w on C_σ can be eliminated by introducing a characteristic magnitude M_Y so that

$$Y = C_\sigma 10^{M_Y/2}. \quad (5.73)$$

Using this quantity, G_w can be expressed (at least, approximately) in a modified form $G_{wm}(M - M_Y)$, which is directly obtained from G_w by the corresponding substitution of the argument. For example, the most important for practical situations is the lower bound of $G_w(X)$. This bound is given by

$$G_{wl}(X) = W_{vol}(X) \approx W_{sp}(X/\gamma). \quad (5.74)$$

In turn,

$$W_{sp}(X/\gamma) = W_{sp}\left(10^{(M-M_Y)/2}\right) = G_{wm}(M - M_Y), \quad (5.75)$$

where $Y = \gamma$.

Therefore, substituting the function $G_w(C_\sigma 10^{M/2})$ by the function $G_{wm}(M - M_Y)$ in equation (5.72) we obtain:

$$W_{\geq M} = 10^{a_d} \int_M^\infty 10^{-bM} G_{wm}(M - M_Y) dM, \qquad (5.76)$$

with the proportionality coefficient

$$10^{a_d} = 1.151 A_X \int_0^\infty C_\sigma^{-2b} f_C(C_\sigma) dC_\sigma = 2.303 b 10^a. \qquad (5.77)$$

Generally, the magnitude M_Y is an unknown quantity effectively representing the range of induced magnitudes. It is defined by the condition of equivalence of equations (5.72) and (5.76). If $C_\sigma^{-2b} f_C(C_\sigma)$ tends to a narrow, δ-function-like distribution around a representative value C_σ then, in accordance with equation (5.61), M_Y will be directly given by (5.73). In reality C_σ is restricted to a limited range between approximately 1 and 1000. Fitting equation (5.76) to a real frequency–magnitude distribution of an induced seismicity yields estimates not only of the a- and b-values but also of the magnitude M_Y. By using equation (5.73) one can estimate the representative value of C_σ and compute the corresponding estimate of the stress drop. Note that, in the case of lower-bound magnitude probability (i.e. $G_w = G_{wl}$), owing to the vanishing probability W_{vol} for $X > Y$, the magnitude M_Y is a limiting value of a largest possible magnitude of an induced earthquake. From (5.59) and (5.73) we obtain:

$$M_Y = \lg Y^2 + (\lg \Delta\sigma - \lg C_0)/1.5 - 6.07. \qquad (5.78)$$

Equation (5.76) can be also represented in the following form:

$$W_{\geq M} = 10^{a - bM + \Psi(M - M_Y)}, \qquad (5.79)$$

where

$$\Psi(M - M_Y) = \lg \left[2.303 b \int_0^\infty 10^{-bm'} G_{wm}(m' + M - M_Y) dm' \right] \qquad (5.80)$$

is a function correcting the magnitude distribution for the finiteness of the stimulated volume (and m' is an integration variable). This function can be roughly estimated in the following way. The exponential function under the integral is a quicker decreasing function than $G_{wm}(m + M - M_Y)$. Thus, by integration we can roughly assume that the last function is a constant equal to $G_{wm}(M - M_Y)$ and we obtain $\Psi(M - M_Y) \approx \lg G_{wm}(M - M_Y)$. It clearly shows that, if M is significantly smaller than M_Y (so that $M - M_Y \ll -1$), the magnitude distribution will be indistinguishable from the classical Gutenberg–Richter one (because $G_{wm} \to 1$). By $M \to M_Y$ the magnitude distribution will quickly drop down in the case $G_{wm} \rightarrow 0$. This is the case for the lower bound of the function $G_w = G_{wl}$, which is given by equation (5.49).

We use (5.79) to further modify equation (5.20):

$$\lg N_{\geq M}(t) = \lg Q_c(t) + \Sigma + \Psi(M - M_Y) - bM. \qquad (5.81)$$

Figure 5.20 shows theoretical cumulative frequency–magnitude curves (i.e. the quantities $\lg N_{\geq M}$ as functions of M) for a given time that has elapsed since the start of the injection. In the figure, the elapsed time has been involved implicitly only. It defines corresponding geometrical sizes (L_{min}, L_{int}, L_{max}) reached by the growing cloud of the seismicity. It also defines a value of the Gutenberg–Richter quantity $a(t) = \lg Q_c(t) + \Sigma$. For the particular examples shown in Figure 5.20, $a = 4.5$ (e.g. $\Sigma = -0.5$ and $Q_c = 10^5$ m^3) and $b = 1.5$. Shapiro *et al.* (2013) numerically computed different functions $\Psi(M - M_Y)$ by using equation (5.80). The functions $G_{wm}(m + M - M_Y)$ were obtained by using substitution (5.61) into the three functions $G_w(X)$ given by equations (5.49)–(5.51), respectively. To compute the lower bound of the quantity $\lg N_{\geq M}$ for the case of an ellipsoidal stimulated volume, $G_w(X)$ was substituted by the approximating function $W_{el}(X)$ defined by equation (5.87). To compute the lower bound of the quantity $\lg N_{\geq M}$ for the case of a cuboidal stimulated volume, $G_w(X)$ was substituted by the function $W_{cub}(X)$ defined by (5.54). To compute the uppermost bound of $\lg N_{\geq M}$ for the

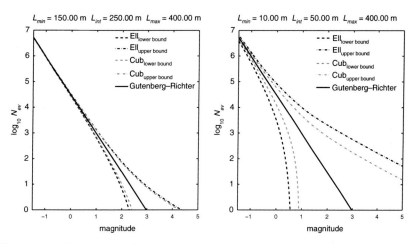

Figure 5.20 Theoretical frequency–magnitude curves. From the lower to the upper curves: the lower bound for the case of an ellipsoidal stimulated volume; the lower bound for the case of a cuboidal stimulated volume; the Gutenberg–Richter distribution; the uppermost bound for the case of a cuboidal stimulated volume; the uppermost bound for the case of an ellipsoidal stimulated volume. On the left: a geothermal-type of the stimulated volume. On the right: a hydraulic-fracturing type of the stimulated volume. The scales of the stimulated volumes shown on Figure 5.18. (After Shapiro *et al.*, 2013.)

case of a cuboidal stimulated volume, the function $G_w(X)$ was substituted by $1/W_{cc}(X)$, which is reciprocal to the one given by equation (5.57). Finally, to compute the uppermost bound of $\lg N_{\geq M}$ for the case of an ellipsoidal stimulated volume, $G_w(X)$ was substituted by the quantity $1/W_c(X)$, which is reciprocal to the approximative function given by equation (5.55) along with the quantity γ_c from equation (5.56). Two situations are represented: geothermal- and hydraulic-fracturing types of stimulated volumes. Both parts of the figure contain the five following curves (from the lowest to the uppermost ones): a lower bound for an ellipsoidal stimulated volume, a lower bound for a cuboidal stimulated volume, a Gutenberg–Richter straight line, an upper bound for a cuboidal stimulated volume and, finally, an upper bound for an ellipsoidal stimulated volume.

Note that the curves for ellipsoidal stimulated volumes are approximations only. By contrast, the curves for cuboidal volumes are exact. However, the equations for cuboidal stimulated volumes assume rupture surfaces inclined under an angle ϕ to the plane of the maximum and intermediate axes. For Figure 5.20, Shapiro *et al.* (2013) accepted $\phi = 30°$. Figure 5.21 shows the curves for different values of ϕ. Finally, Figure 5.22 shows an example of how a sophisticated geometric form of the stimulated volume can influence the frequency–magnitude distribution (a lower bound). Here a situation corresponding to two intersecting ellipsoids has been numerically modeled.

Figures 5.20–5.22 and Figure 5.18 show that, if the seismicity statistics tends to the lower bound, then a fitting of the Gutenberg–Richter straight line will produce a systematically overestimated b-value. This effect will be stronger the smaller is

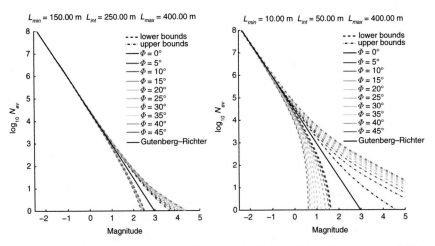

Figure 5.21 The same as Figure 5.20 but cuboidal stimulated volumes and different angles ϕ. (After Shapiro *et al.*, 2013.) A black and white version of this figure will appear in some formats. For the color version, please refer to the plate section.

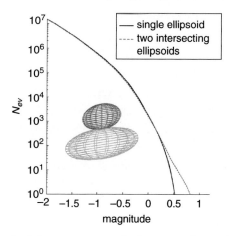

Figure 5.22 Theoretical frequency–magnitude curves (lower bounds) for a stimulated volume in a form of two intersecting ellipsoids.(After Shapiro *et al.*, 2013.)

the size of the stimulated volume. The scale L_{min} is especially important. Thus, the effect will be especially strong for the hydraulic-fracturing type of the geometry. This effect will also be strong for small time periods that elapse from an injection start. For small injection times stimulated volumes are small. Thus, the effect will result in a decrease of b-value estimates with injection times. This effect can be easily understood from equation (5.79). For the lower bound the quantity $\Psi(M - M_Y)$ is negative. Its absolute value is larger the smaller M_Y is and, therefore, the smaller is the size of the stimulated volume. By fitting the Gutenberg–Richter straight line this quantity will directly contribute to the values of the parameters b and a. It will decrease a and increase b. This effect will act in the opposite direction, if the event statistics follow the uppermost bound. It would increase a and decrease b values.

In the following section we compare several observed frequency–magnitude distributions to the theoretical bounds. The differences between the theoretical curves for ellipsoidal and cuboidal volumes are not significant. Moreover, the angle ϕ is not known. In spite of the fact that $\phi = 30°$ seems to be a reasonable approximation, in reality the angle can be broadly distributed. Thus, following Shapiro *et al.* (2013), we will compare real data to the theoretical approximations for ellipsoidal volumes. In practice the stimulated volume can be satisfactory represented by an approximate outer ellipsoidal envelope of the cloud of hypocenters of induced seismicity (see Figure 5.16).

5.2.6 Case studies of magnitude distributions and stress drop

Shapiro *et al.* (2013) compared equation (5.81) to frequency–magnitude distributions in several case studies that are already familiar to us. These are two geothermal locations in crystalline rocks, Basel (Häring *et al.*, 2008) and Soultz 1993 (Baria *et al.*, 1999). There is also the Paradox Valley data set obtained by an injection of a saline water into deep carbonate rocks (Ake *et al.*, 2005). Finally, they have also considered three hydrocarbon locations: a hydraulic-fracturing Stage A in a gas shale from Canada, a hydraulic fracturing Stage B (called stage 2 in the previous chapter) of a tight gas reservoir at the Cotton Valley (Rutledge and Phillips, 2003), and a stage of a hydraulic fracturing from the Barnett Shale (Maxwell *et al.*, 2009).

The corresponding fitting results are shown in Figures 5.23–5.28. For their analysis Shapiro *et al.* (2013) took microseismic clouds at the injection-termination time t_0. In the first step they fitted the real data by a standard Gutenberg–Richter cumulative distribution. They called the resulting straight line and its parameter

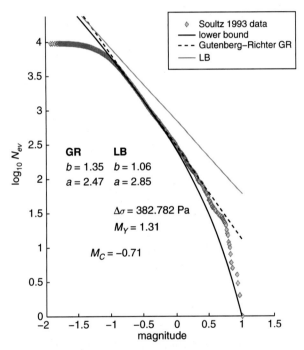

Figure 5.23 Fitting the frequency–magnitude distribution of the seismicity induced by the Soultz 1993 injection. The axes of the ellipsoid representing the stimulated volume are 440 m, 1400 m and 1740 m. The effective sphere scale γ is approximately 630 m. Results of the fitting of the apparent Gutenberg–Richter distribution (dashed line) and of the lower bound G_{wl} (solid line) are shown on the plot below the acronyms GR and LB, respectively. The thin upper straight line shows the Gutenberg–Richter distribution with parameters a and b estimated from the lower-bound curve. M_c, M_Y and $\Delta\sigma$ are estimated values of the completeness magnitude, the maximum induced magnitude and the stress drop, respectively. (After Shapiro *et al.*, 2013.)

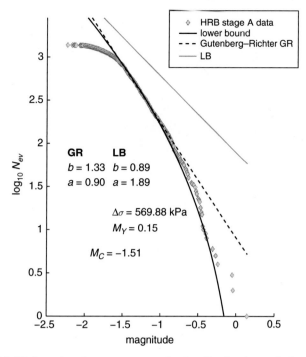

Figure 5.24 Fitting the frequency–magnitude distribution of the seismicity induced by a hydraulic fracturing stage A in a gas-shale deposit at Horn River in Canada. Notations are explained in Figure 5.23. The axes of the ellipsoid are 10 m, 100 m and 650 m. The effective sphere scale γ is approximately 15 m. (After Shapiro *et al.*, 2013.)

apparent parameters of the Gutenberg–Richter distribution. Then they fitted the data by the theoretical curve of the lower bound. Using the a- and b-values from this fit they attempted to reconstruct "real" Gutenberg–Richter cumulative distribution. The reconstructed Gutenberg–Richter distribution describes seismicity in an infinite medium with tectonic conditions of the corresponding injection site. The re-estimated b-values are systematically lower than the parameters obtained by the apparent Gutenberg–Richter fit. We observe that the real frequency–magnitude distributions are usually restricted between the lower bound and the fitting straight line corresponding to the apparent Gutenberg–Richter distribution. Note that this line is located lower than the magnitude–frequency distribution given by the function G_{wu0} (see equation (5.50)) and corresponds to the reconstructed Gutenberg–Richter distribution in the infinite medium. Moreover, nearly all data sets show a tendency of the seismicity to be better represented by the lower bound.

The fitting of the lower bound also yields estimates of maximum induced magnitudes and of an averaged stress drop. The stress drop is computed from equation (5.78) by substituting the estimates of characteristic length γ and of the maximum

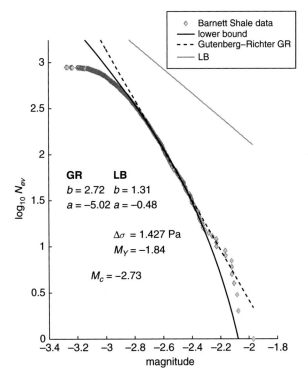

Figure 5.25 Fitting the frequency–magnitude distribution of the seismicity induced by hydraulic fracturing at one of locations in Barnett Shale. Notations are explained in Figure 5.23. The axes of the ellipsoid are 70 m, 340 m and 1000 m. The effective sphere scale γ was approximately 100 m. (After Shapiro *et al.*, 2013.)

induced magnitude M_Y. In reality the values of the stress drop of induced events can be distributed in a broad range. For example, Goertz-Allmann *et al.* (2011) estimated the stress drop of 1000 selected events from the Basel injection experiment mentioned above. Their figure 2 shows values in the range of 0.1–10^2 MPa. Jost *et al.* (1998) estimated stress drops of events induced by the 1994 KTB fluid injection experiment. Their table 1 gives values distributed in the range 5×10^{-3}–6 MPa. On the other hand, our lower-bound-based stress-drop estimate represents an average value in the sense of the equivalence of equations (5.72) and (5.76). It depends on a real distribution of the stress drop and will be dominated by the most probable stress-drop values. Contributions of numerous small-magnitude events can become enhanced. Such events can have rather small stress drops. However, corresponding stress-drop values are frequently not seen because such events are usually not analyzed owing to low signal–noise relations. The presence of such events in the magnitude statistics can lead to even smaller averaged estimates of

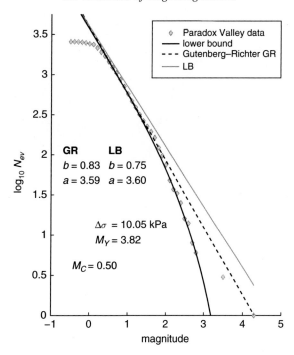

Figure 5.26 Fitting the frequency–magnitude distribution of the seismicity induced by the Paradox Valley injection. Notations are explained in Figure 5.23. The axes of the ellipsoid are 3000 m, 4000 m and 7000 m. The effective sphere scale γ was approximately 3800 m. (After Shapiro *et al.*, 2013.)

stress drops than those measured at the KTB site. This tendency can be further enhanced by the fact that the quantity γ is usually overestimated due to event location errors. Still, because of its independence of any rupture model and any estimate of the corner frequency, the lower-bound approach to estimating the stress drop can yield a reasonable additional constraint of this poorly understood quantity.

The data sets permitting rather simple interpretation are shown in Figures 5.23–5.25. They correspond to a stimulation of a geothermal system in Soultz, and to two hydraulic fracturing stages in the gas shale at Horn River (Canada) and of Barnett Shale (USA).

The estimated values of the stress drop shown in the figures and containing many digits were listed from the fitting algorithm directly. Realistically, these estimates show only the order of magnitude of the stress drop. The Barnett Shale case study (Figure 5.25) is distinguished by an especially low estimate of the stress drop. In addition to the reasons already mentioned above, one more reason for this can be the following. The minimum principal size of the stimulated volume in this particular case study represents the total thickness of the shale reservoir. In reality the induced seismicity is concentrated in several layers of the thickness of 10 m

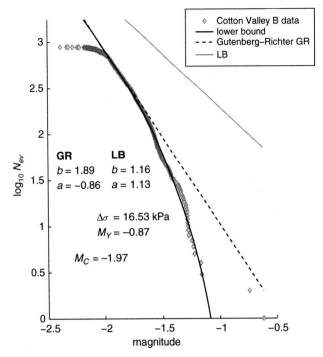

Figure 5.27 Fitting the frequency–magnitude distribution of the seismicity induced by the Stage B of hydraulic fracturing at Cotton Valley. Notations are explained in Figure 5.23. The axes of the ellipsoid approximating the stimulated volume are 10 m, 40 m and 480 m. The effective sphere scale γ is approximately 15 m. (After Shapiro *et al.*, 2013.)

each (see Figure 4.13). Thus, in reality the quantity γ may be less by a factor of 10 than the one used for the estimate of the stress drop. This would yield a stress drop of the order of 2500 Pa.

The complete data sets shown in Figures 5.23–5.25 are described well by the lower-bound approximation. Thus the inducing of events seems to require pore-pressure perturbations involving nearly total rupture surfaces. The fact that the lower-bound curve somewhat underestimates the number of events in the intermediate- to high-magnitude range can be explained by a too-rough analytical approximation of the real rupture statistics. The influence of the geometry, which is more complex than just an ellipsoid (see Figure 5.22), or a rather restricted angular spectrum of the rupture orientations (see Figure 5.21) could also contribute to this effect.

A somewhat more-sophisticated interpretation seems to be required by the data sets shown in Figures 5.26–5.28. These are data from the Paradox Valley, Cotton Valley (Stage B) and Basel case studies.

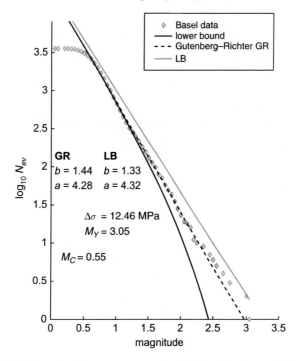

Figure 5.28 Fitting the frequency–magnitude distribution of the seismicity induced by the fluid injection at the Basel EGS site. Notations are explained in Figure 5.23. The axes of the stimulated volume are approximately 100 m, 760 m and 920 m. The effective sphere scale γ is approximately 150 m. (After Shapiro *et al.*, 2013.)

In Figures 5.26 and 5.27 we again observe that a dominant majority of events follow the lower-bound approximation well. However, several data points corresponding to high-magnitude events tend to return backward to the classical Gutenberg–Richter distribution. Thus apparently, the corresponding few large-magnitude events were triggered merely by a pore-pressure-related perturbation of nucleation spots on their rupture surfaces. Thus, possibly the rupture surfaces of these events were not completely included in the stimulated volume. However, for their triggering an excitation of rather large nucleation domains was necessary. We conclude this from the fact that corresponding data points are still located below the Gutenberg–Richter distribution. This indicates that excitation of just a nucleation spot was not sufficient for their occurrence. Note also that we cannot exclude the possibility that these few large-magnitude events correspond to rare statistical fluctuations with high stress drops and/or spatial orientations permitting large-scale (significantly larger than γ) rupture surfaces.

One more example of a similar situation is given by the Basel data set (Figure 5.28). A conventional Gutenberg–Richter fitting yields $a = 4.3$ and $b = 1.4$.

The bound G_{wl} yields close results: $a = 4.3$ and $b = 1.3$. Additionally, $\Delta\sigma = 12.5$ MPa. The maximum magnitude defining the lower bound is found to be $M_Y = 3.05$. It seems that, for the triggering of a majority of events in Basel, stimulation of their nucleation spots was sufficient.

5.2.7 Induced and triggered events

The previous discussion of case studies shows that magnitude distributions on Figures 5.23–5.25 are described well by the lower-bound approximation. The situations shown in Figures 5.26–5.28 seem to be more complicated. A comparison of these two groups of case studies indicates the possibility of distinguishing between triggered and induced events. McGarr *et al.* (2002) defined induced and triggered seismicity in respect to the stress impact of a stimulation. If this impact is of the order of the ambient shear stress they speak about induced seismicity. If this impact is significantly smaller than the ambient shear stress they speak about triggered seismicity. In this terminology, nearly all the fluid-induced seismicity from our case studies could be considered as the triggered ones. Here we use another definition, which is more specific in respect to the geometry of the stimulated volume. It is close to the one used by Dahm *et al.* (2013). We define as "induced" those events resulting from perturbing their nearly complete rupture surfaces. Then their statistics should follow the lower bound of magnitude distributions. We define as "triggered" those events resulting from perturbing nucleation spots of their rupture surfaces only. Their statistic should follow the reconstructed Gutenberg–Richter distribution. It seems that some geothermal-reservoir case studies include triggered events (e.g. the Basel case study). This is rare but cannot be excluded for hydraulic fracturing of hydrocarbon reservoirs. Note also that, for an event triggering, a perturbation of an arbitrary element of its rupture surface is not sufficient. This would correspond to the uppermost bound (5.51). This bound strictly contradicts our observations. The data show that triggering requires a perturbation of a significant part of the rupture surface necessarily including the nucleation domain (which we model by the rupture center). It also seems that the induced events are much more common than triggered events.

5.2.8 Maximum magnitude and sizes of stimulated volume

In Section 5.2.3 we saw that the probability $W_{vol}(X)$ quickly decreases with increasing size X of the rupture surface. As soon as the size X approaches or exceeds a characteristic scale Y of the stimulated volume the probability of events becomes nearly vanishing. Our numerical simulations of W_{vol} show that the size of the stimulated volume defines the scale Y. In comparison to other geometric parameters, the length L_{min} gives a dominant contribution to the magnitude order of Y (see the quantity γ given by equation (5.53)).

The conclusion on a dominant role of L_{min} in the parameter Y is supported by real data. Shapiro *et al.* (2011) considered the largest observed magnitudes M_{max} of induced earthquakes for different case studies. They assumed M_{max} to be a proxy of M_Y and showed that the values of M_{max} are much better described as functions of $\lg L_{min}^2$ than as functions of $\lg L_{int}^2$ or $\lg L_{max}^2$ (see their figure 7).

Equation (5.81) shows that the frequency–magnitude statistic of induced earthquakes is time dependent. This is not only due to the term containing $Q_c(t)$. Another time-dependent factor is the characteristic magnitude M_Y. This magnitude is a function of a geometrical scale of the stimulated volume. This scale is, in turn, a function of time (or, equivalently, of the injected volume). This means that both quantities a and b of a conventionally fitted Gutenberg–Richter frequency–magnitude distribution will be time dependent. We have already mentioned that one should expect a systematic increase of the a-values and decrease of the b-values with increasing time that has elapsed since the start of the injection. However, information on large-magnitude events at intermediate time moments can be included into a consideration of the maximum observed magnitudes as a function of the scale of stimulated volumes.

Figure 5.29 (reproduced from Shapiro *et al.*, 2013) shows the observed maximum magnitudes as function of the minimum axis of the stimulated volume. The figure includes maximum magnitudes for the injection termination moment t_0 as well as maximum magnitudes for the time moments $t_{2/3}$ defined so, that the injected volume $Q_c(t_{2/3})$ numerically satisfies the following condition: $\lg Q_c(t_{2/3}) = (2/3)\lg Q_c(t_0)$ (all volumes are measured in m^3). We see that the data support the general trend shown by the straight line. The error bars on the figure show the possible impact of errors in magnitudes and event locations. For magnitudes, Shapiro *et al.* (2013) assumed an error of the order 0.5. This roughly corresponds to possible differences between local and moment magnitudes (see a thorough study of this subject by Grünthal and Wahlström, 2003), which were taken from different literature sources (see the citations below). For the principal axes, Shapiro *et al.* (2013) assumed the error bars of the order of seismicity location errors: 10 m for hydraulic-fracturing sites Barnet Shale (Maxwell *et al.*, 2009) and Cotton Valley (Rutledge and Phillips, 2003), 50 m for geothermic sites: Basel (Häring *et al.*, 2008), four experiments of Soultz (the data courtesy of Andrew Jupe, EGS Energy), two experiments of Cooper Basin (Baisch *et al.*, 2009), Fenton Hill (Phillips *et al.*, 1997), Berlin (Bommer *et al.*, 2006), and 100 m for the Paradox Valley (Ake *et al.*, 2005). The star corresponds to the largest event of the Basel injection.

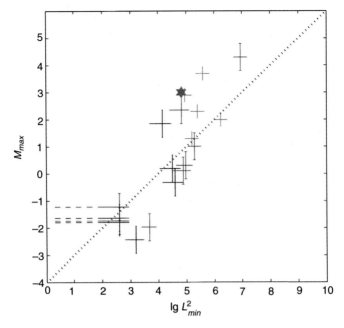

Figure 5.29 Largest observed magnitudes of induced earthquakes as a function of the minimum principal axes of corresponding stimulated rock volumes for different case studies. The length of L_{min} is given in meters. The error bars show the possible impact of errors in magnitudes and event locations. The star corresponds to the largest event of the Basel injection. The crosses correspond to various data sets including also seismicity clouds at the time moments $t_{2/3}$ such that the cumulative injected volume $Q_c(t_{2/3})$ at these times numerically satisfies lg $Q_c(t_{2/3}) = (2/3)$ lg $Q_c(t_0)$, where t_0 is the termination time of the stimulation and all volumes are measured in m³. Injection sites, 2 lg L_{min} and maximum magnitudes for the data catalogs taken till the moment t_0 of injection termination (the latter two quantities are given as a coordinate pair in the coordinate system of the figure) are: Barnett Shale (3.69; −2.0), Cotton Valley A (2.6; −1.2), Cotton Valley C (2.6; −1.7), Basel (4.83; 3.05), Paradox Valley (6.95; 4.3), Soultz 1993 (5.29; 1.0), Soultz 1995 (4.89; 0.1), Soultz 1996 (4.98; 0.3), Soultz 2000 (5.4; 2.3), Berlin (Bommer *et al.*, 2006) (6.23; 2.0), Cooper Basin 2003 (5.59; 3.7), Cooper Basin (4.95; 2.9) and Fenton Hill (5.2; 1.3). (After Shapiro *et al.*, 2013.)

We observe good agreement of the data points with equation (5.78) for $Y = L_{min}$. The corresponding values of $\Delta\sigma$ are of the order of 10^{-4}–10 MPa. Substituting a highest probable limit of stress drops of the order of 10 MPa into equation (5.78) we obtain an approximate estimate of the maximum probable magnitude limit (it would correspond to an upper envelope on the Figure 5.29) of an induced earthquake for a given location:

$$max\{M_{max}\} \sim 2 \cdot \lg L_{min} - 1. \qquad (5.82)$$

This result explains the fact that high-magnitude events are more probable at geothermal sites than via hydraulic fracturing in hydrocarbon reservoirs. L_{min} is much smaller in the latter case than in the former one. By hydraulic fracturing, a typical zone of water penetration behind the fracture walls is of the order of 1–10 m (owing to localization errors, microseismic clouds of hydraulic fractures can have greater L_{min}; examples of a high-precision localization of hydraulic-fracture induced microseismicity can be found in Rutledge and Phillips, 2003; some of them have been shown in the previous chapter). Thus, the largest magnitudes are 0–1. By geothermal stimulations, L_{min} is of the order of 300 m or less. In such a case the largest magnitudes are around 4. Because L_{min} can increase with time, it leads to an enhancement of the large-magnitude probability at the end of injection operation or shortly after it (until the rate of induced seismicity has not dropped significantly). This has been indeed observed in reality (Majer *et al.*, 2007; Baisch *et al.*, 2009).

Result (5.82) can be useful for estimating and constraining an induced seismic hazard. For example, to restrict the hazard one could attempt to keep the minimal principal axis of the stimulated volume restricted by terminating the injection if this size achieves a planned critical value. However, because this result mainly addresses induced events, its application requires a careful analysis of the seismotectonic and geological situation in each particular case. By such an analysis, the probability of triggered events should be carefully constrained. Factors such as scales of faults intersecting the stimulated volume, tendencies of the faults to influence the shape and scales of the stimulated volume, and other seismotectonic parameters (e.g. the seismogenic index, b-values, stress states of significant faults and modifications of these parameters during the stimulation) are of importance for this task.

The formalism of the bounds of magnitude probability is not restricted to fluid-induced seismicity. We hypothesize that it is applicable for any type of seismicity induced in a restricted rock volume, for example an aftershock series of tectonic events.

5.3 Appendix 1. Probability of discs within a sphere

Here we consider the probability of an arbitrary oriented disc, which has its center inside a spherical volume, to completely belong to this volume. As a result we will derive equation (5.52). We follow the derivation of Shapiro *et al.* (2011).

Let us first find possible locations of centers P of such discs. We consider a spherical stimulated volume of a radius R with a center at a point O. We take a disc with a radius $r < R$ and a center at a point P. P is located inside of the stimulated volume. If the distance $y = |OP|$ is less than $R - r$, then the volume will include the disc independently of its orientation. Thus, the volume $V_1 = 4\pi(R-r)^3/3$ will contribute to a possible location of P completely. Let us consider the spherical shell with the inner radius equal to $y_{min} = R - r$ and the outer radius $y_{max} = \sqrt{R^2 - r^2}$. The discs belonging to the volume cannot have centers P outside the outer sphere of the radius y_{max}.

If P is located inside the shell then the disc cannot be arbitrary oriented. An orientation of the disc is given by the orientation of a normal to its plane at the point P. Now we must consider possible orientations of the discs. Let us consider a sphere S1 of radius r with the center at P. This sphere intersects the surface of the stimulated volume along a circle. We consider further a plane including the straight line OP. Such a plane intersects this circle in two points A1 and A2 located symmetrically to the line OP. The sine of angle α between the line PA1 (or PA2) and the normal to the line OP at the point P belonging to the same plane (given by points O, P, A1 and A2) can be computed as $\sin\alpha = (R^2 - r^2)/(2yr) - y/(2r)$. In order to belong to the volume a disc must have its normal located inside the cone with the symmetry axes OP and the limiting angle α. The sphere S1 and this cone define a spherical segment of the height $h = r(1 - \cos\alpha)$ and the surface $2\pi rh$. The probability of a disc having an orientation necessary to belong to the stimulated volume is equal to the ratio of this surface to the surface of the half of the sphere S1, i.e. h/r. This probability W_1 is a function of the variable y.

The probability we are looking for is given by the sum of the contribution of the volume V_1 and of the integral of the probability W_1 over the spherical shell introduced above:

$$W_{sp}(r/R) = \left(3 \int_{y_{min}}^{y_{max}} y^2 \frac{h}{r} dy + (R - r)^3\right)/R^3, \tag{5.83}$$

with $y_{min} = R - r$ and $y_{max} = \sqrt{R^2 - r^2}$. This provides equation (5.52).

5.4 Appendix 2. Probability of discs within an ellipsoid

Here we consider the probability that an arbitrary oriented disc with the center inside an ellipsoidal stimulated volume completely belongs to the volume. We follow the corresponding derivation of Shapiro *et al.* (2013). In contrast to the exact results (5.52) and (5.54), we propose an approximation of the required probability $W_{el}(X)$ (where X is the diameter of the rupture). We consider an ellipsoid with the principal axes $L_{min} < L_{int} < L_{max}$. If the axes are close to each other, then a good approximation of $W_{el}(X)$ will be $W_{sp}(X/\gamma)$, where γ is given by equation (5.53). Let us consider another quite a realistic geometry of stimulated volumes: $L_{min} \ll L_{int} < L_{max}$.

We consider a rupture surface in a form of a plane disc of diameter X of an arbitrary orientation with a center at a point P inside such an ellipsoidal stimulated volume. We will concentrate first on large discs with $X > L_{min}$. The centers of large discs completely belonging to the ellipsoid are approximately located inside the following ellipsoidal volume:

$$V_{large} = \frac{\pi}{6} L_{min}(L_{int} - X)(L_{max} - X).$$ (5.84)

We consider a sphere S1 defined by normals of the length $X/2$ (the sphere's radius) at the point P for all possible orientations of the disc. If $X > L_{min}$ (i.e. large discs) then such a sphere will always intersect the surface of the ellipsoid. We estimate approximately a part of the surface of this sphere (in relation to the spheres complete surface), where the disc's normal can have its end point for a disc belonging to the stimulated volume. For this we consider such a sphere intersecting with a "side surfaces" of the volume. The side surfaces are two ellipsoidal surface halves spanned on the axes L_{int} and L_{max}.

Further, we approximate these surfaces just by planes separated by L_{min} (we call them side planes). We consider the sphere S1 of radius $X/2$ with the center P at a distance y from the closest side plane. This sphere intersects the side plane along a circle. In order to belong to the volume, a disc must have a normal located inside a cone with the symmetry axes coinciding with the the normal from P to the side plane. The sphere S1 and this cone define a spherical segment of the height $h = (X - \sqrt{X^2 - 4y^2})/2$ and the surface $\pi X h$. A probability of a disc having an orientation necessary for belonging to the stimulated volume is equal to the ratio of this surface to the surface of the half of the sphere S1, i.e. $1 - \sqrt{1 - 4y^2/X^2}$. Note that the effect of a possible intersecting of the sphere S1 with the second (farthest) side plane is automatically taken into account. Indeed, the largest intersection is of importance only because of its symmetric effect on permitted orientations of the discs. The probability W_L of large discs belonging to the stimulated volume is given by the integration over y:

$$W_L(X) = \frac{2(L_{int} - X)(L_{max} - X)}{L_{min} L_{int} L_{max}} \int_0^{L_{min}/2} \left(1 - \sqrt{1 - \frac{4y^2}{X^2}}\right) dy.$$ (5.85)

The integration yields:

$$W_L(X) = \left(1 - \frac{1}{2}\sqrt{1 - \frac{L_{min}^2}{X^2}} - \frac{X}{2L_{min}} \arcsin \frac{X}{L_{min}}\right)\left(1 - \frac{X}{L_{int}}\right)\left(1 - \frac{X}{L_{max}}\right).$$ (5.86)

From the derivation it is clear that, with increasing $X \leq L_{int}$, the estimate $W_L(X)$ will adequately decrease to zero. However, for small X the function $W_L(X)$

becomes inadequate. At the point $X = L_{min}$ it must be reasonably combined with the function $W_{sp}(X/\gamma)$. Thus, we propose the following approximation of $W_{el}(X)$:

$$W_{el}(X) \approx W_{sp}(X/\gamma), \quad \text{if } X \leq L_{min},$$

$$W_{el}(X) \approx W_L(X)\frac{W_{sp}(L_{min}/\gamma)}{W_L(L_{min})}, \quad \text{if } L_{min} \leq X \leq L_{int}. \qquad (5.87)$$

5.5 Appendix 3. Probability of discs with centers inside a sphere

Here we consider the probability that an arbitrarily-oriented disc that intersects a sphere has its center within this sphere. We follow the corresponding derivation of Shapiro *et al.* (2013).

We consider a spherical volume (a stimulated volume) of the radius $R = L/2$ with a center at the point O (see Figure 5.30). To calculate a number N_s of discs of the radius $r = X/2$ intersecting this volume, we must locate their centers (points P) inside or outside the stimulated volume. Such discs can belong to two different groups.

The first group consists of discs with centers P belonging to the stimulated volume. Their number is is given by $4\pi N R^3/3$, where N is the bulk concentration of the discs (i.e. the bulk concentration of the potential rupture surfaces).

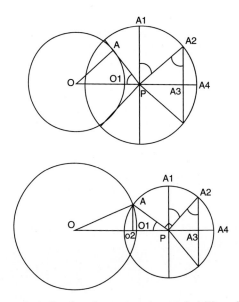

Figure 5.30 Geometrical sketches for computing probability of an arbitrarily oriented disc intersecting a spherical volume having its center inside the volume. (After Shapiro *et al.*, 2013.)

The second group consists of discs with centers outside the stimulated volume. Let us consider a sphere S1 of radius r with the center at P. An orientation of the disc is given by the orientation of the normal to the disc at the point P. A necessary condition for the disc to intersect (or to touch) the stimulated volume is that the sphere S1 completely includes the stimulated volume or intersects the surface of the stimulated volume along a circle (or just touch the volume). We introduce a variable $s = |O1P| = |OP| - R$ (see Figure 5.30) and consider further a plane including the straight line OP. In this plane there are exist two radii of S1 which (or continuations of which) are tangential to the spherical stimulated volume. We denote one of the touching points as A (see the top part of Figure 5.30). There are two such points located symmetrically to the line OP. From the triangle OAP we have $|AP| = \tau = \sqrt{(R+s)^2 - R^2}$. We consider then two possible situations: (1) $\tau \le r$ (shown in Figure 5.30, on the top), and (2) $\tau > r$ (shown in Figure 5.30 on the bottom; note that here A denotes an intersection point of the surface of S1 and of the surface of the stimulated volume).

In case (1) the limiting orientations of a disc intersecting the volume are given by the tangential positions of the radii of S1 (Figure 5.30, on the top). Note that this yields the following restrictions for the variable s: $0 \le s \le \sqrt{r^2 + R^2} - R$. A disc will intersect the stimulated volume if the disc's normal is located within the spherical sector defined by rotation of the plane section A1PA2 around the line PA4. Therefore, such a normal can intersect the surface of the right-hand (upper) half-sphere of S1 everywhere excluding the surface of the spherical segment covered by the rotation of the semi-arc A2A4 around the point A4. Taking into account that $\sin OPA = \sin A1PA2 = \sin PA2A3 = R/(R+s)$ we obtain $|A3A4| = rs/(R+s)$. For the segment surface and the sector surface this yields $2\pi r^2 s/(R+s)$ and $2\pi r^2 R/(R+s)$, respectively. The probability of a disc having an intersection with the volume is given by the relation of the spherical sector surface to the surface of the half sphere. This ratio is equal to $R/(R+S)$. The number of discs intersecting the stimulated volume is then given by the following integral:

$$4\pi N \int_0^{\sqrt{r^2+R^2}-R} (R+s)R\,ds = 2\pi Rr^2 N. \qquad (5.88)$$

An additional number of rupture surfaces intersecting the volume is given by case (2): $\tau > r$. This situation corresponds to the bottom part of Figure 5.30. Here $\sqrt{r^2 + R^2} - R \le s \le r$. The straight line PA is no longer a tangent to the stimulated volume. The height of the excluded spherical segment (defined by the rotation of the semi-arc A2A4) is given by $r - |PA3|$. The length $h_1 = |PA3| = |AO2|$ can be found from the triangle OAP. The probability of a disc intersecting the stimulated volume is given by

$$h_1/r = \sqrt{2(R+s)^2(R^2+r^2) - (R+s)^4 - (R^2-r^2)^2}/(2r(R+s)). \quad (5.89)$$

The number of rupture surfaces intersecting the stimulated volume is then given by the following integral:

$$4\pi N \int_{\sqrt{r^2+R^2}-R}^{r} (R+s)^2(h_1/r)ds = \pi^2 R^2 r N. \quad (5.90)$$

Finally, the sought-after probability $W_c(X)$ is given by the following ratio, taking into account all the contributions discussed above:

$$W_c(X) = \frac{4\pi N R^3/3}{\pi^2 R^2 r N + 2\pi R r^2 N + 4\pi N R^3/3} = \left(1 + \frac{3}{2}\frac{X^2}{L^2} + \frac{3\pi}{4}\frac{X}{L}\right)^{-1}. \quad (5.91)$$

5.6 Appendix 4. Probability notations used in this chapter

symbol	its meaning
$W_n(\nu_I, t)$	probability of n events in the time interval $[0, t]$; ν_I is the event rate
$W_{>\Delta t, <\Delta t}$	probability of waiting time being longer (shorter) than Δt
$W_{n,M}(t)$	probability of n events of magnitude not less than M in the time interval $[0, t]$
W_{ev}	probability of a point-like event
$W_{\geq M}$	probability of an earthquake with a magnitude not less than M
$W_E(X)$	probability of an event with rupture-surface diameter X
$W_f(X)$	probability of a potential rupture surface with diameter X, in unbounded medium
$W_c(X)$	probability of a rupture surface with the center within the stimulated volume
$W_s(X)$	probability of a rupture surface being sufficiently stimulated to produce an event
$W_{vol}(X)$	probability of a rupture surface being completely inside the stimulated volume
$W_{e1}(X)$	probability of a seismic event along a given rupture surface under the condition that the center of the surface is inside of the stimulated volume
$W_{e2}(X)$	probability of a seismic event along a given rupture surface under the condition that the surface is intersecting the stimulated volume but its center is outside of the volume
$W_{sl}(X)$	lower bound of $W_s(X)$: rupture surface is inside the stimulated volume
$W_{su}(X)$	upper bound of $W_s(X)$: arbitrary intersection of the rupture surface and stimulated volume triggers the event
$W_{su0}(X)$	upper bound of $W_s(X)$: center of the rupture surface has to be inside the stimulated volume to trigger the event
W_{sp}	W_{vol} for a sphere
W_{el}	approximation of W_{vol} for an ellipsoid
W_{cub}	W_{vol} for a cuboid
W_{cc}	W_c for a cuboid

References

Ake, J., Mahrer, K., OConnell, D., and Block, L. 2005. Deep-injection and closely monitored induced seismicity at Paradox Valley, Colorado. *Bulletin of the Seismological Society of America*, **95**, 664–683.

Aki, K., and Richards, P. G. 2002. *Quantitative Seismology*, 2nd edn. Sausalito, CA: University Science Books.

Al-Wardy, W., and Zimmerman, R. W. 2004. Effective stress law for the permeability of clay-rich sandstones. *Journal of Geophysical Research: Solid Earth*, **109**(B4), doi:10.1029/2003JB002836.

Altmann, J. B., Müller, T. M., Müller, B. I. R., Tingay, M. R. P., and Heidbach, O. 2010. Poroelastic contribution to the reservoir stress path. *International Journal of Rock Mechanics and Mining Science*, **47**, 1104–1113.

Amenzade, Yu. A. 1976. *Theory of Elasticity*. Moscow: Visshaja Shkola, pp. 210–211 (in Russian).

Auld, B. A. 1990. *Acoustic Fields and Waves in Solids*, 2nd edn, vol. 1. Malabar, Florida: R. E. Krieger.

Avouac, J.-P. 2012. Earthquakes: human-induced shaking. *Nature Geoscience*, **5**, 763–764, doi:10.1038/ngeo1609.

Backus, G., and Mulcahy, M. 1976a. Moment tensors and other phenomenological descriptions of seismic sources I. Continuous Displacements. *Geophysical Journal of the Royal Astronomical Society*, **46**(2), 341–361, doi:10.1111/j.1365-246X.1976.tb04162.x.

Backus, G., and Mulcahy, M. 1976b. Moment tensors and other phenomenological descriptions of seismic sources II. Discontinuous displacements. *Geophysical Journal of the Royal Astronomical Society*, **47**(2), 301–329, doi:10.1111/j.1365-246X.1976.tb01275.x.

Baisch, S., Bohnhoff, M., Ceranna, L., Tu, Y., and Harjes, H.-P. 2002. Probing the crust to 9 km depth: fluid injection experiments and induced seismicity at the KTB superdeep drilling hole, Germany. *Bulletin of Seismological Society of America*, **92**(6), 2369–2380.

Baisch, S., Voros, R., Weidler, R., and Wyborn, D. 2009. Investigation of fault mechanisms during geothermal reservoir stimulation experiments in the Cooper Basin, Australia. *Bulletin of the Seismological Society of America*, **99**(1), 148–158.

Barenblatt, G. I. 1996. *Scaling, Self-similarity, and Intermediate Asymptotics*. New York: Cambridge University Press. ISBN-10 0-521-43522-6.

Barenblatt, G. I., Entov, V. M., and Ryzhik, V. M. 1990. *Flow of Fluids Through Natural Rocks*. Dordrecht: Kluwer Academic Publishers.

Baria, R., Baumgartner, J., Gerard, A., Jung, R., and Garnish, J. 1999. European HDR research programme at Soultz-sous-Florets, France 1987–1996. *Geothermics*, **28**, doi: 10.1016/S0375–6505(99)00036-x.

Becker, K., Shapiro, S. A., Stanchits, S., Dresen, G., and Vinciguerra, S. 2007. Stress induced elastic anisotropy of the Etnean Basalt: theoretical and laboratory examination. *Geophysical Research Letters*, **34**, L11307.

Beresnev, I. A. 2001. What we can and cannot learn about earthquake sources from the spectra of seismic waves. *Bulletin of the Seismological Society of America*, **91**, 397–400, doi:10.1002/jgrb.50362.

Berryman, J. G. 1992. Effective stress for transport properties of inhomogeneous porous rock. *Journal of Geophysical Research*, **97**, 17409–17424.

Berryman, J. G. 1993. Effective stress rules for pore-fluid transport in rocks containing two minerals. *International Journal of Rock Mechanics and Mining Sciences & Geomechanics Abstracts*, **30**, 1165–1168.

Biot, M. A. 1956. Theory of propagation of elastic waves in a fluid-saturated porous solid. I. Low-frequency range. *Journal of the Acoustical Society of America*, **28**, 168–178.

Biot, M. A. 1962. Mechanics of deformation and acoustic propagation in porous media. *Journal of Applied Physics*, **33**, 1482–1498.

Bommer, J. J., Oates, S., Cepeda, J. M., *et al.* 2006. Control of hazard due to seismicity induced by a hot fractured rock geothermal project. *Engineering Geology*, **83**(4), 287–306.

Bouchbinder, E., Goldman, T., and Fineberg, J. 2014. The dynamics of rapid fracture: instabilities, nonlinearities and length scales. *Reports on Progress in Physics*, **77**(4), 1–30, doi:10.1088/0034–4885/77/4/046501.

Brown, R. J. S., and Korringa, J. 1975. On the dependence of the elastic properties of a porous rock on the compressibility of the pore fluid. *Geophysics*, **40**, 608–616.

Brune, J. N. 1970. Tectonic stress and the spectra of seismic shear waves from earthquakes. *Journal of Geophysical Research*, **75**(26), 4997–5009, doi:10.1029/JB07 5i026p04997.

Buske, S. 1999. 3-D prestack Kirchhoff migration of the ISO89-3D data set. *Pure and Applied Geophysics*, D. Gajewski and W. Rabbel (eds). Basel Burkhäuser, pp. 157–171.

Carcione, J. M. 2007. *Wave Fields in Real Media: Wave Propagation in Anisotropic, Anelastic, Porous and Electromagnetic Media*. Oxford: Elsevier.

Carcione, J. M., and Tinivella, U. 2001. The seismic response to overpressure: a modelling study based on laboratory, well and seismic data. *Geophysical Prospecting*, **49**, 523–539.

Carslaw, H. S., and Jaeger, J. C. 1973. *Conduction of Heat in Solids*. Oxford: Clarendon Press.

Carter, R. D. 1957. Derivation of the general equation for estimating the extent of the fractured area. Appendix I of *Optimum Fluid Characteristics for Fracture Extension*, G. C. Howard and C. R. Fant (eds.). New York: American Petroleum Institute, pp. 261–269.

Cerveny, V. 2005. *Seismic Ray Theory*. Cambridge University Press.

Chandler, R. N., and Johnson, D. L. 1981. The equivalence of quasistatic flow in fluid-saturated porous media and Biot's slow wave in the limit of zero frequency. *Journal of Applied Physics*, **52**(5), 3391–3395.

Cheng, A. H.-D. 1997. Material coefficients of anisotropic poroelasticity. *International Journal of Rock Mechanics and Mining Sciences*, **34**(2), 199–205, doi:10.1016/S0148–9062(96)00055-1.

Ciz, R., and Shapiro, S. A. 2007. Generalization of Gassmann's equations for porous media saturated with a solid material. *Geophysics*, **72**, A75–A79.

Ciz, R., and Shapiro, S. A. 2009. Stress-dependent anisotropy in transversely isotropic rocks: Comparison between theory and laboratory experiment on shale. *Geophysics*, **74**, D7–D12.

Cleary, M. P. 1977. Fundamental solutions for a fluid-saturated porous solid. *International Journal of Solids and Structures*, **13**, 785–806.

Cornet, F. 2000. Comment on "Large-scale in situ permeability tensor of rocks from induced microseismicity" by S. A. Shapiro, P. Audigane and J.-J. Royer. *Geophysical Journal International*, **140**, 465–469.

Cornet, F. H., Helm, J., Poitrenaud, H., and Etchecopar, A. 1997. Seismic and aseismic slip induced by large scale fluid-injections. *PAGEOPH*, **150**, 563–583.

Coussy, O. 2004. *Poromechanics*. Chichester: Wiley.

Crank, J. 1975. *The Mathematics of Diffusion*. 2nd edn. Oxford: Clarendon Press.

Dahm, T., Becker, D., Bischoff, M., *et al.* 2013. Recommendation for the discrimination of human-related and natural seismicity. *Journal of Seismology*, **17**, 197–202, doi:10.1007/s10950–012–9295–6.

Deichmann, N., and Giardini, D. 2009. Earthquakes induced by the stimulation of an enhanced geothermal system below Basel (Switzerland). *Seismological Research Letters*, **80**(5), 784–798, doi:10.1785/gssrl.80.5.784.

Detournay, E., and Cheng, A. H.-D. 1993. Fundamentals of poroelasticity. Chapter 5, pp. 113–171 of: Hudson, J. A. (ed.), *Comprehensive Rock Engineering: Principles, Practice and Projects*. Oxford: Pergamon Press.

Dieterich, J. 1978. Pre-seismic fault slip and earthquake prediction. *Journal of Geophysical Research*, **83**, 3940–3948.

Dieterich, J. 1994. A constitutive law for rate of earthquake production and its application to earthquake clustering. *Journal of Geophysical Research*, **99**, 2601–2618.

Dinske, C., and Shapiro, S. A. 2013. Seismotectonic state of reservoirs inferred from magnitude distributions of fluid-induced seismicity. *Journal of Seismology*, **17**, 13–25, doi:10.1007/s10950–012–9292–9.

Dinske, C., Shapiro, S. A., and Rutledge, J. T. 2010. Interpretation of microseismicity resulting from gel and water fracturing of tight gas reservoirs. *Pure and Applied Geophysics*, **167**(1–2), 169–182, doi:10.1007/s00024–009–0003–6.

Duffy, J., and Mindlin, R. D. 1957. Stress–strain relations and vibrations of a granular medium. *Journal of Applied Mechanics*, **24**, 585–593.

Duncan, P., and Eisner, L. 2010. Reservoir characterization using surface microseismic monitoring. *Geophysics*, **75**(5), 75A139–75A146, doi:10.1190/1.3467760.

Dyer, B., Jupe, A., Jones, R. H., *et al.* 1994. *Microseismic Results from the European HDR Geothermal Project at Soultz-sous-Forets, Alsace, France*. CSM Associated Ltd, IR03/24.

Dyer, B. C., Schanz, U., Ladner, F., Häring, M., and Spillman, T. 2008. Microseismic imaging of a geothermal reservoir stimulation. *The Leading Edge*, **27**(7), 856–869, doi:10.1190/1.2954024.

Eberhart-Phillips, D., Han, D.-H., and Zoback, M. D. 1989. Empirical relationships among seismic velocity, effective pressure, porosity and clay content in sandstone. *Geophysics*, **54**, 82–89.

Economides, M. J., and Nolte, K. G. (eds.) 2003. *Reservoir Stimulation*, 3rd edn. Chichester: Wiley, pp. 5-1–5-14.

Edelman, I. Y., and Shapiro, S. A. 2004. An analytical approach to the description of fluid-injection induced microseismicity in porous rock. *Doklady Earth Sciences*, **399**, 1108–1112.

Emmermann, R., and Lauterjung, J. 1997. The German Continental Deep Drilling Program KTB: overview and major results. *Journal of Geophysical Research: Solid Earth*, **102**(B8), 18 179–18 201, doi:10.1029/96JB03945.

Erzinger, J., and Stober, I. 2005. Introduction to special issue: long-term fluid production in the KTB pilot hole. *Geofluids*, **5**, 1–7.

Fehler, M., House, L., Phillips, W. S., and Potter, R. 1998. A method to allow temporal variation of velocity in travel-time tomography using microearthquakes induced during hydraulic fracturing. *Tectonophysics*, **289**, 189–202.

Ferreira, J. M., Oliveira, R. T. De, Assumpcao, M., *et al.* 1995. Correlation of seismicity and water level in the Acu Reservoir – an example from northeast Brazil. *Bulletin of the Seismological Society of America*, **85**(5), 1483–1489.

Fischer, T., Hainzl, S., Eisner, L., Shapiro, S. A., and Le Calvez, J. 2008. Microseismic signatures of hydraulic fracture growth in sediment formations: observations and modeling. *Journal of Geophysical Research*, **113**(B02307), doi:10.1029/2007JB005070.

Fisher, M. K., Davidson, B. M., Goodwin, A. K., *et al.* 2002. Integrating fracture mapping technologies to optimize stimulations in the Barnett Shale. *Paper SPE 77411*.

Fisher, M. K., Heinze, J. R., Harris, C. D, *et al.* 2004. Optimizing horizontal completion techniques in the Barnett Shale using microseismic fracture mapping. *Paper SPE 90051*.

Fletcher, J. B., and Sykes, L. R. 1977. Earthquakes related to hydraulic mining and natural seismic activity in Western New York State. *Journal of Geophysical Research*, **82**, 3767–3780.

Frenkel, J. 2005. On the theory of seismic and seismoelectric phenomena in a moist soil. *Journal of Engineering Mechanics*, **131**, 879–887.

Freund, D. 1992. Ultrasonic compressional and shear velocities in dry clastic rocks as a function of porosity, clay content, and confining pressure. *Geophysical Journal International*, **108**, 125–135.

Gangi, A. F. 1978. Variation of whole and fractured porous rock permeability with confining stress. *International Journal of Rock Mechanics and Mining Science*, **15**, 249–257.

Gangi, A. F., and Carlson, R. L. 1996. An asperity-deformation model for effective pressure. *Tectonophysics*, **256**, 241–251.

Gassmann, F. 1951. Über die Elastizität poröser Medien. *Vierteljahresschrift der Naturforschenden Gesellschaft in Zurich*, **96**, 1–23.

Gavrilenko, P., Singh, Chandrani, and Chadha, R. K. 2010. Modelling the hydromechanical response in the vicinity of the Koyna reservoir (India): results for the initial filling period. *Geophysical Journal International*, **183**, 461–477, doi:10.1111/j.1365–246X.2010.04752.x.

Gelinsky, S., and Shapiro, S. A. 1997. Dynamic-equivalent medium approach for thinly layered saturated sediments. *Geophysical Journal International*, **128**, F1–F4.

Giardini, D. 2009. Geothermal quake risks must be faced. *Nature*, **462**(7275), 848–849, doi:10.1038/462848a.

Giardini, D., Grünthal, G., Shedlock, K. M., and Zhang, P. 2003. The GSHAP Global Seismic Hazard Map. *International Handbook of Earthquake & Engineering Seismology*. International Geophysics Series, vol. 81, no. B. Amsterdam: Academic Press, pp. 1233–1239.

Goertz-Allmann, B. P., Goertz, A., and S., Wiemer. 2011. Stress drop variations of induced earthquakes at the Basel geothermal site. *Geophysical Research Letters*, **38**, L09308, doi:10.1029/2011GL047498.

Goulty, N. R. 1998. Relationship between porosity and effective stress in shales. *First Break*, **16**, 413–419.

Gräsle, W., Kessels, W., Kümpel, H.-J., and Li, X. 2006. Hydraulic observations from a 1 year fluid production test in the 4000 m deep KTB pilot borehole. *Geofluids*, **6**, doi:10.1111/j.1468–8123.2006.00124.x.

Grechka, V. 2009. *Applications of Seismic Anisotropy in the Oil and Gas Industry*. Houten: EAGE Publications.

Grechka, V., and Yaskevich, S. 2013. Inversion of microseismic data for triclinic velocity models. *Geophysical Prospecting*, **61**(6), 1159–1170, doi:10.1111/1365–2478.12042.

Griffith, A. A. 1921. The phenomena of rupture and flow in solids. *Philosophical Transactions of the Royal Society of London*, **A221**, 163–198.

Griffith, A. A. 1924. The theory of rupture. *Proceedings of 1st International Congress on Applied Mechanics, Delft*, pp. 55–63.

Grünthal, G., and Wahlström, R. 2003. An M_w based earthquake catalogue for central, northern and northwestern Europe using a hierarchy of magnitude conversions. *Journal of Seismology*, **7**, 507–531.

Guha, S. K. 2000. *Induced Earthquakes*. London: Kluwer Academic, Dordrecht.

Gupta, H. K. 2002. A review of recent studies of triggered earthquakes by artificial water reservoirs with special emphasis on earthquakes in Koyna, India. *Earth-Science Reviews*, **58**, 279–310.

Gurevich, B. 2004. A simple derivation of the effective stress coefficient for seismic velocities in porous rocks. *Geophysics*, **69**, 393–397, doi:10.1190/1.1707058.

Gurevich, B., and Lopatnikov, S. 1995. Velocity and attenuation of elastic waves in finely layered porous rocks. *Geophysical Journal International*, **121**, 933–947.

Gurevich, B., Makarynska, D., de P. O. Bastos, and Pervukhina, M. 2010. A simple model for squirt-flow dispersion and attenuation in fluid-saturated granular rocks. *Geophysics*, **75**(6), N109–N120, doi:10.1190/1.3509782.

Gutenberg, B., and Richter, C. F. 1954. *Seismicity of Earth and Associated Phenomenon*. Princeton: Princeton University Press.

Hainzl, S., Fischer, T., and Dahm, T. 2012. Seismicity-based estimation of the driving fluid pressure in the case of swarm activity in Western Bohemia. *Geophysical Journal International*, **191**(1), 271–281, doi:10.1111/j.1365–246X.2012.05610.x.

Häring, M., Schanz, U., Ladner, F., and Dyer, B. 2008. Characterisation of the Basel 1 enhanced geothermal system. *Geothermics*, **37**(5), 469–495, doi:10.1016/j.geoth ermics.2008.06.002.

Harjes, H.-P., Bram, K., Dürbaum, H.-J., *et al.* 1997. Origin and nature of crustal reflections: Results from integrated seismic measurements at the KTB superdeep drilling site. *Journal of Geophysical Research*, **102**(B8), 18 267–18 288.

Hashin, Z., and Strikman, S. 1963. A variational approach to the elastic behavior of multiphase materials. *Journal of the Mechanics and Physics of Solids*, **11**, 127–140.

Hirschmann, G., and Lapp, M. 1995. Evaluation of the structural geology of the KTB Hauptbohrung (KTB-Oberpfalz HB). *KTB Report 94-1*. Niedersächsisches Landesamt für Bodenforschung, pp. 285–308.

House, L. 1987. Locating microearthquakes induced by hydraulic fracturing in crystalline rocks. *Geophysical Research Letters*, **14**, 919–921.

Hudson, D. J. 1964. *Lectures on Elementary Statistics and Probability*. Geneva: CERN, European Organization for Nuclear Research.

Hudson, J. A., Pointer, T., and Liu, E. 2001. Effective-medium theories for fluid-saturated materials with aligned cracks. *Geophysical Prospecting*, **49**, 509–522.

Hummel, N., and Müller, T. M. 2009. Microseismic signatures of non-linear pore-fluid pressure diffusion. *Geophysical Journal International*, **179**(3), 1558–1565, doi:10.1111/j.1365–246X.2009.04373.x.

Hummel, N., and Shapiro, S. A. 2012. Microseismic estimates of hydraulic diffusivity in case of non-linear fluid-rock interaction. *Geophysical Journal International*, **188**(3), 1441–1453, doi:10.1111/j.1365–246X.2011.05346.x.

Hummel, N., and Shapiro, S. 2013. Nonlinear diffusion-based interpretation of induced microseismicity: a Barnett Shale hydraulic fracturing case study. *Geophysics*, **78**(5), B211–B226, doi:10.1190/geo2012–0242.1.

Jaeger, J. C., Cook, N. G. W., and Zimmerman, R. W. 2007. *Fundamentals of Rock Mechanics*. Blackwell Publishing.

Jones, S. M. 1995. Velocities and quality factors of sedimentary rocks at low and high effective pressures. *Geophysical Journal International*, **123**, 774–780.

Jost, M. L., Büßelberg, T., Jost, Ö., and Harjes, H.-P. 1998. Source parameters of injection-induced microearthquakes at 9 km depth at the KTB Deep Drilling site, Germany. *Bulletin of the Seismological Society of America*, **88**(3), 815–832.

Jung, R., Cornet, F., Rummel, F., and Willis-Richard, J. 1996. Hydraulic stimulation results 1992/1993. Pages 31–41 of: Baria, R., Baumgärtner, J., and Gérard, A. (eds.), *European Hot Dry Rock Programme, 1992–1995. Extended Summary of the Final Report to European Community (DG XII), Contract N JOU2-CT92-0115 (Date of issue: 28 March 1996, Version 1)*.

Kaieda, H., and Sasaki, S. 1998. *Development of fracture evaluation methods for Hot Dry Rock geothermal power – Ogachi reservoir evaluation by the AE method*. CRIEPI report U97107 (in Japanese with English abstract).

Kaieda, H., Kiho, K., and Motojima, I. 1993. Multiple fracture creation for hot dry rock development. *Trends in Geophysical Research*, **2**, 127–139.

Kalpna, and Chander, R. 2000. Greens function based stress diffusion solutions in the porous elastic half space for time varying finite reservoir. *Physics of the Earth and Planetary Interiors*, **120**, 93–101, doi:10.1002/jgrb.50362.

Kanamori, H. 1977. The energy release in great earthquakes. *Journal of Geophysical Research*, **82**(20), 2981–2987, doi:10.1029/JB082i020p02981.

Kanamori, H., and Brodsky, E. E. 2004. The physics of earthquakes. *Reports on Progress in Physics*, **67**, 1429–1496.

Karpfinger, F., Müller, T. M., and Gurevich, B. 2009. Green's functions and radiation patterns in poroelastic solids revisited. *Geophysical Journal International*, **178**(1), 327–337, doi:10.1111/j.1365–246X.2009.04116.x.

Kaselow, A., and Shapiro, S. A. 2004. Stress sensitivity of elastic moduli and electrical resistivity in porous rocks. *Journal of Geopysics and Engineering*, **1**, 1–11.

Kaselow, A., Becker, K., and Shapiro, S. A. 2006. Stress sensitivity of seismic and electric rock properties of the upper continental crust at the KTB. *PAGEOPH*, **163**, 1021–1029, doi:10.1007/s00024–006–0063–9.

Kern, H., and Schmidt, R. 1990. Physical properties of the KTB core samples at simulated in situ conditions. *Scientific Drilling*, **1**, 217–223.

Kern, H., Popp, T., and Schmidt, R. 1994. The effect of deviatoric stress on the rock properties: an experimental study simulating the in-situ stress field at the KTB drilling site, Germany. *Surveys in Geophysics*, **15**, 467–479.

Khaksar, A., Griffiths, C. M., and McCann, C. 1999. Compressional- and shear-wave velocities as a function of confining stress in dry sandstones. *Geophysical Prospecting*, **47**, 487–508.

Kirstetter, O., and MacBeth, C. 2001. Compliance-based interpretation of dry frame pressure sensitivity in shallow marine sandstone. *Expanded Abstracts*. Society of Exploration Geophysicists, San Antonio, pp. 2132–2135.

Klee, G., and Rummel, F. 1993. Hydrofrac stress data for the european HDR research project test site Soultz-sous-Forêts. *International Journal of Rock Mechanics and Mining Sciences & Geomechanics Abstracts*, **30**(7), 973–976.

Kummerow, J. 2010. Using the value of the crosscorrelation coefficient to locate micro-seismic events. *Geophysics*, **75**(4), MA47–MA52.

Kummerow, J. 2013. Joint arrival time optimization for microseismic events recorded by seismic borehole arrays. doi:10.3997/2214–4609.20130401 *Extended Abstracts of 75th EAGE Conference & Exhibition, Session: Microseismic Event Characterisation*.

Kummerow, J., Reshetnikow, A., Häring, M., and Asanuma, H. 2012. Distribution of the Vp/Vs ratio within the Basel 1 Geothermal Reservoir from microseismic data. In: *Extended Abstracts of 74th EAGE Conference & Exhibition, Session: Microseismic Interpretation*.

Kümpel., H.-J, Erzinger, J., and Shapiro, S. A. 2006. Two massive hydraulic tests completed in deep KTB pilot hole. *Scientific Drilling*, (3) pp. 40–42.

Landau, L. D., and Lifshitz, E. M. 1987. *Theory of elasticity* (in Russian). Moscow: Nauka, Glavnaja Redaktsija Phys.-Math. Lit.

Landau, L. D., and Lifshitz, E. M. 1991. *Lehrbuch der theoretischen Physik* (in German), 5th edn. Vol. VI. Moscow: Acadamie, Berlin, Germany.

Langenbruch, C., and Shapiro, S. A. 2010. Decay rate of fluid-induced seismicity after termination of reservoir stimulations. *Geophysics*, **75**, doi:10.1190/1.3506005.

Langenbruch, C., and Shapiro, S. A. 2014. Gutenberg–Richter relation originates from Coulomb stress fluctuations caused by elastic rock heterogeneity. *Journal of Geophysical Research: Solid Earth*, **119**(2), 1220–1234, doi:10.1002/2013JB010282.

Langenbruch, C., Dinske, C., and Shapiro, S. A. 2011. Inter event times of fluid induced earthquakes suggest their Poisson nature. *Geophysical Research Letters*, **38**(21), doi:10.1029/2011GL049474.

Lay, T., and Wallace, T. C. 1995. *Modern Global Seismology*. Academic Press.

Li, M., Bernabé, Y., Xiao, W.-I., Chen, Z.-Y., and Liu, Z.-Q. 2009. Effective pressure law for permeability of E-bei sandstones. *Journal of Geophysical Research: Solid Earth*, **114**(B7), doi:10.1029/2009JB006373.

Lin, G., and Shearer, P. 2007. Estimating local V_p/V_s ratios within similar earthquake clusters. *Bulletin of the Seismological Society of America*, **97**(2), 379–388.

Lomax, A., Michelini, A., and Curtis, A. 2009. Earthquake location, direct, global-search methods. Pages 2449–2473, doi:10.1007/978–0–387–30440–3–150 Meyers, R. A. (ed.), *Encyclopedia of Complexity and Systems Science*. New York: Springer.

Lopatnikov, S. L., and Cheng, A. H.-D. 2005. If you ask a physicist from any country: a tribute to Yacov Il'ich Frenkel. *Journal of Engineering Mechanics*, **131**, 875–878.

Madariaga, R. 1976. Dynamics of an expanding circular fault. *Bulletin of the Seismological Society of America*, **66**(3), 639–666.

Majer, E. L., Baria, R., Stark, M., *et al.* 2007. Induced seismicity associated with Enhanced Geothermal Systems. *Geothermics*, **36**, 185–222, doi:10.1016/j.geothermics.2007.03.003.

Mandelis, A. 2000. Diffusion waves and their uses. *Physics Today*, **53**(8), 29–34.

Mavko, G., and Jizba, D. 1991. Estimating grain-scale fluid effects on velocity dispersion in rocks. *Geophysics*, **12**, 1940–1949.

Mavko, G., Mukerji, T., and Dvorkin, J. 1998. *The Rock Physics Handbook: Tools for Seismic Analysis in Porous Media*. Cambridge University Press.

Maxwell, S. 2014. *Microseismic Imaging of Hydraulic Fracturing: Improved Engineering of Unconventional Shale Reservoirs*. SEG, Tulsa.

Maxwell, S. C., Waltman, C., Warpinski, N. R., Mayerhofer, M. J., and Boroumand, N. 2009. Integrating fracture-mapping technologies to improve stimulations in the Barnett Shale. *SPE Reservoir Evaluation & Engineering*, **12**(1), 48–52, doi:10.2118/102801–PA.

Maxwell, S. C., Waltman, C. K., Warpinski, N. R., Mayerhofer, M. J., and Boroumand, N. 2006. Imaging seismic deformation induced by hydraulic fracture complexity. *Paper SPE 102801.*

McClintock, F. A., and Walsh, J. B. 1962. Friction on Griffith cracks in rocks under pressure. Pages 1015–1021 of: *Proceedings of 4th US National Congress on Applied Mechanics.*

McGarr, A. 1976. Seismic moments and volume changes. *Journal of Geophysical Research,* **81**(8), 1487–1494, doi:10.1029/JB081i008p01487.

McGarr, A., Simpson, D., and Seeber, L. 2002. Case histories of induced and triggered seismicity. Chapter 40, pages 647–665 of: Lee, W. H. K. *et al.* (eds.), *International Handbook of Earthquake and Engineering Seismology, International Geophysics Series, vol. 81A.* New York: Elsevier.

Merkel, R. H., Barree, R. D., and Towle, G. 2001. Seismic response of Gulf of Mexico reservoir rocks with variations in pressure and water saturation. *The Leading Edge,* **20**, 290–299.

Millich, E., Neugebauer, H. J., Huenges, E., and Nover, G. 1998. Pressure dependence of permeability and Earth tide induced fluid flow. *Geophysical Research Letters,* **25**(6), 809–812, doi:10.1029/98GL00395.

Mindlin, R. D. 1949. Compliance of elastic bodies in contact. *Journal of Applied Mechanics,* **16**, 259–268.

Mukuhira, Y., Asanuma, H., Niitsuma, H., Schanz, U., and Häring, M. 2008. Characterization of microseismic events with larger magnitude collected at Basel, Switzerland in 2006. *GRC Transactions,* **32**, 87–93.

National Research Council. 2013. *Induced Seismicity Potential in Energy Technologies.* Washington, DC: The National Academies Press.

Norris, A. 1989. Stoneley-wave attenuation and dispersion in permeable formations. *Geophysics,* **54**, 330–341.

Norris, A. 1993. Low-frequency dispersion and attenuation in partially saturated rocks. *Journal of the Acoustical Society of America,* **94**, 359–370.

Nur, A., and Booker, J. 1972. Aftershocks caused by pore fluid flow? *Science,* **175**, 885–887.

Oelke, A., Alexandrov, D., Abakumov, I., *et al.* 2013. Seismic reflectivity of hydraulic fractures approximated by thin fluid layers. *Geophysics,* **78**(4), T79–T87, doi:10.1190/geo2012–0269.1.

Ohtake, M. 1974. Seismic activity induced by water injection at Matsushiro, Japan. *Journal of Physics of the Earth,* **22**, 163–176.

Parotidis, M., Rothert, E., and Shapiro, S. A. 2003. Pore-pressure diffusion: A possible triggering mechanism for the earthquake swarms 2000 in Vogtland/NW-Bohemia, central Europe. *Geophysical Research Letters,* **30**(20), 2075, doi:10.1029/2003GL018110.

Parotidis, M., Shapiro, S. A., and Rothert, E. 2004. Back front of seismicity induced after termination of borehole fluid injection. *Geophysical Research Letters,* **31**, doi:10.1029/2003GL018987.

Parotidis, M., Shapiro, S. A., and Rothert, E. 2005. Evidence for triggering of the Vogtland swarms 2000 by pore pressure diffusion. *Journal of Geophysical Research,* **110**(B05S10), 1–12, doi:10.1029/2004JB003267.

Pavlis, L. 1986. Appraising earthquake hypocenter location errors: a complete practical approach for single-event locations. *Bulletin of the Seismological Society of America,* **76**, 1699–1717.

Pearson, C. 1981. The relationship between microseismicity and high pore pressures during hydraulic stimulation experiments in low permeability granitic rocks. *Journal of Geophysical Research,* **86**, 7855–7864.

Peirce, A., and Detournay, E. 2008. An implicit level set method for modeling hydrauli-cally driven fractures. *Computer Methods in Applied Mechanics and Engineering*, **197**(3340), 2858–2885, doi:10.1016/j.cma.2008.01.013.

Pervukhina, M., Gurevich, B., Dewhurst, D. N., and Siggins, A. F. 2010. Applicability of velocity–stress relationships based on the dual porosity concept to isotropic porous rocks. *Geophysical Journal International*, **181**, 1473–1479, doi:10.1111/j.1365–246X.2010.04535.x.

Phillips, W. S., House, L. S., and Fehler, M. C. 1997. Detailed joint structure in a geothermal reservoir from studies of induced microearthquake clusters. *Journal of Geophysical Research*, **102**(B6), 11 745–11 763, doi:10.1029/97JB00762.

Prasad, M., and Manghnani, M. H. 1997. Effects of pore and differential pressure on compressional wave velocity and quality factor in Berea and Michigan sandstones. *Geophysics*, **62**, 1163–1176.

Pride, S. R., Berryman, J. G., and Harris, J. M. 2004. Seismic attenuation due to wave-induced flow. *Journal of Geophysical Research: Solid Earth*, **109**(B1), doi:10.1029/2003JB002639.

Rentsch, S., Buske, S., Lüth, S., and Shapiro, S. A. 2007. Fast location of seismicity: a migration-type approach with application to hydraulic-fracturing data. *Geophysics*, **72**, S33–S40, doi:10.1190/1.2401139.

Rentsch, S., Buske, S., Gutjahr, S., Kummerow, J., and Shapiro, S. A. 2010. Migration-based location of seismicity recorded with an array installed in the main hole of the San Andreas Fault Observatory at Depth (SAFOD). *Geophysical Journal International*, **182**(1), 477–492, doi:10.1111/j.1365–246X.2010.04638.x.

Reshetnikov, A., Buske, S., and Shapiro, S. A. 2010. Seismic imaging using microseismic events: results from the San Andreas Fault System at SAFOD. *Journal of Geophysical Research: Solid Earth*, **115**(B12), doi:10.1029/2009JB007049.

Reshetnikov, A., Kummerow, J., Shapiro, S. A., Asanuma, H., and Häring, M. 2013. Microseismic reflection imaging of stimulated reservoirs and fracture zones. doi:10.3997/2214–4609.20131279 *Extended Abstracts of 75th EAGE Conference & Exhibition – Workshops Session: WS16 – Micro Seismicity – What Now? What Next?*.

Rice, J. R. 1980. The mechanics of earthquake rupture, course 78, 1979, pages 556–649 of: Dziewonski, A. M., and Boschi, E. (eds.), *Physics of the Earth's Interior (Proc. International School of Physics "Enrico Fermi")*. Italian Physical Society and North-Holland Publ. Co.

Rice, J. R., and Cleary, M. P. 1976. Some basic stress diffusion solutions for fluid-saturated elastic porous media with compressible constituents. *Reviews of Geophysics and Space Physics*, **14**, 227–241.

Richards, P. G., Waldhauser, F., Schaff, D., and Kim, W.-Y. 2006. The applicability of modern methods of earthquake location. *Pure and Applied Geophysics*, **163**(2–3), 351–372, doi:10.1007/s00024–005–0019–5.

Roeloffs, E. A. 1988. Fault stability changes induced beneath a reservoir with cyclic variations in water level. *Journal of Geophysical Research: Solid Earth*, **93**(B3), 2107–2124, doi:10.1029/JB093iB03p02107.

Rothert, E., and Shapiro, S. A. 2003. Microseismic monitoring of borehole fluid injections: data modeling and inversion for hydraulic properties of rocks. *Geophysics*, **68**(2), 685–689, doi:10.1190/1.1567239.

Rothert, E., and Shapiro, S. A. 2007. Statistics of fracture strength and fluid-induced microseismicity. *Journal of Geophysical Research*, **112**, B04309, doi:10.1029/2005JB003959.

Rudnicki, J. W. 1986. Fluid mass sources and point forces in linear elastic diffusive solids. *Mechanics of Materials*, **5**, 383–393.

Ruina, A. 1983. Slip instability and state variable friction laws. *Journal of Geophysical Research*, **88**, 359–370.

Rutledge, J. T., and Phillips, W. S. 2003. Hydraulic stimulation of natural fractures as revealed by induced microearthquakes, Carthage Cotton Valley gas field, east Texas. *Geophysics*, **68**(2), 441–452, doi:10.1190/1.1567214.

Rutledge, J. T., Phillips, W. S., and Mayerhofer, M. J. 2004. Faulting induced by forced fluid injection and fluid flow forced by faulting: An interpretation of hydraulic-fracture microseismicity, Carthage Cotton Valley gas field, Texas. *Bulletin of the Seismological Society of America*, **94**, 1817–1830.

Rytov, S. M., Kravtsov, Yu. A., and Tatarskii, V. I. 1989. *Wave Propagation Through Random Media*. Principles of statistical radiophysics, vol. 4. Berlin: Springer Verlag.

Schoenball, M., Müller, T. M., Müller, B. I. R., and Heidbach, O. 2010. Fluid-induced microseismicity in pre-stressed rock masses. *Geophysical Journal International*, **180**(2), doi:10.1111/j.1365–246X.2009.04443.x.

Scholz, C. H. 2002. *The Mechanics of Earthquakes and Faulting*, 2nd edn. Cambridge: Cambridge University Press.

Scholze, M., Stürmer, K., Kummerow, J., and Shapiro, S. A. 2010. An approach to analyse microseismic event similarity. In: *Extended Abstracts of 72nd EAGE Conference & Exhibition, Session: Passive & Microseismic Methods (EAGE)*.

Segall, P. 1989. Earthquakes triggered by fluid extraction. *Geology*, **17**(10), 942–946, doi:10.1130/0091–7613(1989).

Segall, P. 2010. *Earthquake and Volcano Deformation*. Princeton: Princeton University Press.

Segall, P., and Fitzgerald, S. 1998. A note on induced stress changes in hydrocarbon and geothermal reservoirs. *Tectonophysics*, **289**, 117–128.

Segall, P., and Rice, J. R. 1995. Dilatancy, compaction, and slip instability of a fluid-infiltrated fault. *Journal of Geophysical Research*, **100**, 22 155–22 172.

Segall, P., Desmarais, E., Shelly, D., Miklius, A., and Cervelli, P. 2006. Earthquakes triggered by silent slip events on Kilauea volcano, Hawaii. *Nature*, **442**, 71–74.

Shapiro, S. A. 1992. Elastic waves scattering and radiation by fractal inhomogeneity of a medium. *Geophysical Journal International*, **110**, 591–600.

Shapiro, S. A. 2000. An inversion for fluid transport properties of three-dimensional heterogeneous rocks using induced microseismicity. *Geophysical Journal International*, **143**, 931–936.

Shapiro, S. A. 2003. Elastic piezosensitivity of porous and fractured rocks. *Geophysics*, **68**, 482–486, doi: 10.1190/1.1567215.

Shapiro, S. A. 2008. *Microseismicity: a Tool for Reservoir Characterization*. Houten, the Netherlands: EAGE Publications.

Shapiro, S. A., and Dinske, C. 2009a. Fluid-induced seismicity: pressure diffusion and hydraulic fracturing. *Geophysical Prospecting*, **57**, 301–310, doi:10.1111/j.1365–2478.2008.00770.xi.

Shapiro, S. A., and Dinske, C. 2009b. Scaling of seismicity induced by nonlinear fluid-rock interaction. *Journal of Geophysical Research*, **114**(B9), B09307, doi:10.1029/2008JB006145.

Shapiro, S. A., and Fayzullin, I. S. 1992. Fractal properties of fault systems by scattering of body seismic waves. *Tectonophysics*, **202**, 177–181.

Shapiro, S. A., and Hubral, P. 1999. *Elastic Waves in Random Media. Fundamentals of Seismic Stratigraphic Filtering*. Berlin: Springer Verlag.

Shapiro, S. A., and Kaselow, A. 2005. Porosity and elastic anisotropy of rocks under tectonic stress and pore-pressure changes. *Geophysics*, **70**, N27–N38, doi:10.1190/1.2073884.

Shapiro, S. A., Huenges, E., and Borm, G. 1997. Estimating the permeability from fluid-injection-induced seismic emissions at the KTB site. *Geophysical Journal International*, **131**, F15–F18.

Shapiro, S. A., Audigane, P., and Royer, J.-J. 1999. Large-scale in situ permeability tensor of rocks from induced microseismicity. *Geophysical Journal International*, **137**, 207–213.

Shapiro, S. A., Rothert, E., Rath, V., and Rindschwentner, J. 2002. Characterization of fluid transport properties of reservoirs using induced microseismicity. *Geophysics*, **67**, 212–220, doi:10.1190/1.1451597.

Shapiro, S. A., Patzig, R., Rothert, E., and Rindschwentner, J. 2003. Triggering of microseismicity due to pore-pressure perturbation: permeability related signatures of the phenomenon. *PAGEOPH*, **160**, 1051–1066.

Shapiro, S. A., Rentsch, S., and Rothert, E. 2005a. Characterization of hydraulic properties of rocks using probability of fluid-induced microearthquakes. *Geophysics*, **70**, F27–F34, doi:10.1190/1.1897030.

Shapiro, S. A., Rentsch, S., and Rothert, E. 2005b. Fluid-induced seismicity: theory, modeling, and applications. *Journal of Engineering Mechanics*, **131**, 947–952.

Shapiro, S. A., Kummerow, J., Dinske, C., *et al.* 2006a. Fluid induced seismicity guided by a continental fault: injection experiment of 2004/2005 at the German Deep Drilling Site (KTB). *Geophysical Research Letters*, **33**(1), L01309, doi:10.1029/2005GL024659.

Shapiro, S. A., Dinske, C., and Rothert, E. 2006b. Hydraulic-fracturing controlled dynamics of microseismic clouds. *Geophysical Research Letters*, **33**, L14312, doi:10.1029/2006GL026365.

Shapiro, S. A., Dinske, C., and Kummerow, J. 2007. Probability of a given-magnitude earthquake induced by a fluid injection. *Geophysical Research Letters*, **34**, L22314, doi:10.1029/2007GL031615.

Shapiro, S. A., Dinske, C., Langenbruch, C., and Wenzel, F. 2010. Seismogenic index and magnitude probability of earthquakes induced during reservoir fluid stimulations. *The Leading Edge*, **29**(3), 304–309, doi:10.1190/1.3353727.

Shapiro, S.A., Krüger, O. S., Dinske, C., and Langenbruch, C. 2011. Magnitudes of induced earthquakes and geometric scales of fluid-stimulated rock volumes. *Geophysics*, **76**, WC53–WC61, doi:10.1190/GEO2010–0349.1.

Shapiro, S. A., Krüger, O. S., and Dinske, C. 2013. Probability of inducing given-magnitude earthquakes by perturbing finite volumes of rocks. *Journal of Geophysical Research: Solid Earth*, **118**(7), 3557–3575, doi:10.1002/jgrb.50264.

Shearer, P. M. 1994. Global seismic event detection using a matched filter on long-period seismograms. *Journal of Geophysical Research: Solid Earth*, **99**(B7), 13 713–13 725, doi:10.1029/94JB00498.

Shearer, P. M. 2009. *Introduction to Seismology*. Cambridge: Cambridge University Press.

Shelly, D. R., Hill, D. P., Massin, F., *et al.* 2013. A fluid-driven earthquake swarm on the margin of the Yellowstone caldera. *Journal of Geophysical Research*, **188**, 4872–4886, doi:10.1002/jgrb.50362.

Simon, M., Gebrande, H., and Bopp, M. 1996. Pre-stack migration and true-amplitude processing of DEKORP near-normal incidence and wideangle reflection measurements. *Tectonophysics*, **264**, 381–392.

N., Asanuma, H., Kaieda, H., *et al.* 2004. On site mapping of microseis-
ꞏcity at Cooper Basin, Australia HDR project by the Japanese team. In: *Proceedings, Twenty-ninth Workshop on Geothermal Reservoir Engineering, Stanford University.*

Stober, I., and Bucher, K. 2005. The upper continental crust, an aquifer and its fluid: hydraulic and chemical data from 4 km depth in fractured crystalline basement rocks at the KTB test site. *Geofluids*, **5**(1), 8–19, doi:10.1111/j.1468–8123.2004.00106.x.

Talwani, P. 1997. On the nature of reservoir-induced seismicity. *PAGEOPH*, **150**, 473–492.

Talwani, P., and Acree, S. 1985. Pore pressure diffusion and the mechanism of reservoir-induced seismicity. *PAGEOPH*, **122**, 947–965.

Terzaghi, K. 1936. The shear resistance of saturated soils. *Proceedings for the 1st International Conference on Soil Mechanics and Foundation Engineering (Cambridge, MA)*, vol. 1, pp. 54–56.

Thomsen, L. 1986. Weak elastic anisotropy. *Geophysics*, **51**(10), 1954–1966.

Thomsen, L. 1995. Elastic anisotropy due to aligned cracks in porous rocks. *Geophysical Prospecting*, **43**, 805–830.

Thurber, H. C., and Rabinowitz, N. 2000. *Advances in Seismic Event Location*. Kluwer Academic Publishers.

Tsvankin, I. 2005. *Seismic Signatures and Analysis of Reflection Data in Anisotropic Media, Volume 29 (Handbook of Geophysical Exploration: Seismic Exploration)*. Oxford: Elsevier Science.

Turcotte, D. L., Holliday, J. R., and Rundle, J. B. 2007. BASS, an alternative to ETAS. *Geophysical Research Letters*, **34**, L12303, doi:10.1029/2007GL029696.

Urbancic, T., and Baig, A. 2013. Validating engineering objectives of hydraulic fracture stimulations using microseismicity. *First Break*, **31**(7), 73–90.

Van der Kamp, G., and Gale, J. E. 1983. Theory of Earth tide and barometric effects in porous formations with compressible grains. *Water Resources Research*, **19**, 538–544.

Waldhauser, F., and Ellsworth, W.L. 2000. A double-difference earthquake location algorithm: method and application to the northern Hayward fault, California. *Bulletin of the Seismological Society of America*, **90**, 1353–1368.

Walsh, J. B. 1981. Effect of pore pressure and confining pressure on fracture permeability. *International Journal of Rock Mechanics and Mining Sciences & Geomechanics Abstracts*, **18**(5), 429–435.

Wang, H. F. 2000. *Theory of Linear Poroelasticity with Applications to Geomechanics and Hydrogeology*. Princeton: Princeton University Press.

Warpinski, N., Wolhart, S., and Wright, C. 2001. Analysis and prediction of microseismicity induced by hydraulic fracturing. *Paper SPE 71649.*

White, J. E. 1983. *Underground Sound: Application of Seismic Waves*. Amsterdam: Elsevier.

Yu, C. P., Reshetnikov, A., and Shapiro, S.A. 2013. Simultaneous inversion of anisotropic velocity model and microseismic event location – synthetic and real data examples. doi:10.3997/2214–4609.20130407 *Extended Abstracts of 75th EAGE Conference & Exhibition, Session: Microseismic Event Characterisation.*

Zarembo, L. K., and Krasilnikov, V. A. 1966. *Introduction into Non-linear Acoustics* (in Russian). Moscow: Nauka, Glavnaja Redaktsija Phys.-Math. Lit.

Zimmerman, R. W., Somerton, W. H., and King, M. S. 1986. Compressibility of porous rocks. *Journal of Geophysical Research*, **91**, 12765–12777.

Zoback, M. D. 2010. *Reservoir Geomechanics*. Cambridge: Cambridge University Press.

Zoback, M. D., and Gorelick, S. M. 2012. Earthquake triggering and large-scale geologic storage of carbon dioxide. *Proceedings of the National Academy of Sciences USA*, **109**, 10 164–10 168, doi: 10.1073/pnas.1202473109.

Zoback, M. D., and Harjes, H.-P. 1997. Injection-induced earthquakes and crustal stress at 9 km depth at the KTB deep drilling site, Germany. *Journal of Geophysical Research*, **102**(B8), 18 477–18 491.

Index

272

Index

k, 9, 109, 111, 113, 132, 166
nonic plane wave, 11, 67
torques, 3
tortuosity, 65
total moment tensor, 33
total stress, 49, 86
traction, 2–4, 15, 16, 49, 95, 117
transformation strain, 30
transverse isotropy, 8, 43–45, 60
treatment fluid, 165, 166, 168, 172, 177, 178
triclinic media, 7
triggering front, 118, 127–149, 152, 158, 161–166, 176–200

undrained rock, 55, 59–62, 66, 69, 83, 88–90, 94, 108
undrained system, 54, 55, 59, 61, 79
uniaxial strain, 87
uniaxial tensile strength, 22, 165
unjacketed sample, 50, 52

vector of fluid flux, 62, 146
velocity strengthening, 21
velocity weakening, 21
viscosity of a fluid, 63, 64, 145, 177
Voigt's notations, 6
volume averaged strain, 27, 50, 53
volume balance, 164, 165, 168, 181, 187
volume of a hydraulic fracture, 166
volumetric hydraulic fracturing, 177, 180, 183
volumetric strain, 8

wave number, 68–71, 73, 142
wave vector, 11, 67, 68, 70, 71
wavelength, 13, 32, 33, 71, 128
weak anisotropy, 44
well-head pressure, 119, 120, 155, 157
width of fracture, 173

Printed in the United States
By Bookmasters